Antimicrobial Peptides: An Emerging Hope in the Era of New Infections and Resistance

Antimicrobial Peptides: An Emerging Hope in the Era of New Infections and Resistance

Guest Editors

Piyush Baindara
Marisa Di Pietro

MDPI

Basel • Beijing • Wuhan • Barcelona • Belgrade • Novi Sad • Cluj • Manchester

Guest Editors
Piyush Baindara
Animal Science Research Center
University of Missouri
Columbia
United States

Marisa Di Pietro
Dipartimento di Sanità Pubblica e Malattie Infettive
Università degli Studi di Roma La Sapienza
Rome
Italy

Editorial Office
MDPI AG
Grosspeteranlage 5
4052 Basel, Switzerland

This is a reprint of the Special Issue, published open access by the journal *Antibiotics* (ISSN 2079-6382), freely accessible at: www.mdpi.com/journal/antibiotics/special_issues/Era_Peptides.

For citation purposes, cite each article independently as indicated on the article page online and as indicated below:

Lastname, A.A.; Lastname, B.B. Article Title. *Journal Name* **Year**, *Volume Number*, Page Range.

ISBN 978-3-7258-3960-5 (Hbk)
ISBN 978-3-7258-3959-9 (PDF)
https://doi.org/10.3390/books978-3-7258-3959-9

Contents

About the Editors

Piyush Baindara

Piyush Baindara is currently working as a senior research associate at the Somatic Cell Genome Editing program, Animal Science Research Center, University of Missouri, Columbia, USA. He received his Ph.D. in microbiology from the CSIR-Institute of Microbial Technology, Chandigarh, India. His Ph.D. research focus was on bacterial antimicrobial peptides. Subsequently, he pursued his post-doctoral research at the University of Arkansas for Medical Sciences, Little Rock, and the University of Missouri, Columbia, USA. He has published more than 50 research articles in reputed peer-reviewed journals. His research interests include the discovery and characterization of novel antimicrobial peptides that includes bacteriocins, lanthipeptides, and human defensins. Additionally, he is fascinated about antimicrobial drug resistance, peptide-based vaccine development, infectious agents and disease pathogenesis, anticancer therapeutics, NETs (Neutrophil Extracellular Traps) in infectious and autoimmune diseases. His recent interest is developing towards AMPs in gene delivery strategies and generation of transgenic animals to fight against rapid evolving drug-resistance.

Marisa Di Pietro

Marisa Di Pietro works at the Dipartimento di Sanità Pubblica e Malattie Infettive, Università degli Studi di Roma La Sapienza, Rome, Italy. Her research interests include clinical microbiology, chlamydia, bacterial pathogenesis, pathogen–host interactions, infection, cervicovaginal microbiota, biofilm, in vitro susceptibility testing, and antibacterial agents.

antibiotics

Editorial

Antimicrobial Peptides: An Emerging Hope in the Era of New Infections and Resistance

Piyush Baindara

Animal Science Research Center, Division of Animal Sciences, University of Missouri, Columbia, MO 65211, USA; piyush.baindara@gmail.com

Received: 8 May 2025
Accepted: 22 May 2025
Published: 27 May 2025

Citation: Baindara, P. Antimicrobial Peptides: An Emerging Hope in the Era of New Infections and Resistance. *Antibiotics* **2025**, *14*, 546. https://doi.org/10.3390/antibiotics14060546

Recently, antimicrobial peptides (AMPs) have garnered significant attention as a viable alternative to traditional antibiotics. AMPs are naturally occurring important elements of the host defense system that are functional across all biological domains, from prokaryotes to eukaryotes [1]. AMPs are powerful antimicrobial agents exhibiting a wide range of biological activity against infectious and pathogenic entities, including bacteria, fungi, viruses, and parasites [2]. Furthermore, the immuno-modulatory and potential anticancer properties of AMPs have been well reported [3].

Interestingly, AMPs are extremely diverse, thus leading to endless possibilities for new variants. The human gut microbiota is one such example of a complex system that is vital to human health due to its diversity and dynamic competition [4]. Notably, gut AMPs work synergistically with other gut microbiota and antimicrobials to maintain gut homeostasis. Additionally, gut AMPs are evolving under complicated and highly synergistic co-evolutionary pressure developed by interactions between various competitive microbiota and their respective AMPs [5]. The synergistic actions of gut AMPs with conventional antibiotics have been suggested as a key weapon to fight against multi-antibiotic-resistant bacteria [6]. In a recent study, Pandey et al. reported multi-antibiotic resistance in different Vibrio species recovered from environmental water samples [7]. This suggests that antibiotic disposal in the environment is also triggering the emergence of drug resistance not only competition or direct antibiotic intake.

AMPs have been considered one of the potential alternatives to fight against multidrug-resistant bacteria, and several are undergoing clinical trials [8]. Similarly, a new AMP, NNS5-6, produced by mangrove bacteria *Paenibacillus thiaminolyticus* NNS5-6, has displayed antimicrobial activity against drug-resistant *Pseudomonas aeruginosa* and *Klebsiella pneumonia* [9]. Furthermore, recombinant mussel adhesive proteins fused with functional peptides (MAP-FPs) have been characterized and exhibit specific activity against Gram-negative bacteria such as *Escherichia coli*, *Salmonella typhimurium*, and *K. pneumonia* [9]. Notably, MAP-FPs were found to be nontoxic against mammalian cell lines, suggesting they are suitable candidates for therapeutic applications.

Natural AMP-inspired synthetic peptides are one strategy to achieve highly efficient AMPs with low toxicity to combat drug-resistant pathogens in therapeutic settings [10]. Additionally, synthetic peptides and their truncated forms facilitate the easy examination and characterization of antimicrobial efficacy to design better AMPs against drug-resistant pathogens [11]. In a recent study, Meier et al. examined and demonstrated the antimicrobial potential of a synthetic peptide, C18G, and its several truncated forms using model lipid membranes and vesicles. The findings suggest that peptide length and ensuing hydrophobic matching are critical factors to consider in the evolution and design of membrane-disrupting AMPs [12]. Moreover, Keeratikunakorn et al. synthesized a natural AMP,

BiF2_5K7K, using amino acid substitution based on residue composition and distributic BiF2_5K7K displayed superior activity against both Gram-positive and Gram-negati bacteria isolated from boar semen and sow vaginal discharge. Interestingly, BiF2_5K7 treatment achieved a superior pregnancy and farrowing rate in an artificial inseminati test at a pig farm [13]. Similarly, synthetic AMPs A-11 and AP19 were reported to restri the growth of Gram-negative bacteria isolated from fresh and stored boar semen at 18 ° while not harming sperm motility, acrosomal integrity, and viability [14]. Overall, sy thetic AMPs, BiF2_5K7K, A-11, and AP19 could be potential alternatives to convention antibiotics for use in boar semen extenders.

Other than direct antimicrobial activity, AMPs engage in immunomodulatory acti ities, thus shaping the outcomes of antimicrobial therapies. Finkina et al. reported t immunomodulatory effects of a tobacco defensin, NaD1, on human macrophages, dendrit cells, and bold monocytes. Also, NaD1 could induce both pro-inflammatory and an inflammatory cytokines, suggesting its possible potential for therapeutic applications [1 Another synthetic AMP, MV6, showed synergistic interactions and an aminoglycosid netilmicin. Interestingly, MV6 lacks intrinsic antimicrobial activity; however, it reduc mutant prevention concentration of an aminoglycoside netilmicin against *Acinetobacter ba mannii* when used synergistically [16]. Conclusively, the development of synthetic AM based on natural AMPs or by using special algorithms has the potential to generate mo efficient and specific AMPs to fight the battle against drug resistance. Slavokhotova et developed an algorithm for the prediction of α-hairpins based on characteristic motifs co taining four or six cysteines. There were more than 2000 putative α-hairpins predicted, an the authors concluded that AMPs containing six cysteines had more potent antimicrobi activity than the AMPs with four cysteines [16]. Interestingly, AMPs containing cystei motifs are evolutionarily conserved in all domains of life, ranging from prokaryotes t eukaryotes, which also suggests a potential evolutionary link and role with antimicrobi efficacy [17–19].

In conclusion, AMPs are promising drug candidates to combat drug-resistan pathogens. Notably, 12 peptide-based drugs with potential antimicrobial or antifung properties have been approved by the US Food and Drug Administration (FDA) sinc 1955. As of now, several AMPs are undergoing clinical trials targeting drug-resistan pathogens [8]. Interestingly, rezafungin (a novel systemic antifungal), an echinocandi and a class of cyclic lipopeptides, was approved by the FDA in March 2023 [20]. Overal AMPs are an emerging hope in the era of new infections and drug resistance (Figure 1 AMPs provide endless opportunities to discover new peptides, especially in the case o bacterial AMPs [21,22]. As bacterial diversity is wide, there are significant possibilities t discover new AMPs; in fact, more than 99% of bacterial diversity is still unexplored [23 Additionally, complex biological systems such as skin, the lungs, and gut microbiota ar highly competitive environments that have favorable conditions for the production o new AMPs [24–26]. Also, recent scientific advances along with artificial intelligence (A have contributed exceptionally to the discovery of novel unexplored AMPs from differen domains of life, including the extinct ones [27,28]. Recently, AMPs have been extensivel explored as an alternative to conventional antibiotics for therapeutic applications; howeve there is a long way to go before AMPs can be fully adopted in clinical settings, as there ar many unanswered questions about AMPs that require further detailed study [29].

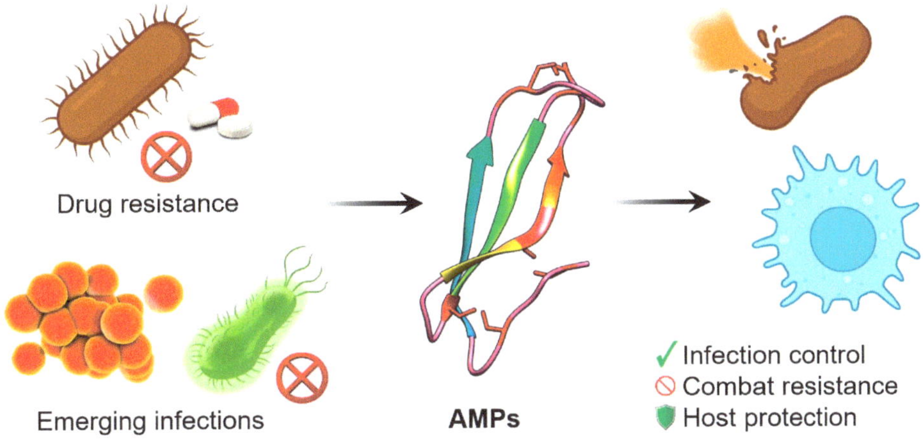

Figure 1. AMPs combating drug resistance and infection.

- How do AMPs selectively target microbes over host cells? Can this selectivity be manipulated synthetically?
- How do the AMP producers maintain self-immunity?
- Is it possible to develop resistance against AMPs? If so, what are the AMP resistance genes and what is the global prevalence? Can it be transferred horizontally like conventional antibiotics?
- How do repeated AMP treatments influence microbial diversity?
- How does AMP treatment influence pro- and anti-inflammatory immune response?
- What signaling pathways are activated upon AMP–host cell interactions?

Acknowledgments: P.B. is thankful to the Animal Science Research Center, Division of Animal Sciences, University of Missouri, Columbia, USA, for providing the necessary space and facilities to conduct this work.

Conflicts of Interest: The author declares no conflicts of interest.

References

1. Dijksteel, G.S.; Ulrich, M.M.W.; Middelkoop, E.; Boekema, B.K.H.L. Review: Lessons Learned from Clinical Trials Using Antimicrobial Peptides (AMPs). *Front. Microbiol.* **2021**, *12*, 616979. [CrossRef] [PubMed]
2. Yang, M.; Liu, S.; Zhang, C. Antimicrobial peptides with antiviral and anticancer properties and their modification and nanodelivery systems. *Curr. Res. Biotechnol.* **2023**, *5*, 100121. [CrossRef]
3. Duarte-Mata, D.I.; Salinas-Carmona, M.C. Antimicrobial peptides' immune modulation role in intracellular bacterial infection. *Front. Immunol.* **2023**, *14*, 1119574. [CrossRef]
4. Ma, Y.; Guo, Z.; Xia, B.; Zhang, Y.; Liu, X.; Yu, Y.; Tang, N.; Tong, X.; Wang, M.; Ye, X.; et al. Identification of antimicrobial peptides from the human gut microbiome using deep learning. *Nat. Biotechnol.* **2022**, *40*, 921–931. [CrossRef]
5. Ostaff, M.J.; Stange, E.F.; Wehkamp, J. Antimicrobial peptides and gut microbiota in homeostasis and pathology. *EMBO Mol. Med.* **2013**, *5*, 1465–1483. [CrossRef]
6. Baindara, P.; Mandal, S.M. Gut-Antimicrobial Peptides: Synergistic Co-Evolution with Antibiotics to Combat Multi-Antibiotic Resistance. *Antibiotics* **2023**, *12*, 1732. [CrossRef] [PubMed]
7. Pandey, R.; Sharma, S.; Sinha, K.K. Evidence of Antibiotic Resistance and Virulence Factors in Environmental Isolates of Vibrio Species. *Antibiotics* **2023**, *12*, 1062. [CrossRef]
8. Baindara, P.; Kumari, S.; Dinata, R.; Mandal, S.M. Antimicrobial peptides: Evolving soldiers in the battle against drug-resistant superbugs. *Mol. Biol. Rep.* **2025**, *52*, 432. [CrossRef]
9. Sermkaew, N.; Atipairin, A.; Krobthong, S.; Aonbangkhen, C.; Yingchutrakul, Y.; Uchiyama, J.; Songnaka, N. Unveiling a New Antimicrobial Peptide with Efficacy against *P. aeruginosa* and *K. pneumoniae* from Mangrove-Derived Paenibacillus thiaminolyticus NNS5-6 and Genomic Analysis. *Antibiotics* **2024**, *13*, 846. [CrossRef]

10. Yang, C.-L.; Wang, P.-P.; Zhou, Z.-Y.; Wu, X.-W.; Hua, Y.; Chen, J.-W.; Wang, H.; Wei, B. Discovery of naturally inspire antimicrobial peptides using deep learning. *Bioorganic Chem.* **2025**, *160*, 108444. [CrossRef]
11. Klubthawee, N.; Adisakwattana, P.; Hanpithakpong, W.; Somsri, S.; Aunpad, R. A novel, rationally designed, hybrid antimicrobi peptide, inspired by cathelicidin and aurein, exhibits membrane-active mechanisms against *Pseudomonas aeruginosa*. *Sci. Re* **2020**, *10*, 9117. [CrossRef] [PubMed]
12. Meier, S.; Ridgway, Z.M.; Picciano, A.L.; Caputo, G.A. Impacts of Hydrophobic Mismatch on Antimicrobial Peptide Efficacy ar Bilayer Permeabilization. *Antibiotics* **2023**, *12*, 1624. [CrossRef] [PubMed]
13. Keeratikunakorn, K.; Chanapiwat, P.; Aunpad, R.; Ngamwongsatit, N.; Kaeoket, K. Effect of Antimicrobial Peptide BiF2_5K7K c Contaminated Bacteria Isolated from Boar Semen and Semen Qualities during Preservation and Subsequent Fertility Test on P Farm. *Antibiotics* **2024**, *13*, 579. [CrossRef] [PubMed]
14. Keeratikunakorn, K.; Chanapiwat, P.; Aunpad, R.; Ngamwongsatit, N.; Kaeoket, K. The Effects of Different Antimicrobial Peptid (A-11 and AP19) on Isolated Bacteria from Fresh Boar Semen and Semen Quality during Storage at 18 °C. *Antibiotics* **2024**, *13*, 48 [CrossRef]
15. Finkina, E.I.; Bogdanov, I.V.; Shevchenko, O.V.; Fateeva, S.I.; Ignatova, A.A.; Balandin, S.V.; Ovchinnikova, T.V. Immunomodul tory Effects of the Tobacco Defensin NaD1. *Antibiotics* **2024**, *13*, 1101. [CrossRef]
16. Roson-Calero, N.; Lucas, J.; Gomis-Font, M.A.; de Pedro-Jové, R.; Oliver, A.; Ballesté-Delpierre, C.; Vila, J. Cyclic Peptide MV6, a Aminoglycoside Efficacy Enhancer Against Acinetobacter baumannii. *Antibiotics* **2024**, *13*, 1147. [CrossRef]
17. Yount, N.Y.; Yeaman, M.R. Multidimensional signatures in antimicrobial peptides. *Proc. Natl. Acad. Sci. USA* **2004**, *101*, 7363–736 [CrossRef]
18. Baindara, P.; Singh, N.; Ranjan, M.; Nallabelli, N.; Chaudhry, V.; Pathania, G.L.; Sharma, N.; Kumar, A.; Patil, P.B.; Korpole, Laterosporulin10: A novel defensin like class iid bacteriocin from *Brevibacillus* sp. strain SKDU10 with inhibitory activity again microbial pathogens. *Microbiology* **2016**, *162*, 1286–1299. [CrossRef]
19. Cardoso, M.H.; de Lima, L.R.; Pires, A.S.; Maximiano, M.R.; Harvey, P.J.; Freitas, C.G.; Costa, R.A.; Fensterseifer, I.C.M.; Rigueira P.O.; Migliolo, L.; et al. Discovery of Five Classes of Bacterial Defensins: Ancestral Precursors of Defensins from Eukarya? *AC Omega* **2024**, *9*, 45297–45308. [CrossRef]
20. Al Musaimi, O. FDA-Approved Antibacterials and Echinocandins. *Antibiotics* **2025**, *14*, 166. [CrossRef]
21. Cotter, P.D.; Ross, R.P.; Hill, C. Bacteriocins-a viable alternative to antibiotics? *Nat. Rev. Microbiol.* **2013**, *11*, 95–105. [CrossRef] [PubMed]
22. Telhig, S.; Ben Said, L.; Zirah, S.; Fliss, I.; Rebuffat, S. Bacteriocins to Thwart Bacterial Resistance in Gram Negative Bacteria. *Fron Microbiol.* **2020**, *11*, 586433. [CrossRef]
23. Louca, S.; Mazel, F.; Doebeli, M.; Parfrey, L.W. A census-based estimate of earth's bacterial and archaeal diversity. *PLoS Biol.* **201** *17*, e3000106. [CrossRef]
24. Baindara, P.; Mandal, S.M. The interplay of gut-microbiome between infection and inflammation. *Front. Cell. Infect. Microbio* **2024**, *14*, 1413473. [CrossRef]
25. Zeeuwen, P.L.J.M.; Grice, E.A. Skin microbiome and antimicrobial peptides. *Exp. Dermatol.* **2021**, *30*, 1362–1365. [CrossRef]
26. Di, Y.P.; Kuhn, J.M.; Mangoni, M.L. Lung antimicrobial proteins and peptides: From host defense to therapeutic strategies. *Physio Rev.* **2024**, *104*, 1643–1677. [CrossRef] [PubMed]
27. Wan, F.; Torres, M.D.T.; Peng, J.; de la Fuente-Nunez, C. Deep-learning-enabled antibiotic discovery through molecular de extinction. *Nat. Biomed. Eng.* **2024**, *8*, 854–871. [CrossRef]
28. Santos-Júnior, C.D.; Torres, M.D.T.; Duan, Y.; Rodríguez del Río, Á.; Schmidt, T.S.B.; Chong, H.; Fullam, A.; Kuhn, M.; Zhu C.; Houseman, A.; et al. Discovery of antimicrobial peptides in the global microbiome with machine learning. *Cell* **2024** *187*, 3761–3778.e16. [CrossRef]
29. Botelho Sampaio de Oliveira, K.; Lopes Leite, M.; Albuquerque Cunha, V.; Brito da Cunha, N.; Luiz Franco, O. Challenges and advances in antimicrobial peptide development. *Drug Discov. Today* **2023**, *28*, 103629. [CrossRef]

antibiotics

Review

FDA-Approved Antibacterials and Echinocandins

Othman Al Musaimi [1,2]

[1] School of Pharmacy, Newcastle University, Newcastle upon Tyne NE1 7RU, UK; othman.almusaimi@newcastle.ac.uk
[2] Department of Chemical Engineering, Imperial College London, London SW7 2AZ, UK

Abstract: Since 1955, a total of 12 peptide-based drugs with antimicrobial or antifungal properties have received approval from the Food and Drug Administration (FDA). Peptides present a promising opportunity to address serious infections that may be challenging to manage through other means. Peptides exhibit the capability to leverage various mechanisms, and in some cases, multiple mechanisms are employed for this purpose. Despite the initial approval dating back to 1955, the FDA recently approved an echinocandin peptide just last year. The ongoing approvals underscore the significance of peptides in addressing ongoing medical challenges. Approximately 22 peptide therapeutics with an antibacterial and antifungal spectrum are currently undergoing various phases of clinical trials, showing promising results. In this review, antimicrobial and antifungal peptides are analyzed in terms of their chemical structure, indication, mode of action, and development journey, concluding with their arrival in the pharmaceutical market.

Keywords: peptides; pharmaceuticals; FDA; antimicrobial; echinocandin; bacteria; fungus

Academic Editors: Piyush Baindara and Marisa Di Pietro

Received: 21 January 2025
Revised: 1 February 2025
Accepted: 6 February 2025
Published: 7 February 2025

Citation: Al Musaimi, O. FDA-Approved Antibacterials and Echinocandins. *Antibiotics* **2025**, *14*, 166. https://doi.org/10.3390/antibiotics14020166

1. Introduction

The extraordinary growth in the global pharmaceutical industry has now spread to include peptides [1]. Thanks to their tolerable safety profile, biocompatibility, biodegradability, and specificity, they are considered an appealing class of drugs. They have been incorporated in various medical fields, including cardiology [2], oncology [3], wound treatment, and others [4,5]. Therapeutic peptides are well suited to address existing unmet medical challenges, making them excellent complements and sometimes more preferable alternatives to small molecules and very large biologics [6]. With their medium size (500–5000 Da) [7], peptides possess a substantial binding footprint for interacting with therapeutic targets, often outperforming small molecules. This allows them to bind to cell surface receptors and act through agonism/antagonism mechanisms [8]. Additionally, peptides can exert their biological effects through direct translocation into cells, surpassing the capabilities of large biologics [9,10]. Peptides are utilized as targeted therapy to deliver a range of payloads to their therapeutic targets, encompassing small molecules, peptides, and large proteins [3,7,11,12]. A well-known example of a commonly administered peptide is insulin [13]. At present, there are around 120 peptide drugs on the global market, and research into new peptide therapeutics continues at a steady pace, with more than 150 peptides in clinical development and another 400–600 peptides undergoing preclinical studies [14]. These new medicines require pharmaceutical companies to adopt new synthetic strategies to deliver these peptide therapeutics more effectively and in a green and sustainable fashion wherever possible [15].

The UK pharmaceutical industry, valued at USD 53 billion annually, accounts for 2.6% of the global pharmaceutical market. Peptides, monoclonal antibodies, and oligonucleotides constitute the three largest and fastest-growing classes of new biotherapeut products under development. Projections indicate that the peptide therapeutic market w expand from USD 38 billion to USD 106 billion between 2023 and 2033, with a compour annual growth rate (CAGR) of 10.8% [16]. Over the last five years, the FDA has approve 23 peptides for a range of applications, including therapeutic, diagnostic, and, in son instances, theranostic uses, representing almost half of the approvals seen in the previo decade [1,17]. This indicates a growing prominence of peptides, with expectations to secu a significant share by 2030, potentially surpassing the threshold of 60 approvals.

Antimicrobial peptides (AMPs) typically comprise 10–50 amino acid residues, an are found in a variety of organisms, including plants (e.g., *Brassica species*) [18], insec (e.g., fruit fly *Drosophila melanogaster*) [19], amphibians (e.g., African clawed frog *Xenop laevis*) [20], and mammals (e.g., humans *Homo sapiens*) [21]. They exhibit a wide range effectiveness against various microbes including bacteria, viruses, and fungi [22–24]. Th existing antibiotic regimen has been excessively utilized and is now confronted with th challenge of microbial resistance [22]. The surge in multidrug-resistant (MDR) bacteria ha resulted in increased infections and higher mortality rates, and poses a significant threat global health [25,26]. Hence, the advancement of various strategies to counter infectior is steadily underway. Antimicrobial peptides (AMPs) have undergone extensive researc and demonstrate promising potential as antibiotics [27,28]. The primary advantage AMPs lies in their distinct mode of action, specifically their ability to interact with microbi membranes and distinguish between host and pathogenic cells [29]. Their antimicrobi properties and selectivity enable them to effectively target and eliminate bacteria withi mixed cultures. A key reason for their antimicrobial efficacy and safety is that mammalia cells typically feature neutral zwitterionic phospholipid-containing bilayer membrane which lack electrostatic attraction to AMPs. Additionally, the higher cholesterol content i mammalian cells provides another basis for differentiation from bacterial cells [30]. Lipid and proteins form the phospholipid bilayer, the core structure of cell membranes. Pho phatidylcholine (PC) and phosphatidylethanolamine (PE) are typically uncharged, whi phosphatidylserine (PS), cardiolipin (CL), and phosphatidylglycerol (PG) are negativel charged. PS, CL, and PG are common in bacterial pathogens but rare in mammalian men branes, where PC and PE dominate. Eukaryotic membranes contain sterols like cholester (mammals) and ergosterol (fungi), which are scarce in prokaryotes. Gram-negative bacteri have lipopolysaccharide (LPS), and Gram-positive bacteria have lipoteichoic acid, bot adding negative charges. In fungi, negative charges come from phosphomannan an components like phosphatidylinositol (PI), PS, and diphosphatidylglycerol (DPG) [31,32]

Resistance mechanisms observed against traditional antibiotics like β-lactams, aminc glycosides, quinolones, and fluoroquinolones are ineffective against AMPs. Consequentl AMPs hold promise for treating infections caused by antibiotic-resistant microbes [33,34 Moreover, the mode of action of AMPs on microbial membranes suggests that microbia resistance to AMPs is challenging to develop. Microbes would potentially need to underg significant restructuring of their cell membranes to acquire resistance. Given that suc reorganization would require changes to enzymes across various pathways, the likelihoo of rapid adaptation leading to resistance is low.

This review aims to provide insights into FDA-approved analogs of antimicrobial pep tides and echinocandins. The analysis will encompass their chemical structures, indications mode of action, administration routes, development journey, and adverse effects.

2. Antimicrobial Peptides (AMPs)

The FDA has granted approval for seven antimicrobial peptides, categorized as com mon peptide structures, glycopeptides (containing glycosylated cyclic or polycyclic pep tides), lipopeptides (containing a lipid (fat) component), and lipoglycopeptides (combining features of both lipids (fats) and glycopeptides) (Table 1).

Table 1. FDA-approved peptide-based antibiotics.

Peptide (Trade Name)	Indication	Therapeutic Target	Route	FDA Approval Year
	Common peptide structure			
Gramicidin D (Neocidin)	To treat infected surface wounds, and eye, nose, and throat infections	Bacterial membranes	Lotion or ointment	1955
	Glycopeptides			
Vancomycin (Vancocin)	IV: to treat septicemia, infective endocarditis, skin and skin structure infections, bone infections, and lower respiratory tract infections Orally: to treat clostridioides difficile-associated diarrhea and enterocolitis caused by Staphylococcus aureus (including methicillin-resistant strains)	D-alanyl-D-alanine moieties	IV and orally	1958
	Lipopeptides			
Colistin (Coly-Mycin M)	To treat infections due to MDR Gram-negative bacteria	Lipopolysaccharide (LPS)	IV	1959
Daptomycin (Cubicin)	For skin and skin structure infections caused by Gram-positive infections	Bacterial membranes	IV	2003
	Lipoglycopeptides			
Telavancin (Vibativ)	To treat the following infections in adult patients: complicated skin and skin structure infections	D-alanyl-D-alanine moieties	IV	2013
Dalbavancin (Dalvance)	To treat acute bacterial skin and skin structure infections (ABSSSIs)	D-alanyl-D-alanine moieties	IV	2014
Oritavancin (Kimyrsa)	To treat adult patients with acute bacterial skin and skin structure infections	D-alanyl-D-alanine moieties	IV	2015

IV, intravenous; LPS, lipopolysaccharide; MDR, multidrug-resistant.

Glycopeptides consist of a peptide and a sugar unit, with varied classes determined by substituents and the type of residues at positions 1 and 3 of the heptapeptide [35]. These peptides are used in the treatment of severe infections caused by methicillin-resistant *Staphylococcus aureus* (MRSA) or *Enterococcus* bacteria resistant to β-lactams and other antibiotics. Notably, they operate independently of penicillin-binding proteins (PBPs), enabling them to overcome the PBP mutations responsible for MRSA's resistance to β-lactams.

Several antimicrobial agents are currently undergoing different phases of clinical trials (Figure 1) [36]. Cresti and colleagues have provided a concise overview of AMPs progressing through various clinical phases [37].

AMPs can be categorized as membrane-acting or non-membrane-acting. Membrane-acting AMPs primarily target anionic microbial membranes, leading to disruptions in the membrane structure, whereas non-membrane-acting peptides facilitate membrane translocation without causing membrane destruction (Figure 2) [38]. The membrane-targeting mechanisms of AMPs can be explained using various models, such as the carpet model and the pore model. The pore model can be further categorized into the toroidal pore and barrel-stave models. AMPs enter cells through direct penetration or endocytosis.

Once inside the cytoplasm, they identify and interact with specific targets. Based on these targets, AMPs can be categorized into distinct groups [39]. Enninful et al. has extensively detailed the major mechanisms of AMP–membrane interactions [40].

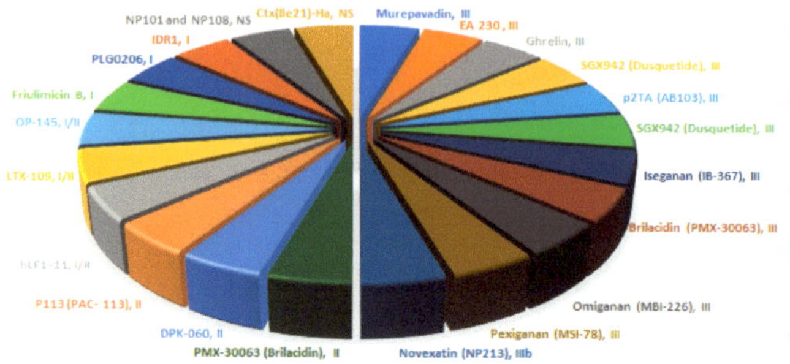

Figure 1. AMPs undergoing clinical trials and their corresponding phase.

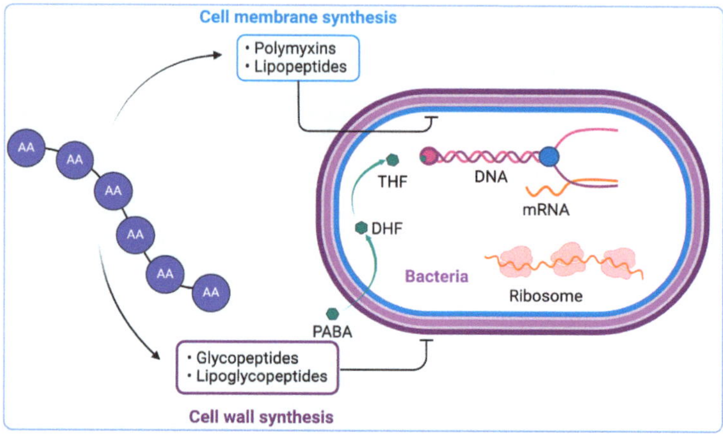

Figure 2. General mechanism of action of AMPs. Lipopeptides disrupt the bacterial cell membrane through electrostatic interactions between their positively charged amino acid residues and the negatively charged membrane components. Glycopeptides and lipoglycopeptides inhibit cell wall synthesis by binding to the tetrapeptide motif, preventing its involvement in peptidoglycan synthesis. Created with www.biorender.com (accessed on 19 January 2025).

Glycopeptides exert their inhibitory effect on cell wall biosynthesis by binding to tetrapeptide chains, hindering their linkage by the PBP enzyme. Specifically, they bind to the D-Ala-D-Ala terminus of *N*-acetylmuramic acid (NAM) and *N*-acetylglucosamine (NAG) peptides, preventing their incorporation into peptidoglycan, a crucial component of the cell wall (Figure 2).

2.1. *Gramicidin D (Neocidin)*

Gramicidin D is a 15-mer linear acid peptide, and it is composed of a heterogeneous mixture of three pore-forming peptides, A (80%), B (5%), and C (15%) (Figure 3) [41,42]. Gramicidin D is used to treat infected surface wounds, as well as eye, nose, and throat infections.

Gramicidin D exhibits a robust binding affinity towards cell membranes, particularly targeting the membranes of Gram-positive bacteria. This interaction disrupts and permeabilizes the membrane, serving as a channel. Consequently, a cascade of detrimental effects ensues, including (i) depletion of intracellular solutes such as K+ and amino acids; (ii) dissipation of the trans-membrane potential; (iii) inhibition of respiration; (iv) reduction in ATP pools; and (v) interference with DNA, RNA, and protein synthesis, ultimately leading to cell death (refer to the colistin mechanism of action, Section 2.3).

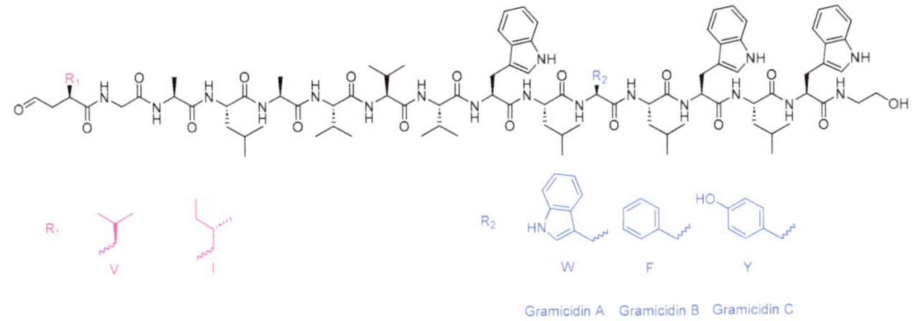

Figure 3. Chemical structure of gramicidin D.

In 1939, gramicidin was initially discovered in the soil bacterium Bacillus brevis [43]. In 1964, the sequence of gramicidin A was elucidated by Reinhard Sarges and Bernhard Witkop [44,45]. Gramicidin was approved by the FDA in 1955. Gramicidin D is not administered intravenously/systemically due to the risk of hemolysis, where significant intake may lead to the rupture of red blood cells. Therefore, it is typically used in the form of a lotion or ointment. Side effects may include redness, burning, stinging, or itching of the eye or ear, as well as blurred vision.

2.2. Vancomycin (Vancocin)

Vancomycin is a tricyclic glycopeptide antibiotic originally derived from the organism *Streptococcus orientalis* (Figure 4). When administered intravenously, it is used to treat conditions such as septicemia, infective endocarditis, skin and skin structure infections, bone infections, and lower respiratory tract infections. When administered orally, it is used to treat clostridioides difficile-associated diarrhea and enterocolitis caused by *Staphylococcus aureus*, including strains that are methicillin-resistant [46].

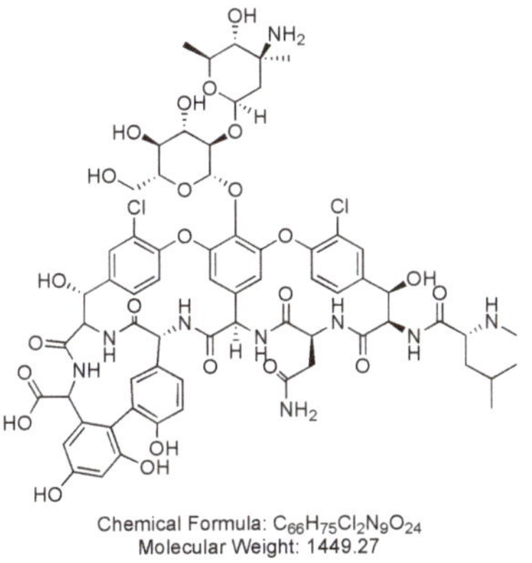

Figure 4. Chemical structure of vancomycin. CID. 14969.

Vancomycin forms hydrogen bonds with the terminal D-alanyl-D-alanine moieties present in the NAM and NAG peptide subunits (Figure 5). This interaction hinders their integration into the peptidoglycan matrix, the primary structural element of Gram-positive cell walls. As a result, the antibiotic disrupts cell wall synthesis, causing changes bacterial–cell–membrane permeability and impeding RNA synthesis [46].

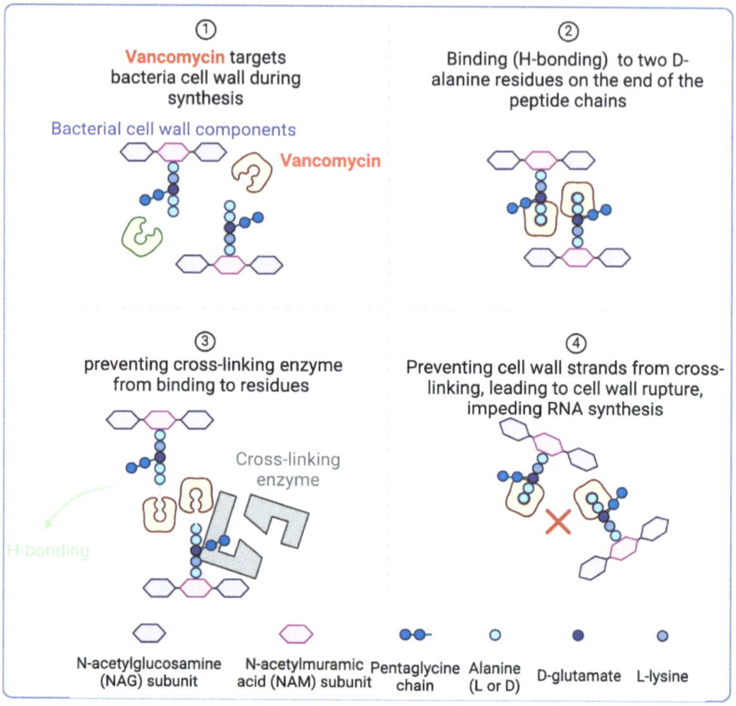

Figure 5. Vancomycin mechanism of action. Vancomycin binds to D-alanyl-D-alanine residue blocking peptidoglycan synthesis in Gram-positive bacteria. This disrupts cell wall formation, alte membrane permeability, and inhibits RNA synthesis, leading to bacterial cell death. Created wit www.biorender.com (accessed on 19 January 2025).

Zhejiang Novus Pharmaceuticals developed vancomycin, and it received FDA ap proval in 1958 [46]. Vancomycin can be administered both orally and intravenously, wit associated side effects such as acute kidney injury, hearing loss, neutropenia, anaphylaxis and vancomycin infusion reaction. The predominant adverse reactions observed with ora administration include nausea, abdominal pain, and hypokalemia [46].

2.3. Colistin (Coly-Mycin M)

Colistin is a 10-mer lipopeptide, a polymyxin antibiotic. It comprises three parts hydrophobic acyl tail, linear tripeptide, and hydrophilic heptapeptide (Figure 6) [47 Colistin is used for the treatment of infections caused by MDR Gram-negative bacteria [48

The mechanism entails the interaction between the cationic cyclic segment of colistin and the anionic lipopolysaccharide (LPS) molecules, resulting in the displacement of Mg^{2} and Ca^{2+} ions from the outer cell membrane of Gram-negative bacteria. Consequently, this interaction induces permeability changes in the cell envelope, ultimately leading to the leakage of cellular contents (Figure 7) [49].

Colistin was developed by JHP Pharmaceuticals and received FDA approval in 1959 [49]. Colistin is administered intravenously and may entail side effects such as gastrointestinal upset, tingling of extremities and tongue, slurred speech, dizziness, vertigo paresthesia, generalized itching, urticaria, rash, and fever [48].

Polymyxin B and colistin both belong to the polymyxin class, and they are used for similar indications. Polymyxin B received FDA approval in 1964 [50]. Polymyxin B can be administered via several routes including intramuscular, intravenous drip, intrathecal, or ophthalmic use. However, it is associated with various adverse effects such as nephrotoxicity (kidney damage), neurotoxicity (nerve damage), drug fever, urticarial rash (hives), and pain at injection sites. Thrombophlebitis (inflammation of veins) can also occur at intravenous injection sites [50].

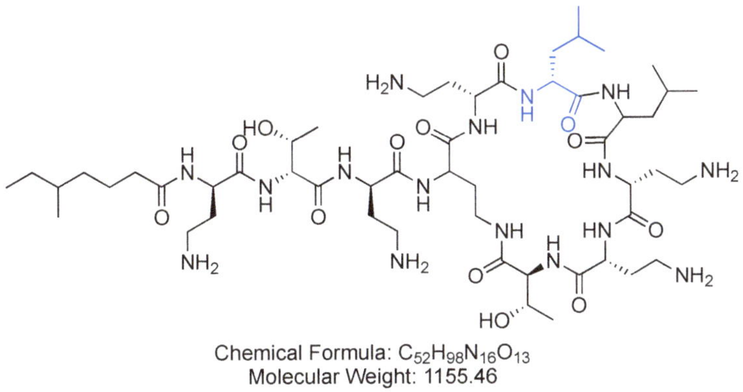

Figure 6. Chemical structure of colistin. Blue—in polymyxin b, a D-Phe residue replaces the D-Leu residue. CID. 5311054.

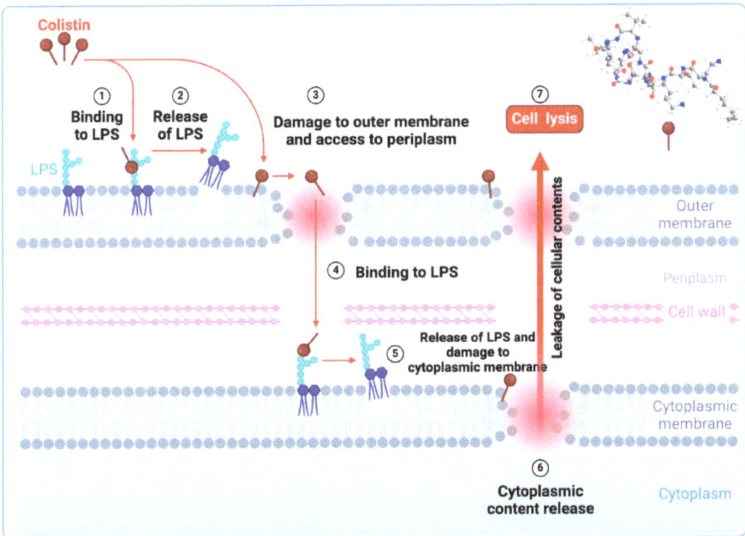

Figure 7. Colistin mechanism of action. Colistin binds LPS, displacing Mg^{2+} and Ca^{2+}, disrupting membrane integrity, and causing cell leakage in Gram-negative bacteria. Created with www.biorender.com (accessed on 19 January 2025).

2.4. Daptomycin (Cubicin)

Daptomycin is a cyclic lipopeptide, derived from the fermentation of *Streptomyces roseosporus* (Figure 8). Daptomycin is used for the treatment of skin and skin structure infections caused by Gram-positive bacteria, *S. aureus* bacteremia, and right-sided *S. aureus* endocarditis.

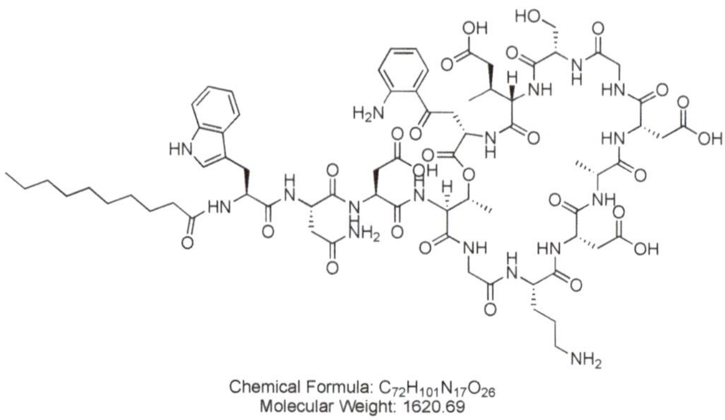

Figure 8. Chemical structure of daptomycin. CID. 21585658.

Daptomycin operates through a distinct mechanism of action by binding to bacterial membranes, inducing rapid depolarization of their potential through ion leakage [51]. This action leads to the inhibition of protein, DNA, and RNA synthesis, ultimately resulting in bacterial death (Figure 9) [52,53]. Huang conducted a more in-depth exploration to elucidate its distinctive mechanism [54]. Daptomycin forms a $Dap_2Ca_3PG_2$ complex in the membrane, acting as a transient ionophore. It binds and releases ions at the membrane boundary, facilitating ion transport from high to low concentrations. Mobility is key to its function, with only small dimeric complexes likely acting as ionophores. Over time, daptomycin aggregates grow, eventually exiting the membrane in a lipid-extracting effect [55], supporting its transient ionophore role [54].

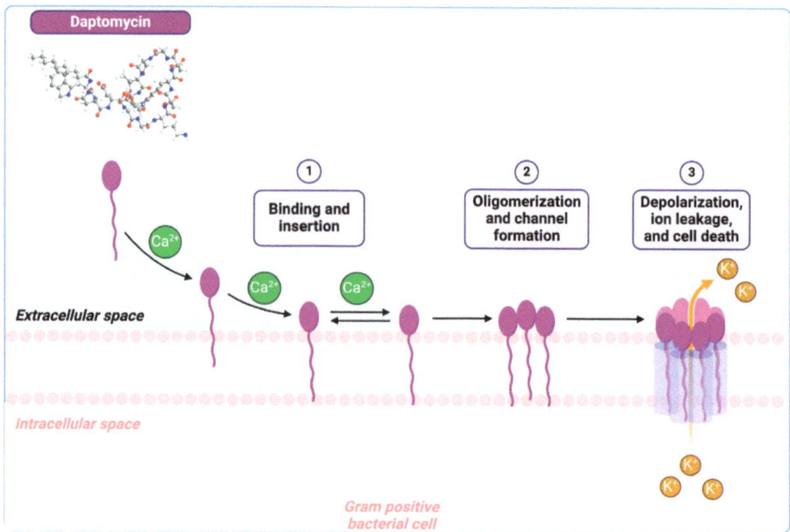

Figure 9. Daptomycin mechanism of action. Daptomycin binds bacterial membranes, causing depolarization through ion leakage, which inhibits protein, DNA, and RNA synthesis, leading to cell death. Created with www.biorender.com (accessed on 19 January 2025).

Daptomycin was first discovered by in 1980s by researchers at Eli Lilly and company in soil samples from Mount Ararat in Turkey [53]. Subsequently, it was developed by Cubist Pharmaceuticals and obtained FDA approval in 2003 [56]. Daptomycin is administered intravenously and may elicit adverse effects such as anemia, anxiety, asthenia, constipation, diarrhea, dizziness, fever, flatulence, gastrointestinal discomfort, headache, hypertension, hypotension, increased risk of infection, insomnia, nausea, pain, skin reactions, and vomiting [52].

2.5. Telavancin (Vibativ)

Telavancin is a 7-mer lipoglycopeptide, derived from vancomycin. It consists of a lipophilic side chain (decylaminoethyl) attached to an amino sugar, and a hydrophilic moiety (phosphonomethyl aminomethyl) at the 4′-position of amino acid 7, anchored to a vancosamine sugar (Figure 10). Telavancin is used in the treatment of various infections in adult patients, including complicated skin and skin structure infections (CSSSIs), as well as hospital-acquired and ventilator-associated bacterial pneumonia (HABP/VABP) caused by *Staphylococcus aureus* [57].

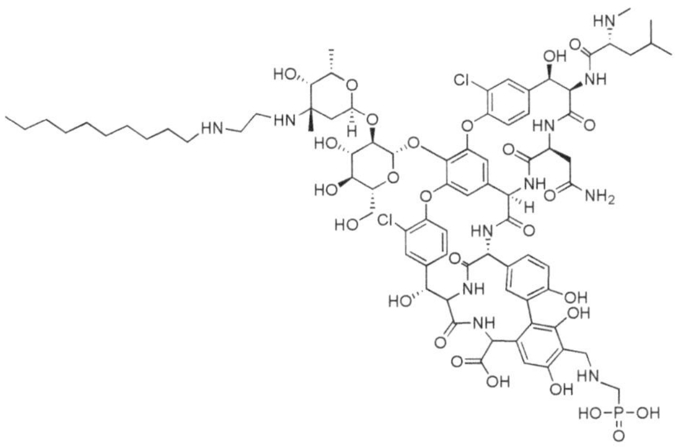

Chemical Formula: $C_{80}H_{106}Cl_2N_{11}O_{27}P$
Molecular Weight: 1755.65

Figure 10. Chemical structure of telavancin. CID. 3081362.

Telavancin binds to late-stage peptidoglycan precursors, including lipid II, thereby preventing the polymerization of NAM and NAG, as well as the cross-linking of peptidoglycan by binding to D-Ala-D-Ala. This dual action ultimately leads to the inhibition of cell wall synthesis. Additionally, telavancin binds to the cell membrane, disrupting its function as a barrier [57,58].

Telavancin was developed by Theravance and received FDA approval in 2013 [59]. Telavancin is administered intravenously and is associated with adverse effects such as diarrhea, taste disturbance, nausea, vomiting, and discoloration of urine [57].

2.6. Dalbavancin (Dalvance)

Dalbavancin is a semisynthetic lipoglycopeptide consisting of a mixture of five closely related active homologs (A0, A1, B0, B1, and B2). Among these, component B0 is the major constituent of dalbavancin. The variations lie in the fatty acid side chain of the *N*-acylaminoglucuronic acid moiety (R1) structure and/or the presence of an additional methyl group (R2) on the terminal amino group (Figure 11). Dalbavancin is used in the treatment of acute bacterial skin and skin structure infections (ABSSSIs) caused by designated susceptible strains of Gram-positive microorganisms [60].

Dalbavancin has a mechanism of action similar to vancomycin (Figure 5). It disrupts cell wall synthesis by binding to the D-alanyl-D-alanine terminus of the stem pentapeptide in nascent cell wall peptidoglycan. This interaction prevents cross-linking, ultimately halting the process of cell wall synthesis [60,61].

Dalbavancin was developed by Durata Therapeutics and received FDA approval 2014 [62]. Dalbavancin is administered intravenously and may be associated with adver effects such as nausea, headache, and diarrhea [60].

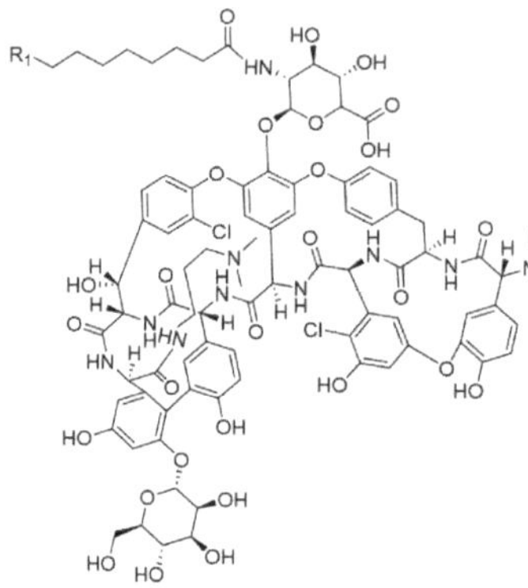

Homolog	R1	R2
A0	$CH(CH_3)_2$	H
A1	$CH_2CH_2CH_3$	H
B0	$CH_2CH(CH_3)_2$	H
B1	$CH_2CH_2CH_2CH_3$	H
B2	$CH_2CH(CH_3)_2$	CH_3

Figure 11. Chemical structure of dalbavancin.

2.7. *Oritavancin (Kimyrsa)*

Oritavancin is a lipoglycopeptide antibacterial drug that distinguishes itself fro vancomycin through the incorporation of an aromatic lipophilic side chain and an unsu stituted sugar (Figure 12). Oritavancin is used in the treatment of adult patients with acu bacterial skin and skin structure infections caused or suspected to be caused by susceptib isolates of designated Gram-positive microorganisms [63].

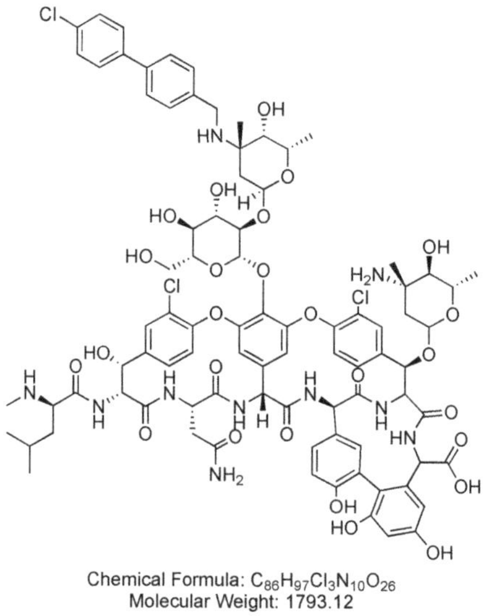

Figure 12. Chemical structure of oritavancin. CID. 16136912.

Oritavancin exerts its effects through three mechanisms: (i) inhibition of the transg lycosylation (polymerisation) step in cell wall synthesis by binding to the stem D-alany D-alanine peptide of peptidoglycan precursors, (ii) inhibition of the transpeptidatio (cross-linking) step in cell wall biosynthesis by binding to the peptide bridging segments o the cell wall, and (iii) disruption of bacterial membrane integrity, resulting in depolarization permeabilization, and ultimately, cell death [63,64].

Oritavancin was developed by Melinta Therapeutics and obtained FDA approval in 2015 [65]. Oritavancin is administered intravenously and may be associated with side effects such as headache, nausea, vomiting, limb and subcutaneous abscesses, and diarrhea [63].

3. Echinocandins Analogs

Echinocandins analogs play a crucial role in targeting and inhibiting *β*-(1,3)-D-glucan synthase, specifically at the Fks1p extracellular subdomain [66]. This subdomain is an integral part responsible for constructing the fungal cell wall [67]. The inability of the organism to synthesize *β*-(1,3)-D-glucan results in osmotic instability and eventual cell death (Figure 13) [68–71].

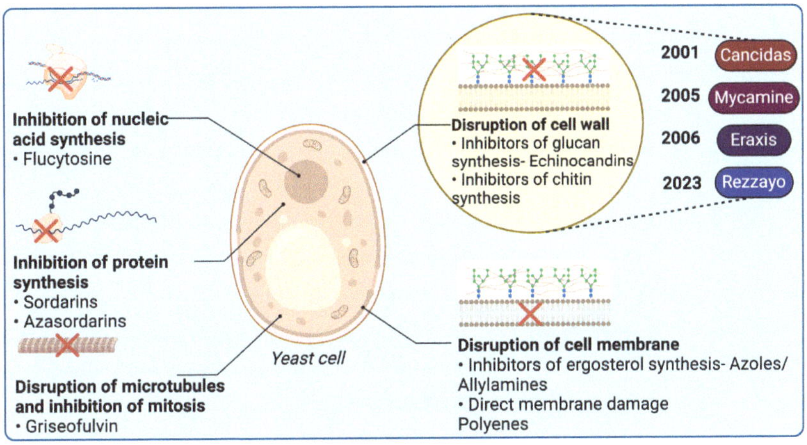

Figure 13. Echinocandins sites of action. Echinocandins inhibit β-(1,3)-D-glucan synthase at the Fks1p subdomain, disrupting fungal cell wall synthesis, leading to osmotic instability and cell death. Created with www.biorender.com (accessed on 19 January 2025).

These analogs are semisynthetic cyclic lipopeptides exhibiting antifungal activity. They are acylated with different fatty acids attached to the α-amino group of dihydroxyornithine, facilitating the attachment of the drug to the cell membrane of the therapeutic target [72].

This class of drugs traces its roots back to 1992 when caspofungin was initially synthesized from pneumocandin B_0 and subsequently approved for clinical trials [73]. Since 2001, the FDA has approved four analogs from this class (Table 2).

Table 2. FDA-approved echinocandin peptide analogs.

Peptide (Trade Name)	Indication	Therapeutic Target	Route	FDA Approval Year
Caspofungin (Cancidas)	To treat serious fungal infections	*β*-1,3-D-glucan synthase	IV	2001
Micafungin (Mycamine)	To help the body overcome serious fungus infections, such as candidemia			2005
Anidulafungin (Eraxis)	To treat patients with esophageal candidiasis			2006
Rezafungin (Rezzayo)	To treat patients 18 years of age or older who have limited or no alternative options for the treatment of candidemia and invasive candidiasis			2023

IV, intravenous.

3.1. *Caspofungin (Cancidas)*

Caspofungin is a cyclic lipopeptide and an echinocandin antifungal agent that fur tions as a β-1,3-D-glucan synthase inhibitor (Figure 14). Caspofungin is used in the tre ment of severe fungal infections, encompassing conditions such as candidemia (fung infection in the blood), esophageal candidiasis (fungal infection of the esophagus), oth candida infections, and aspergillosis (fungal infection in the lungs) [74].

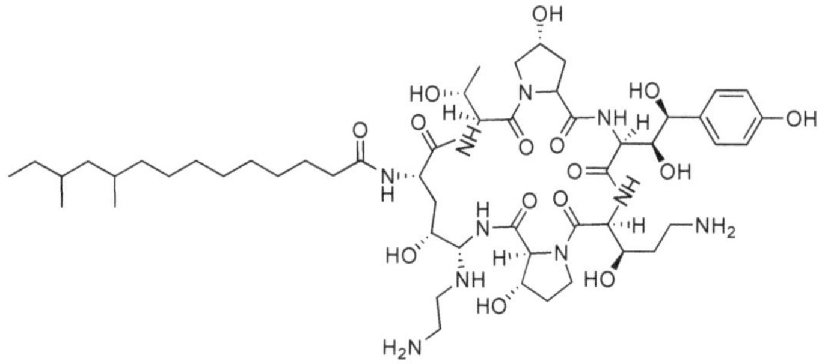

Chemical Formula: $C_{52}H_{88}N_{10}O_{15}$
Molecular Weight: 1093.33

Figure 14. Chemical structure of caspofungin. CID. 16119814.

Caspofungin inhibits β-1,3-D-glucan synthase, thereby halting the production c β-1,3-D-glucan, a crucial component of fungal cell walls. This disruption compromises th structural integrity of the fungal cell walls, ultimately impeding fungal growth and leadin to the control of the infection [75].

Caspofungin was developed by Merck Laboratories and became the first withi the echinocandin class to receive FDA approval in 2001 [76]. Caspofungin is admir istered intravenously and may be associated with adverse effects such as chills, feve phlebitis/thrombophlebitis, tachycardia, nausea, vomiting, rash, abdominal pain, headach and diarrhea [74].

3.2. *Micafungin (Mycamine)*

Micafungin is a cyclic lipopeptide and an echinocandin antifungal agent that acts as β-1,3-D-glucan synthase inhibitor. Micafungin is utilized to assist the body in overcomin severe fungal infections, including conditions such as candidemia (Figure 15) [77].

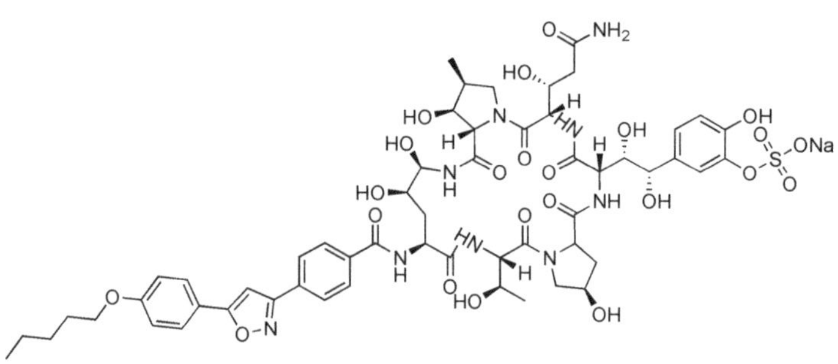

Chemical Formula: $C_{56}H_{70}N_9NaO_{23}S$
Molecular Weight: 1292.27

Figure 15. Chemical structure of micafungin. CID. 477468.

Micafungin functions by inhibiting β-1,3-D-glucan synthase, disrupting the productior of β-1,3-D-glucan, a crucial component of fungal cell walls. This disruption compromises the structural integrity of the fungal cell walls, ultimately leading to the control of the fungal infection [77,78].

It was developed by Fujisawa Healthcare and received FDA approval in 2005 [79]. Micafungin is administered intravenously and may be associated with adverse effects such as anxiety, black or tarry stools, bleeding gums, bloating or swelling of the face, arms, hands, lower legs, or feet, cold sweats, coma, and cool, pale skin. Additionally, micafungin can lead to a decreased frequency or amount of urine [77].

3.3. Anidulafungin (Eraxis)

Anidulafungin is a cyclic lipopeptide and an echinocandin antifungal agent that operates as a β-1,3-D-glucan synthase inhibitor (Figure 16). Anidulafungin is used in the treatment of patients with esophageal candidiasis [75].

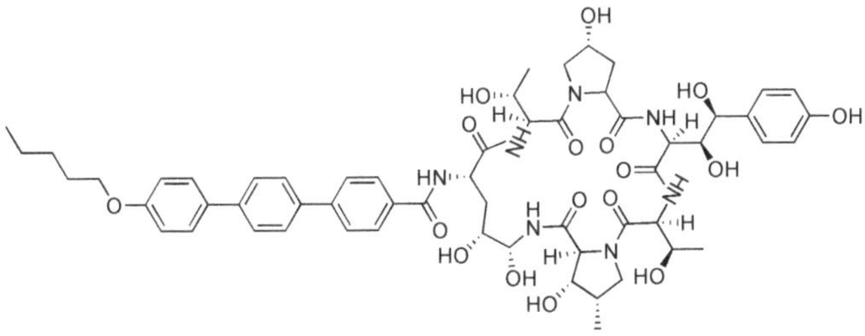

Figure 16. Chemical structure of anidulafungin. CID. 166548.

Anidulafungin inhibits β-1,3-D-glucan synthase, disrupting the production of β-1,3-D-glucan, which is an essential component of fungal cell walls. This interference compromises the structural integrity of the fungal cell walls, leading to the control of the fungal infection [75].

Anidulafungin was developed by Vicuron Pharmaceuticals and received FDA approval in 2006 [80]. Anidulafungin is administered intravenously, may be associated with adverse effects including black or tarry stools, chills, decreased urine, fever, increased thirst, irregular heartbeat, and lower back or side pain. Additionally, anidulafungin might lead to mood or mental changes [81].

3.4. Rezafungin (Rezzayo)

Rezafungin is a semisynthetic echinocandin antifungal lipopeptide designed for the treatment of patients aged 18 years or older, particularly those with limited or no alternative options for managing candidemia and invasive candidiasis. Its structure is analogous to that of anidulafungin, with the exception that an OH group in the latter is replaced by a (trimethylammonio)ethoxy moiety in the former (Red) (Figure 17) [82].

Rezafungin functions by inhibiting the 1,3-β-D-glucan synthase enzyme complex found in fungal cells. This inhibition results in the prevention of the formation of 1,3-β-D-glucan, a crucial component of the fungal cell wall in various fungi, including Candida species. The disruption of 1,3-β-D-glucan synthesis contributes to the compromised integrity of the fungal cell wall, ultimately inhibiting fungal growth and aiding in the treatment of infections [82].

Rezafungin was developed by Cidara Therapeutics Inc., and Melinta Therapeutics LLC obtained an exclusive license to commercialize it. Rezafungin received FDA approval in 2023 [83]. Rezafungin is administered intravenously and has demonstrated several adverse effects, including hypokalemia, pyrexia, diarrhea, anemia, vomiting, nause hypomagnesemia, abdominal pain, constipation, and hypophosphatemia [82].

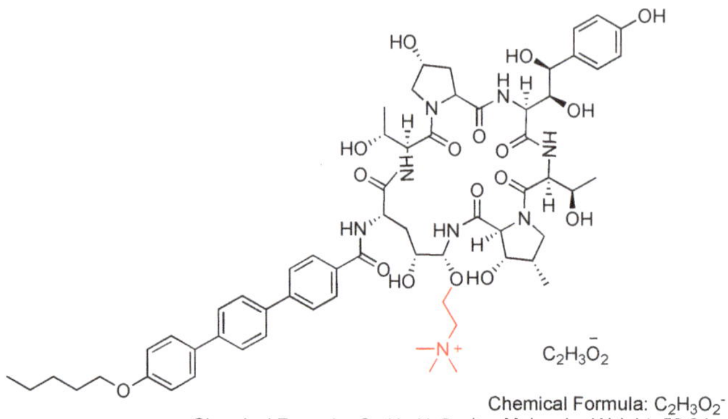

Figure 17. Chemical structure of rezafungin acetate. Difference from anidulafungin is shown in re CID. 78318119.

4. Model Architecture and Methodology

Key experimental databases provide detailed functional activity annotations, suc as the Antimicrobial Peptide Database (APD3) (containing 5099 peptides from divers sources) [83], dbAMP 2.0 (featuring 26,447 AMPs and 2262 antimicrobial proteins fro 3044 organisms) [84], LAMP (with 3904 natural and 1643 synthetic AMPs) [85], and AVPd (including 2683 peptides, 624 of which are modified and tested for antiviral activity). Thes resources are vital for advancing AMP research and discovery.

Computational methods have emerged as a central focus in bioinformatics, wit machine learning-based approaches playing a key role in the identification of AMP Advanced techniques such as deep learning and feature extraction have been leveraged t develop numerous models and algorithms, improving the accuracy and efficiency of AM prediction. Additionally, optimization methods like genetic algorithms, early linguisti models, and QSAR-based models have been employed to generate and optimize AMP further advancing their discovery and design [86].

AI methods like CAMPr3, iAMPpred, AmPEPpy, AntiBP2, and CS-AMPPred ar widely used for antimicrobial peptide (AMP) discovery. Machine learning (ML) an deep learning (DL) techniques, including random forests, neural networks (NNs), Hidde Markov models, and deep-AMPpred—a two-stage AMP predictor—analyze bacterial phe notypic fingerprints to predict mechanisms of action, estimate antibiotic potency, and asses phenotypic changes. These tools also support the creation of AMP discovery database advancing antibiotic research [87,88].

5. Conclusions

Peptides have played a crucial role in the pharmaceutical arena for almost seve decades, showcasing their significance. Antimicrobial peptides, for instance, contribut to combating bacterial infections through diverse mechanisms. Glycopeptides have mad significant strides in treating MRSA by circumventing PBP mutations. Another noteworth class, echinocandins, proves effective in treating serious fungal infections. The primary rol of peptides as antimicrobial agents is attributed to their mechanism of action, positionin them as a last resort in this field. Peptides predominantly target microbial membrane thereby circumventing known resistance pathways. Simultaneously, microbes woul need to undergo a complete restructuring of their membrane to develop resistance to antimicrobial peptides, an occurrence considered unlikely. Even if resistance were to emerge, the process would likely require an extended period for development.

Despite reluctance in developing new antimicrobial agents, the FDA's focus on antibiotics reflects the urgency of addressing serious infections. Zevtera, a cephalosporin-based drug approved in 2024, is now approved for treating bloodstream infections, bacterial skin and soft tissue infections, and community-acquired bacterial pneumonia, emphasizing the critical need for innovative solutions [89].

Peptides hold promise through the continuous isolation of new ones exhibiting potent activity against microbes. Moreover, the diverse structures of peptides enable them to exploit various mechanisms, making them versatile agents for addressing both bacterial and fungal infections. In contrast to their small molecule counterparts, peptides can interact with their targets through a larger interface and exert their activity through multiple synergistic mechanisms. Currently, there are around 22 peptide-based therapeutics that are undergoing clinical trials across different phases. These peptides have shown promising results in combating bacterial and fungal infections with a broad spectrum of activity. Moreover, numerous lasso peptides, such as Microcin J25 (MccJ25) and capistruin, demonstrate promising activity against antibiotic-resistant bacteria and hold potential for various applications [90].

Experimental databases, computational tools, and AI-based toolkits are invaluable resources for identifying, engineering, and delivering AMPs capable of overcoming resistance challenges. In addition, the development of a comprehensive computational tool capable of analyzing multiple aspects and activities of AMPs will significantly enhance the discovery of potential AMPs with broad-spectrum capabilities.

Funding: This research received no external funding.

Institutional Review Board Statement: Not applicable.

Informed Consent Statement: Not applicable.

Data Availability Statement: Data sharing is not applicable.

Conflicts of Interest: The authors declare no conflict of interest.

References

1. Al Shaer, D.; Al Musaimi, O.; Albericio, F.; de la Torre, B.G. 2023 FDA TIDES (Peptides and Oligonucleotides) Harvest. *Pharmaceuticals* **2024**, *17*, 243. [CrossRef]
2. Al Musaimi, O. FDA's stamp of approval: Unveiling peptide breakthroughs in cardiovascular diseases, ACE, HIV, CNS, and beyond. *J. Pept. Sci.* **2024**, *30*, e3627. [CrossRef]
3. Al Musaimi, O. Peptide Therapeutics: Unveiling the Potential against Cancer—A Journey through 1989. *Cancers* **2024**, *16*, 1032. [CrossRef]
4. de la Torre, B.G.; Albericio, F. The Pharmaceutical Industry in 2023: An Analysis of FDA Drug Approvals from the Perspective of Molecules. *Molecules* **2024**, *29*, 585. [CrossRef]
5. Mullard, A. 2023 FDA approvals. *Nat. Rev.* **2024**, *23*, 88–95. [CrossRef]
6. Wang, L.; Wang, N.; Zhang, W.; Cheng, X.; Yan, Z.; Shao, G.; Wang, X.; Wang, R.; Fu, C. Therapeutic peptides: Current applications and future directions. *Signal Transduct. Target Ther.* **2022**, *7*, 48. [CrossRef]
7. Al Musaimi, O.; Lombardi, L.; Williams, D.R.; Albericio, F. Strategies for Improving Peptide Stability and Delivery. *Pharmaceuticals* **2022**, *15*, 1283. [CrossRef]
8. Li, C.M.; Haratipour, P.; Lingeman, R.G.; Perry, J.J.P.; Gu, L.; Hickey, R.J.; Malkas, L.H. Novel Peptide Therapeutic Approaches for Cancer Treatment. *Cells* **2021**, *10*, 2908. [CrossRef]
9. Ruseska, I.; Zimmer, A. Internalization mechanisms of cell-penetrating peptides. *Beilstein J. Nanotechnol.* **2020**, *11*, 101–123. [CrossRef]
10. Lamers, C. Overcoming the shortcomings of peptide-based therapeutics. *Future Drug Discov.* **2022**, *4*, FDD75. [CrossRef]

11. Derakhshankhah, H.; Jafari, S. Cell penetrating peptides: A concise review with emphasis on biomedical applications. *Biome Pharmacother.* **2018**, *108*, 1090–1096. [CrossRef]
12. Al Musaimi, O.; Morse, S.V.; Lombardi, L.; Serban, S.; Basso, A.; Williams, D.R. Successful synthesis of a glial-specific blood–bra barrier shuttle peptide following a fragment condensation approach on a solid-phase resin. *J. Pept. Sci.* **2022**, *29*, e3448. [CrossRe
13. Rosenfeld, L. Insulin: Discovery and controversy. *Clin. Chem.* **2002**, *48*, 2270–2288. [CrossRef]
14. Ferrazzano, L.; Catani, M.; Cavazzini, A.; Martelli, G.; Corbisiero, D.; Cantelmi, P.; Fantoni, T.; Mattellone, A.; De Luca, C.; Fellet S.; et al. Sustainability in peptide chemistry: Current synthesis and purification technologies and future challenges. *Green Che* **2022**, *24*, 975–1020. [CrossRef]
15. Sharma, A.; Kumar, A.; de la Torre, B.G.; Albericio, F. Liquid-Phase Peptide Synthesis (LPPS): A Third Wave for the Preparatio of Peptides. *Chem. Rev.* **2022**, *122*, 13516–13546. [CrossRef]
16. Unveiling the Phenomenal Rise of Peptide Therapeutics: A Whopping 10.8% CAGR Propels Market Value to US$ 106.0 Billio by 2033. 2023. Available online: https://www.pharmiweb.com/press-release/2023-07-25/unveiling-the-phenomenal-rise-o peptide-therapeutics-a-whopping-108-cagr-propels-market-value-to (accessed on 20 January 2025).
17. de la Torre, B.G.; Albericio, F. The Pharmaceutical Industry in 2024: An Analysis of the FDA Drug Approvals from the Perspectiv of Molecules. *Molecules* **2025**, *30*, 482. [CrossRef]
18. Cao, H.; Ke, T.; Liu, R.; Yu, J.; Dong, C.; Cheng, M.; Huang, J.; Liu, S. Identification of a Novel Proline-Rich Antimicrobial Peptid from Brassica napus. *PLoS ONE* **2015**, *10*, e0137414. [CrossRef]
19. Imler, J.L.; Bulet, P. Antimicrobial peptides in Drosophila: Structures, activities and gene regulation. *Chem. Immunol. Allergy* **200** *86*, 1–21. [CrossRef]
20. McMillan, K.A.M.; Coombs, M.R.P. Review: Examining the Natural Role of Amphibian Antimicrobial Peptide Magainin. *Molecul* **2020**, *25*, 5436. [CrossRef]
21. Satchanska, G.; Davidova, S.; Gergova, A. Diversity and Mechanisms of Action of Plant, Animal, and Human Antimicrobi Peptides. *Antibiotics* **2024**, *13*, 202. [CrossRef]
22. Sethi, N. *Antimicrobial Peptides Small Molecules as Antibiotics Substitute, in Emerging Modalities in Mitigation of Antimicrobial Resistanc* Akhtar, N., Singh, K.S., Prerna, Goyal, D., Eds.; Springer International Publishing: Cham, Switzerland, 2022; pp. 261–289.
23. Qureshi, A. A review on current status of antiviral peptides. *Discov. Viruses* **2025**, *2*, 3. [CrossRef]
24. Willyard, C. The drug-resistant bacteria that pose the greatest health threats. *Nature* **2017**, *543*, 15. [CrossRef]
25. Petty, N.K.; Ben Zakour, N.L.; Stanton-Cook, M.; Skippington, E.; Totsika, M.; Forde, B.M.; Phan, M.-D.; Gomes Moriel, D.; Peter K.M.; Davies, M.; et al. Global dissemination of a multidrug resistant Escherichia coli clone. *Proc. Natl. Acad. Sci. USA* **2014**, *11* 5694–5699. [CrossRef]
26. Jaradat, D.s.M.M.; Saleh, K.K.Y.; Za'arir, B.H.M.; Arafat, T.; Alzoubi, K.H.; Al-Taweel, S.A.; Mallah, E.; Haddad, M.A.; Haimu B.A.K. Correction to: Solid-Phase Synthesis and Antibacterial Activity of an Artificial Cyclic Peptide Containing Two Disulfid Bridges. *Int. J. Pept. Res. Ther.* **2019**, *25*, 1103. [CrossRef]
27. Jaradat, D.s.M.M.; Al-Karablieh, N.; Zaarer, B.H.M.; Li, W.; Saleh, K.K.Y.; Rasras, A.J.; Abu-Romman, S.; O'Brien-Simpson, N.M Wade, J.D. Human glucose-dependent insulinotropic polypeptide (GIP) is an antimicrobial adjuvant re-sensitising multidrug resistant Gram-negative bacteria. *Biol. Chem.* **2021**, *402*, 513–524. [CrossRef]
28. Gorr, S.-U.; Abdolhosseini, M. Antimicrobial peptides and periodontal disease. *J. Clin. Periodontol.* **2011**, *38* (Suppl. S11), 126–14 [CrossRef]
29. Glukhov, E.; Stark, M.; Burrows, L.L.; Deber, C.M. Basis for selectivity of cationic antimicrobial peptides for bacterial versu mammalian membranes. *J. Biol. Chem.* **2005**, *280*, 33960–33967. [CrossRef]
30. Ebenhan, T.; Gheysens, O.; Kruger, H.G.; Zeevaart, J.R.; Sathekge, M.M. Antimicrobial Peptides: Their Role as Infection-Selectiv Tracers for Molecular Imaging. *Biomed. Res. Int.* **2014**, *2014*, 867381. [CrossRef]
31. Zhang, Q.-Y.; Yan, Z.-B.; Meng, Y.-M.; Hong, X.-Y.; Shao, G.; Ma, J.-J.; Cheng, X.-R.; Liu, J.; Kang, J.; Fu, C.-Y. Antimicrobia peptides: Mechanism of action, activity and clinical potential. *Mil. Med. Res.* **2021**, *8*, 48. [CrossRef]
32. Tummanapalli, S.S.; Willcox, M.D. Antimicrobial resistance of ocular microbes and the role of antimicrobial peptides. *Clin. Exp Optom.* **2021**, *104*, 295–307. [CrossRef]
33. Bechinger, B.; Gorr, S.U. Antimicrobial Peptides: Mechanisms of Action and Resistance. *J. Dent. Res.* **2017**, *96*, 254–260. [CrossRef
34. Butler, M.S.; Hansford, K.A.; Blaskovich, M.A.T.; Halai, R.; Cooper, M.A. Glycopeptide antibiotics: Back to the future. *J. Antibiot* **2014**, *67*, 631–644. [CrossRef]
35. Asif, F.; Zaman, S.U.; Arnab, M.K.H.; Hasan, M.; Islam, M.M. Antimicrobial peptides as therapeutics: Confronting delivery challenges to optimize efficacy. *Microbe* **2024**, *2*, 100051. [CrossRef]
36. Cresti, L.; Cappello, G.; Pini, A. Antimicrobial Peptides towards Clinical Application-A Long History to Be Concluded. *Int. J. Mo Sci.* **2024**, *25*, 4870. [CrossRef]
37. Hancock, E.W.R.; Patrzykat, A. Clinical Development of Cationic Antimicrobial Peptides: From Natural to Novel Antibiotics *Curr. Drug Targets Infect. Disord.* **2002**, *2*, 79–83. [CrossRef]

28. Huan, Y.; Kong, Q.; Mou, H.; Yi, H. Antimicrobial Peptides: Classification, Design, Application and Research Progress in Multiple Fields. *Front. Microbiol.* **2020**, *11*, 582779. [CrossRef]
29. Enninful, G.N.; Kuppusamy, R.; Tiburu, E.K.; Kumar, N.; Willcox, M.D.P. Non-canonical amino acid bioincorporation into antimicrobial peptides and its challenges. *J. Pept. Sci.* **2024**, *30*, e3560. [CrossRef]
30. Hotchkiss, R.D.; Dubos, R.J. The isolation of Bactericidal Substances from Cultures of Bacillus Brevis. *J. Biol. Chem.* **1941**, *141*, 155–162. [CrossRef]
31. Chen, C.H.; Lu, T.K. Development and Challenges of Antimicrobial Peptides for Therapeutic Applications. *Antibiotics* **2020**, *9*, 24. [CrossRef]
32. Dubos, R.J. Studies on a bactericidal agent extracted from a soil bacillus: I. preparation of the agent. its activity in vitro. *J. Exp. Med.* **1939**, *70*, 1–10. [CrossRef]
33. Sarges, R.; Witkop, B.; Gramicidin, A., IV. Primary Sequence of Valine and Isoleucine Gramicidin A. *J. Am. Chem. Soc.* **1964**, *86*, 1862–1863. [CrossRef]
34. Meikle, T.G.; Conn, C.E.; Separovic, F.; Drummond, C.J. Exploring the structural relationship between encapsulated antimicrobial peptides and the bilayer membrane mimetic lipidic cubic phase: Studies with gramicidin A′. *RSC Adv.* **2016**, *6*, 68685–68694. [CrossRef]
35. Vancomycin Drug Label. Available online: https://www.accessdata.fda.gov/drugsatfda_docs/label/2023/210274s000lbl.pdf (accessed on 20 January 2025).
36. Rhouma, M.; Beaudry, F.; Thériault, W.; Letellier, A. Colistin in Pig Production: Chemistry, Mechanism of Antibacterial Action, Microbial Resistance Emergence, and One Health Perspectives. *Front. Microbiol.* **2016**, *7*, 1789. [CrossRef]
37. Colistin Approval Letter and Drug Label. Available online: https://www.accessdata.fda.gov/drugsatfda_docs/label/2013/050108s030lbl.pdf (accessed on 20 January 2025).
38. El-Sayed Ahmed, M.A.E.; Zhong, L.L.; Shen, C.; Yang, Y.; Doi, Y.; Tian, G.B. Colistin and its role in the Era of antibiotic resistance: An extended review (2000–2019). *Emerg. Microbes Infect.* **2020**, *9*, 868–885. [CrossRef]
39. Polymyxin B Drug Label. Available online: https://www.accessdata.fda.gov/drugsatfda_docs/label/2012/060716s018lbl.pdf (accessed on 20 January 2025).
40. Hover, B.M.; Kim, S.-H.; Katz, M.; Charlop-Powers, Z.; Owen, J.G.; Ternei, M.A.; Maniko, J.; Estrela, A.B.; Molina, H.; Park, S.; et al. Culture-independent discovery of the malacidins as calcium-dependent antibiotics with activity against multidrug-resistant Gram-positive pathogens. *Nat. Microbiol.* **2018**, *3*, 415–422. [CrossRef]
41. Daptomycin Drug Label. Available online: https://www.accessdata.fda.gov/drugsatfda_docs/nda/2003/21-572_Cubicin_Prntlbl.pdf (accessed on 20 January 2025).
42. Karas, J.A.; Carter, G.P.; Howden, B.P.; Turner, A.M.; Paulin, O.K.A.; Swarbrick, J.D.; Baker, M.A.; Li, J.; Velkov, T. Structure-Activity Relationships of Daptomycin Lipopeptides. *J. Med. Chem.* **2020**, *63*, 13266–13290. [CrossRef]
43. Huang, H.W. DAPTOMYCIN, its membrane-active mechanism vs. that of other antimicrobial peptides. *Biochim. Et Biophys. Acta (BBA) Biomembr.* **2020**, *1862*, 183395. [CrossRef]
44. Chen, Y.-F.; Sun, T.-L.; Sun, Y.; Huang, H.W. Interaction of Daptomycin with Lipid Bilayers: A Lipid Extracting Effect. *Biochemistry* **2014**, *53*, 5384–5392. [CrossRef]
45. Daptomycin Approval Letter. Available online: https://www.accessdata.fda.gov/drugsatfda_docs/nda/2003/21-572_Cubicin_Approv.pdf (accessed on 20 January 2025).
46. Telavancin Drug Label. Available online: https://www.accessdata.fda.gov/drugsatfda_docs/nda/2013/022407Orig1s000Lbl.pdf (accessed on 20 January 2025).
47. Damodaran, S.E.; Madhan, S. Telavancin: A novel lipoglycopeptide antibiotic. *J. Pharmacol. Pharmacother.* **2011**, *2*, 135–137. [CrossRef]
48. Telavancin Approval Letter. Available online: https://www.accessdata.fda.gov/drugsatfda_docs/nda/2013/022407Orig1s000Approv.pdf (accessed on 20 January 2025).
49. Dalbavancin Drug Label. Available online: https://www.accessdata.fda.gov/drugsatfda_docs/nda/2014/021883Orig1s000Lbl.pdf (accessed on 20 January 2025).
50. Cercenado, E. Antimicrobial spectrum of dalbavancin. Mechanism of action and in vitro activity against Gram-positive microorganisms. *Enferm. Infecc. Microbiol. Clin.* **2017**, *35* (Suppl. S1), 9–14. [CrossRef]
51. Dalbavancin Approval Letter. Available online: https://www.accessdata.fda.gov/drugsatfda_docs/nda/2014/021883Orig1s000Approv.pdf (accessed on 20 January 2025).
52. Oritavancin Drug Label. Available online: https://www.accessdata.fda.gov/drugsatfda_docs/label/2021/206334s006lbl.pdf (accessed on 20 January 2025).
53. Zhanel, G.G.; Schweizer, F.; Karlowsky, J.A. Oritavancin: Mechanism of action. *Clin. Infect. Dis.* **2012**, *54* (Suppl. S3), S214–S219. [CrossRef]

64. Oritavancin Approval Letter. Available online: https://www.accessdata.fda.gov/drugsatfda_docs/appletter/2021/214155Ori s000ltr.pdf (accessed on 20 January 2025).
65. Loh, B.S.; Ang, W.H. "Illuminating" Echinocandins' Mechanism of Action. *ACS Cent. Sci.* **2020**, *6*, 1651–1653. [CrossRef]
66. Szymański, M.; Chmielewska, S.; Czyżewska, U.; Malinowska, M.; Tylicki, A. Echinocandins—Structure, mechanism of actic and use in antifungal therapy. *Enzym. Inhib. Med. Chem.* **2022**, *37*, 876–894. [CrossRef]
67. Wiederhold, N.P.; Lewis, R.E. The echinocandin antifungals: An overview of the pharmacology, spectrum and clinical effica *Expert Opin. Investig. Drugs* **2003**, *12*, 1313–1333. [CrossRef]
68. Denning, D.W. Echinocandin antifungal drugs. *Lancet* **2003**, *362*, 1142–1151. [CrossRef]
69. Odds, F.C.; Brown, A.J.; Gow, N.A. Antifungal agents: Mechanisms of action. *Trends Microbiol.* **2003**, *11*, 272–279. [CrossRef]
70. Zaas, A.K.; Alexander, B.D. Echinocandins: Role in antifungal therapy, 2005. *Expert Opin. Pharmacother.* **2005**, *6*, 1657–166 [CrossRef]
71. Hüttel, W. Echinocandins: Structural diversity, biosynthesis, and development of antimycotics. *Appl. Microbiol. Biotechnol.* **202** *105*, 55–66. [CrossRef]
72. Balkovec, J.M.; Hughes, D.L.; Masurekar, P.S.; Sable, C.A.; Schwartz, R.E.; Singh, S.B. Discovery and development of first in cla antifungal caspofungin (CANCIDAS®)—A case study. *Nat. Prod. Rep.* **2014**, *31*, 15–34. [CrossRef]
73. Caspofungin Drug Label. Available online: https://www.accessdata.fda.gov/drugsatfda_docs/nda/2001/21227_Cancida prntlbl.pdf (accessed on 20 January 2025).
74. Kofla, G.; Ruhnke, M. Pharmacology and metabolism of anidulafungin, caspofungin and micafungin in the treatment of invasi candidosis: Review of the literature. *Eur. J. Med. Res.* **2011**, *16*, 159–166. [CrossRef]
75. Caspofungin Approval Letter. Available online: https://www.accessdata.fda.gov/drugsatfda_docs/nda/2001/21227_Cancida Approv.pdf (accessed on 20 January 2025).
76. Micafungin Drug Label. Available online: https://www.accessdata.fda.gov/drugsatfda_docs/nda/2005/21-506_Mycamine Prntlbl.pdf (accessed on 20 January 2025).
77. Jarvis, B.; Figgitt, D.P.; Scott, L.J. Micafungin. *Drugs* **2004**, *64*, 969–982, discussion 983–984. [CrossRef]
78. Micafungin Approval Letter. Available online: https://www.accessdata.fda.gov/drugsatfda_docs/nda/2005/21-506 Mycamine_Approv.pdf (accessed on 20 January 2025).
79. Anidulafungin Approval Letter. Available online: https://www.accessdata.fda.gov/drugsatfda_docs/nda/2006/21948s0C _Eraxis_Approv.pdf (accessed on 20 January 2025).
80. Anidulafungin Drug Label. Available online: https://www.accessdata.fda.gov/drugsatfda_docs/nda/2006/21948s000_Eraxis ApprovBle.pdf (accessed on 20 January 2025).
81. Rezzayo Drug Label. Available online: https://www.accessdata.fda.gov/drugsatfda_docs/label/2023/217417s000lbl.pdf (a cessed on 20 January 2025).
82. Syed, Y.Y. Rezafungin: First Approval. *Drugs* **2023**, *83*, 833–840. [CrossRef]
83. Wang, G.; Li, X.; Wang, Z. APD3: The antimicrobial peptide database as a tool for research and education. *Nucleic Acids Res.* **201** *44*, D1087–D1093. [CrossRef]
84. Jhong, J.H.; Yao, L.; Pang, Y.; Li, Z.; Chung, C.R.; Wang, R.; Li, S.; Li, W.; Luo, M.; Ma, R.; et al. dbAMP 2.0: Updated resourc for antimicrobial peptides with an enhanced scanning method for genomic and proteomic data. *Nucleic Acids Res.* **2022**, *5* D460–D470. [CrossRef]
85. Zhao, X.; Wu, H.; Lu, H.; Li, G.; Huang, Q. LAMP: A Database Linking Antimicrobial Peptides. *PLoS ONE* **2013**, *8*, e6655 [CrossRef]
86. Nagarajan, D.; Nagarajan, T.; Roy, N.; Kulkarni, O.; Ravichandran, S.; Mishra, M.; Chakravortty, D.; Chandra, N. Computationa antimicrobial peptide design and evaluation against multidrug-resistant clinical isolates of bacteria. *J. Biol. Chem.* **2018**, *29* 3492–3509. [CrossRef]
87. David, L.; Brata, A.M.; Mogosan, C.; Pop, C.; Czako, Z.; Muresan, L.; Ismaiel, A.; Dumitrascu, D.I.; Leucuta, D.C.; Stanculete, M.F et al. Artificial Intelligence and Antibiotic Discovery. *Antibiotics* **2021**, *10*, 1376. [CrossRef]
88. Zhao, J.; Liu, H.; Kang, L.; Gao, W.; Lu, Q.; Rao, Y.; Yue, Z. deep-AMPpred: A Deep Learning Method for Identifying Antimicrobia Peptides and Their Functional Activities. *J. Chem. Inf. Model.* **2025**, *65*, 997–1008. [CrossRef]
89. Zevtera Drug Label. 2024. Available online: https://www.accessdata.fda.gov/drugsatfda_docs/label/2024/218275s000lbl.pd (accessed on 20 January 2025).
90. Al Musaimi, O. Lasso Peptides Realm: Insights and Applications. *Peptides* **2024**, *182*, 171317. [CrossRef]

antibiotics

rticle

Cyclic Peptide MV6, an Aminoglycoside Efficacy Enhancer Against *Acinetobacter baumannii*

Natalia Roson-Calero [1,2], Jimmy Lucas [1], María A. Gomis-Font [3,4], Roger de Pedro-Jové [1], Antonio Oliver [3,4], Clara Ballesté-Delpierre [3,5] and Jordi Vila [1,2,3,5,*]

1 Barcelona Institute for Global Health (ISGlobal), 08036 Barcelona, Spain; natalia.rc96.nr@gmail.com (N.R.-C.); jimmy.lucas@isglobal.org (J.L.); roger.depedro@isglobal.org (R.d.P.-J.)
2 Department of Basic Clinical Practice, School of Medicine, University of Barcelona, 08036 Barcelona, Spain
3 CIBER de Enfermedades Infecciosas (CIBERINFEC), Instituto Salud Carlos III, 28029 Madrid, Spain; mariaantonia.gomis@ssib.es (M.A.G.-F.); antonio.oliver@ssib.es (A.O.); ccballes@clinic.cat (C.B.-D.)
4 Department of Microbiology, Hospital Universitario Son Espases, Health Research Institute of the Balearic Islands (IdISBa), 07120 Palma de Mallorca, Spain
5 Department of Clinical Microbiology, Biomedical Diagnostic Center, Hospital Clinic, 08036 Barcelona, Spain
* Correspondence: jvila@clinic.cat

Abstract: Background/Objectives: *Acinetobacter baumannii* is a globally emerging pathogen with widespread antimicrobial resistance driven by multiple mechanisms, such as altered expression of efflux pumps like AdeABC, placing it as a priority for research. Driven by the lack of new treatments, alternative approaches are being explored to combat its infections, among which efficacy-enhancing adjuvants can be found. This study presents and characterizes MV6, a synthetic cyclic peptide that boosts aminoglycoside efficacy. **Methods**: MV6's activity was assessed through antimicrobial susceptibility testing in combination with different antibiotic classes against *A. baumannii* strains characterized by PCR and RT-qPCR. PAβN served as a reference efflux pump inhibitor. Synergy was evaluated using checkerboard assays, and spontaneous mutants were generated with netilmicin with/without MV6 (100 mg/L). Whole-genome sequencing and variant calling analysis were then performed. **Results**: MV6 presented low antimicrobial activity in *A. baumannii* with MICs higher than 2048 mg/L. MV6 showed a better boosting effect for aminoglycosides, especially netilmicin, exceeding that of PAβN. Checkerboard assays confirmed a strong synergy between netilmicin and MV6, and a significant correlation was found between netilmicin MIC and *adeB* overexpression, which was mitigated by the presence of MV6. MV6 reduced, by 16-fold, the mutant prevention concentration of netilmicin. Mutations in a TetR-family regulator and ABC-binding proteins were found in both groups, suggesting a direct or indirect implication of these proteins in the resistance acquisition process. **Conclusions**: MV6 lacks intrinsic antimicrobial activity, minimizing selective pressure, yet enhances netilmicin's effectiveness except for strain 210, which lacks the AdeABC efflux pump. Resistant mutants indicate specific aminoglycoside resistance mechanisms involving efflux pump mutations, suggesting synergistic interactions. Further research, including transcriptomic analysis, is essential to elucidate MV6's role in enhancing netilmicin efficacy and its resistance mechanisms.

Keywords: antimicrobial peptides; antimicrobial resistance; *Acinetobacter baumannii*; adjuvants; efflux pump inhibition; aminoglycosides; MV6

Citation: Roson-Calero, N.; Lucas, J.; Gomis-Font, M.A.; de Pedro-Jové, R.; Oliver, A.; Ballesté-Delpierre, C.; Vila, J. Cyclic Peptide MV6, an Aminoglycoside Efficacy Enhancer Against *Acinetobacter baumannii*. *Antibiotics* **2024**, *13*, 1147. https://doi.org/10.3390/antibiotics13121147

Academic Editors: Marisa Di Pietro and Piyush Baindara

Received: 2 November 2024
Revised: 22 November 2024
Accepted: 24 November 2024
Published: 1 December 2024
Corrected: 11 February 2025

1. Introduction

Acinetobacter baumannii is a globally emerging opportunistic pathogen, notable for its broad range of antimicrobial resistance (AMR) mechanisms, which confer resistance to all classes of antimicrobials including last-resort carbapenems, therefore becoming a pan-drug resistant bacteria [1,2]. Its adaptability is driven by two key factors: (i) the acquisition of foreign resistance-conferring elements, such as transposons, plasmids, and resistance islands; and (ii) the regulation of innate resistance mechanisms, allowing for it to survive under selective pressure in the environment [1,3,4]. The threat of AMR in *A. baumannii* h positioned it as a critical priority for research investment and control efforts [5]. It is no recognized as a significant nosocomial pathogen, ranking among the top five pathoger responsible for AMR-related deaths globally, which may be recovered from bloodstrea infections, ventilator-associated pneumonia, wound infections, urinary tract infectior (UTIs), or meningitis [6–9].

A. baumannii can acquire AMR through horizontal gene transfer (HGT) via plasmid insertion sequences (IS), and other mobile genetic elements. Additionally, it can develo resistance through spontaneous mutations that affect endogenous genes associated wit membrane permeability or the expression of key resistance and transport proteins [1(Efflux pumps play a crucial role in the clinically relevant pathogenicity and resistanc profiles of this bacterium, as they actively expel multiple substrates from the cell, inclu ing a diverse array of antimicrobial agents [11]. The resistance-nodulation-cell divisio (RND) superfamily of transporters is notable in *A. baumannii*, with AdeABC and intrins AdeIJK as its main representatives [12,13]. Overproduction of these two is reportedly ass ciated with an increase in resistance that acts synergistically with additional mechanism particularly the ones improving the permeability barrier for efflux pumps' specific sul strates. Therefore, there is an interplay between lower permeability associated with porin(deficiency and overproduction of efflux pumps [14]. Other efflux pumps belonging t different families, such as CraA, AbeM, and TetA/B, also contribute to multidrug resistanc (MDR) [15–17]. Considering the high prevalence of genes encoding resistance enzyme particularly aminoglycoside-modifying enzymes (AMEs), extended-spectrum β-lactamase (ESBLs), and carbapenemases in *A. baumannii*, it is becoming a particularly challengin pathogen to treat [1,18].

The current pipeline for anti-*Acinetobacter* treatments reveals a concerning lack o new-in-class antimicrobial agents nearing market approval [19,20]. During the decade fron 2010 to 2019, only twelve antimicrobials were approved for market entry [21], of which onl a few are viable options for treating *A. baumannii* infections. The approved agents includ Cefiderocol [22], Eravacycline [23,24], Plazomicin [25], and Sulbactam/Durlobactam [26 Recent attention has also been directed toward novel, yet non-approved, compound targeting *A. baumannii*, including carbapenem-resistant strains. Some examples are GT-1, siderophore-cephalosporin [27]; DS-8587, a broad-spectrum quinolone [28]; and AIC499, β-lactam combined with a β-lactamase inhibitor [29]. Among the most promising nove antimicrobials is Abaucin, an antibiotic recently discovered through artificial intelligenc that disrupts lipoprotein trafficking and exhibits a narrow spectrum of activity limited to *A baumannii* [30].

In light of the notable lack of new treatments against this species, alternatives hav arisen. The use of bacteriophages has shown success in mice [31,32], and several successfu case studies involving phages have been reported [29]. Monoclonal antibodies are als studied for pneumonia and sepsis prevention [33,34]. Furthermore, the developmen of agents that reverse resistance mechanisms and restore susceptibility is actively bein investigated. A typical example of this approach is the inhibition of β-lactamases; howeve other strategies, such as efflux pump inhibitors (EPIs), also represent promising avenue for research [35]. Different EPIs, such as phenylalanine-arginine β-naphthylamide (PAβN) carbonyl cyanide-m-chlorophenylhydrazone (CCCP), and reserpine, have been identified However, their use is currently limited to research purposes, as they show toxicity a therapeutic levels [36].

In this study, we present a synthetic cyclic peptide named MV6, which is able tc resensitize bacteria to specific antibiotics. We investigate MV6's potential application by examining its mechanisms of action and the genetic alterations present in spontaneous resistant mutants.

2. Results

2.1. MV6 Structure

The MV6 cyclic peptide was selected for further study against *A. baumannii* from a synthetic library of 28 cyclic peptides following a small-scale "shot in the dark" approach in which each peptide was tested in combination with various antibiotics and bacterial species. Its structure consists of six amino acids, two arginine residues (Arg), two D-proline residues (D-Pro), and two tryptophan residues (Trp), arranged in a cyclic configuration. The final structure is &Arg-D-Pro-Trp-Arg-D-Pro-Trp& (Figure 1).

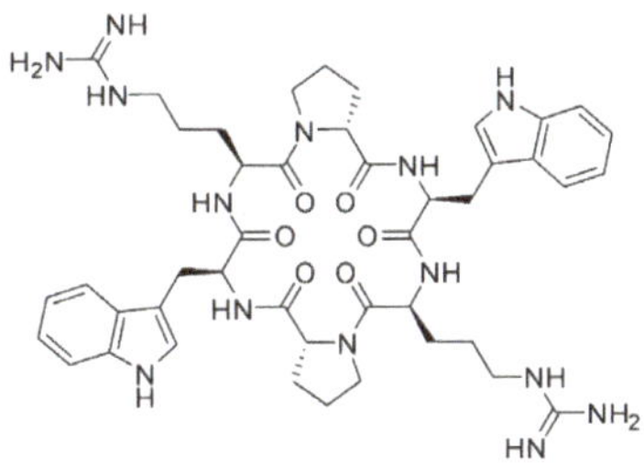

Figure 1. Chemical structure of MV6 cyclic peptide.

2.2. Strains' Resistance Mechanisms

A high prevalence of AME among the selected *A. baumannii* strains was observed. The *aacC1* gene was the most prevalent, found in strains 80, 81, CR17, and CS01, and in some cases, it was the only AME detected. Strain 210 contained the highest number of AME-coding genes, including *aacC2*, *aphA6*, and *aphA1*. The *aadA1* gene was present in strains 80 and 81. Surprisingly, strain 306 did not exhibit any AME-related genes. Regarding efflux pump-related genes, all strains presented *tetB*, but none had *tetA*. The genes *adeJ* and *adeG* were present in all strains and so was *adeB* except for strain 210, which tested negative. The complete PCR results together with the corresponding protein products and expected substrates are listed in Table 1.

Table 1. Results of PCR screening and the corresponding products and substrates.

Strain	Aminoglycoside Modifying Enzymes (AMEs)							Efflux Pumps				
	aacC1	*aacC2*	*aacA4*	*aadA1*	*aadB*	*aphA6*	*aphA1*	*adeB*	*adeJ*	*adeG*	*tetA*	*tetB*
80												
81												
210												
306												
CR17												
CS01												
Product	AAC(3)-I	AAC(3)-II	AAC(6′)-I	ANT(3″)-9	ANT(2″)-I	APH(3′)-VI	APH(3′)-I	AdeABC [1]	AdeIJK [1]	AdeFGH [1]	TetA	TetB
Substrates	GM, TOB, NET	KN, NIT, GM	GM, AK, TOB	STR, SPT	GM, KN AK	AK, KN, NEO	NEO, KN	AMG, FQ, BL, CHL, TMP, TET, E, EtBr	BL, TET, FQ, CHL, TMP, FA, RIF, E, LIN, ACR, NOV, PYO, SDS	TET, TGC, NAL, FQ, SUL, EtBr, E, SDS	TET	TET, MIN
Source	[37–39]	[40]	[41]	[37]	[38,42]	[43]	[38]	[35]	[35]	[35]	[35]	[35]

Color code. Light Green: Positive PCR. White: Negative PCR. [1] Refers to the complete efflux pump to which the AdeB/J/G product belongs. **Abbreviations**. ACR: Acridine; AK: Amikacin; AMG: Aminoglycosides; BL: β-lactams; CHL: Chloramphenicol; E: Erythromycin; EtBr: Ethidium Bromide; FA: Fusidic Acid; FQ: Fluoroquinolones; GM: Gentamicin; KN: Kanamycin; MIN: Minocycline; NAL: Nalidixic Acid; NIT: Nitrofurantoin; NOV: Novobiocin; PYO: Pyonine; RIF: Rifampicin; SDS: Sodium Dodecyl Sulphate; SUL: Sulphonamides; SPT: Spectinomycin; STR: Streptomycin; TET: Tetracycline; TGC: Tigecycline; TMP: Trimethoprim; TOB: Tobramycin.

RT-qPCR revealed a statistically significant overexpression of *adeJ* in all strains (p-value < 0.05), with the highest relative quantification (RQ) values compared to the other genes tested. Although *adeG* was present in all strains, only constitutive expression was detected, with RQ values lower than 1. The expression of *adeB* varied across strains, wi RQ values ranging from 15 to 200. Despite these differences, all strains except for stra 306 exhibited statistically significant overexpression of *adeB*. RT-qPCR also confirmed t absence of *adeB* in *A. baumannii* strain 210. All RQ values for *adeB*, *adeJ*, and *adeG* are show in Figure 2.

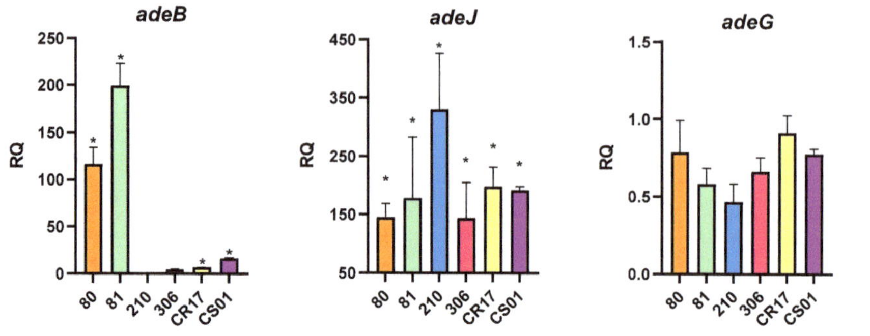

(a) (b) (c)

Figure 2. Expression levels of *adeB* (**a**), *adeJ* (**b**), and *adeG* (**c**) efflux pump genes (RQ) in selected *baumannii* strains. The symbol (*) indicates statistically significant overexpression of the correspondir gene in the specified strain (*p*-value < 0.05). All *p*-values showing statistical significance for *ade* overexpression were <0.0001. *A. baumannii* ATCC 17978 was used as the reference strain, with *rpo* and *gyrB* genes serving as internal controls for basal expression.

2.3. Antimicrobial Susceptibility Testing and Checkerboard Assays

Neither MV6 nor the efflux pump inhibitor PAβN exhibited growth inhibition at th concentrations used in this study. MICs for MV6 and PAβN were in all cases >2048 mg/ and 512 mg/L, respectively. All strains were able to tolerate 12.5% DMSO, making it feasib to use as a solvent, as the maximum concentration reached in the MIC plate was 0.5%. Th results from combined susceptibility testing revealed that the MV6 peptide enhances th activity of aminoglycosides, particularly netilmicin (NET). No activity-boosting effect c MV6 was observed with other classes of antimicrobials. MV6 (100 mg/L) reduced the MI of NET by 8- to 4-fold except for strain 210, in which the reduction was not significan When compared to PAβN, the resensitizing activity of MV6 was slightly superior: strain 80 and 81 had a NET MIC of 256 mg/L, which was reduced to 128–64 mg/L with PAβ and to 32 mg/L in the presence of MV6 (Table 2). When treated with MV6, four out o six strains showed a reduction in NET's MIC below the resistance breakpoint establishe by CLSI, lowering it to the intermediate category. The checkerboard assays revealed FICI value of 0.0097 for strain 80 and of 0.0166 for strain 306, in both cases indicating synergistic effect of MV6 over NET (Figure 3).

Table 2. Minimum inhibitory concentrations (MICs) in mg/L of NET, alone or in combination wit PAβN and MV6, for selected *A. baumannii* strains. Both PAβN and MV6 are used at a constan concentration of 100 mg/L.

Strains *A. baumannii*	NET	NET + MV6	Fold-Change	NET + PAβN	Fold-Change
80	256	32	8-fold	128	2-fold
81	256	32	8-fold	64	4-fold
210	16	8 *	2-fold	8 *	2-fold
306	64	8 *	8-fold	16	4-fold
CR17	32	8 *	4-fold	8 *	4-fold
CS01	32	8 *	4-fold	16	2-fold

* MIC that breaks the CLSI resistance breakpoint of NET (≥16 mg/L) for *A. baumannii*.

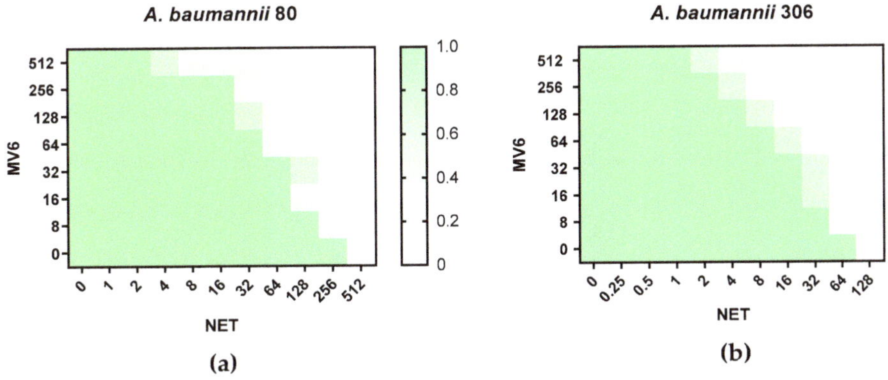

Figure 3. MIC distribution (mg/L) of NET and MV6 on the checkerboard assays. The plate figure ranges from 0, where no growth is observed, to 1.0, where growth was detected in all biological replicates. (**a**) Checkerboard assays for strain 80. (**b**) Checkerboard assays for strain 306.

The analysis of the correlation between the increasing NET MIC values among the studied *A. baumannii* strains and the expression levels of the *adeB* gene indicated that, for both the NET and NET/MV6 datasets, *adeB* expression is significantly related to the log2(MIC) values, as shown in Figure 4. Moreover, the R^2 values, 0.802 and 0.773 for NET and NET/MV6, respectively, suggest a strong relationship between the two parameters. The estimated intercept coefficients of 0.4678436 and 0.225576, respectively, represent the expected *adeB* expression values for a MIC = 0, with an increase in the logarithmic scale of *adeB* expression by 0.0067306 and 0.061134, respectively, for each treatment.

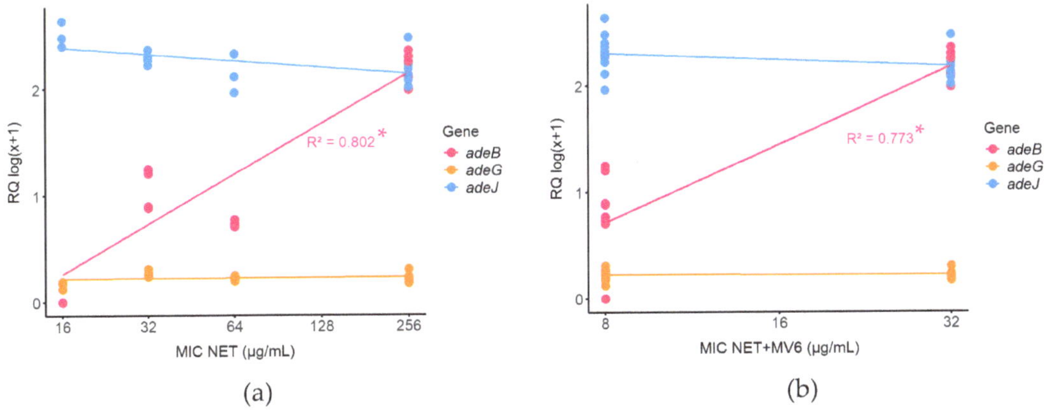

Figure 4. Linear regression analysis of *adeB* expression versus the MIC of NET alone and combined with MV6 peptide. (**a**) Correlation of *adeB*, *adeG*, and *adeJ* expression and NET's MIC (* *p*-value = 5.089×10^{-7}) (**b**) Correlation of *adeB*, *adeG*, and *adeJ* expression and NET's MIC in presence of MV6 100 mg/L (* *p*-value = 1.552×10^{-6}).

2.4. Resistance Profile Characterization of Spontaneous Mutants

Population studies, as illustrated in Figure 5, demonstrate that the mutant preventic concentration (MPC) is significantly reduced when NET is combined with 100 mg/L c MV6. This combination inhibited the emergence of resistant mutants at 64 mg/L, wherea NET alone limited resistant mutant generation at 512 mg/L. Therefore, the presence c MV6 reduces the MPC by 8-fold. For perspective, the MIC for the combination and for NE alone was determined at 16 mg/L and 64 mg/L, respectively. Analysis of the resistanc mechanisms in ten spontaneous mutants obtained from various resistance levels reveale as expected, resistance to NET (Table 3). In the NET/MV6 mutant group, the media MIC to NET stands between 512 and 1024 mg/L, which decreased to 256 mg/L whe MV6 was added. In contrast, in the NET-generated mutants, the median MIC for NE was 2048 mg/L, again dropping to 256 mg/L in the presence of MV6. The addition c MV6 resulted in an average 8-fold reduction in MIC for NET/MV6 compared to NE alone. No significant differences in the degree of MIC reduction were observed betwee the two groups.

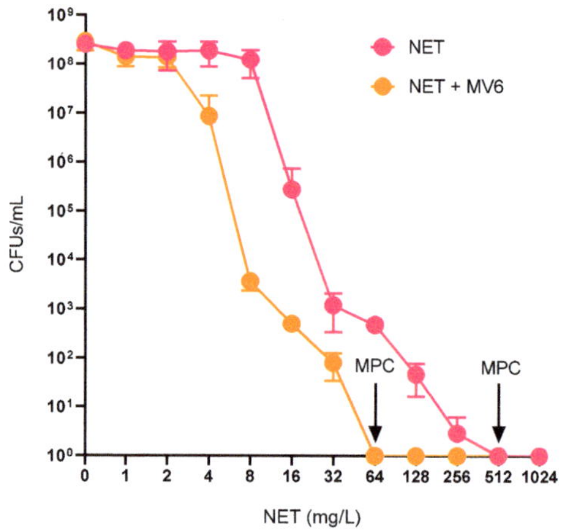

Figure 5. CFU/mL recount of mutant spontaneous generation for NET treatment and NET (+MV 100 mg/L) combination treatment. The first concentration with no CFU recovered determines th mutant prevention concentration (MPC).

The results of microdilution assays using DKNGM Sensititre plates revealed the resis tance profiles of the selected mutants. As expected, NET was not the only aminoglycosid impacted; variations in the MICs for tobramycin, amikacin, and gentamicin were als observed. This confirms that the mutation conferring resistance to NET affected resistanc mechanisms involved in aminoglycoside resistance more broadly. No other significant cros resistances were observed for antibiotics belonging to other classes. The 30-day mutan reversion study demonstrated that the mutations conferring resistance to NET, whethe generated in combination with MV6 or through monotherapy, were largely stable. Only one strain, generated with NET alone (Rev.3), exhibited a reduction in the MIC for NET/MV6 decreasing from 256 mg/L to 64 mg/L, representing a 4-fold reduction in MIC. However no significant changes in the MIC for NET were observed in any of the four mutants testec over the 30-day experiment. Consequently, the original resistance levels were not regained showing stable mutations across time, as represented in Figure 6.

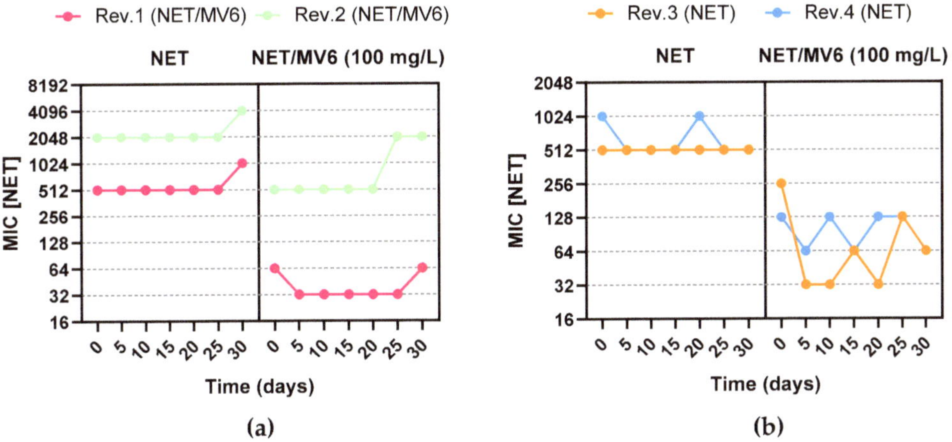

Figure 6. The 30-day analysis of NET MIC variation with and without MV6 for 4 different mutants. MICs for NET and NET/MV6 are represented in a time-lapse of 5 days. (**a**) Rev.1 and Rev.2 are spontaneous mutants generated with NET + MV6 100 mg/L; (**b**) Rev.3 and Rev.4 are spontaneous mutants generated with NET.

2.5. Prediction of Resistance Mechanisms

Whole-genome sequencing (WGS) revealed recurrent mutations across the spontaneous mutants. As expected, mutations were identified in genes associated with resistance to aminoglycosides and regulators, especially those regulating the expression of efflux pumps. In the group of NET-generated mutants, alterations were identified in genes encoding ATP-binding proteins, TetR family regulators, and TetR-AcrR-like regulators as well as in the intergenic region between *adeR* (part of the AdeRS two-component system) and *adeA* (which encodes the membrane fusion protein of the efflux pump) within the *ade* operon responsible for expressing the AdeABC efflux pump. Additionally, mutations in various hypothetical proteins were detected, suggesting a potential, yet unknown, role of these genes in aminoglycoside resistance. Contrary to expectations, mutants generated from exposure to NET/MV6 exhibited similar mutations, including those affecting regulators of efflux pump systems. However, one exception was observed: a mutated tyrosyl-tRNA synthetase found in strain AB21, which was generated in NET at 128 mg/L. This mutation was not detected in any members of the NET/MV6 group. No other known resistance mechanisms were identified.

Most of the mutations exhibited a moderate to high impact, suggesting a potentially significant contribution to the resistance profile of the mutants. The most prevalent type of mutations were single nucleotide polymorphisms (SNPs), although insertions and deletions (INDELs) were also detected. Based on the results, high-impact mutations are typically associated with frameshift variants, moderate-impact mutations are commonly linked to missense variants, and low-impact mutations are mostly found in intergenic regions. All high-impact mutations correspond to a mutated TetR family gene (GBFHJJIP_01215). Table 4 lists the identified mutations for each mutant, their respective nature, and reference.

Table 3. Variation in the resistance profile of the NET and NET/MV6 spontaneous mutants compared to the original parental strain *A. baumannii* 306.

Mutant Strains		MIC (mg/L)		Sensititre (DKNGM)																
		[NET]	[NET]/MV6	MERO	GEN	CIP	AUGC	COL	TGC	TAZ	IMI	AZT	C/T	SXT	P/T4	FOT	CZA	ETP	AMI	TOB
AB1	WT 306	64	8	16	8	>2	>64/2	1	2	>16	>16	32	16/4	>8/152	>32/4	>8	>16/4	>2	8	2
AB2	Ab306_NET8_MV6	1024	32	16	>8	>2	>64/2	1	1	>16	>16	16	16/4	>8/152	>32/4	>8	>16/4	>2	16	2
AB3	Ab306_NET32_MV6	512	64	16	>8	>2	>64/2	1	1	>16	>16	32	>32/4	>8/152	>32/4	>8	>16/4	>2	32	2
AB4	Ab306_NET64_MV6	4096	512	16	>8	>2	>64/2	0.5	>4	>16	16	16	32/4	>8/152	>32/4	>8	>16/4	>2	32	8
AB5	Ab306_NET64_MV6	4096	512	8	>8	>2	>64/2	0.5	4	>16	>16	16	32/4	>8/152	>32/4	>8	>16/4	>2	32	8
AB6	Ab306_NET32_MV6	2048	512	>16	>8	>2	>64/2	0.5	>4	>16	>16	16	>32/4	>8/152	>32/4	>8	>16/4	>2	16	8
AB7	Ab306_NET16_MV6	1024	128	16	>8	>2	>64/2	0.5	>4	>16	16	16	16/4	>8/152	>32/4	>8	>16/4	>2	16	8
AB8	Ab306_NET16_MV6	512	256	8	>8	>2	>64/2	0.5	4	>16	>16	32	32/4	>8/152	>32/4	>8	>16/4	>2	16	8
AB9	Ab306_NET16_MV6	512	256	16	>8	>2	>64/2	0.5	4	>16	16	16	32/4	>8/152	>32/4	>8	>16/4	>2	32	8
AB10	Ab306_NET16_MV6	512	256	16	>8	>2	>64/2	0.5	4	>16	16	8	32/4	>8/152	>32/4	>8	>16/4	>2	32	8
AB11	Ab306_NET16_MV6	512	256	16	>8	>2	>64/2	0.5	4	>16	16	8	16/4	>8/152	>32/4	>8	>16/4	>2	32	8
AB12	Ab306_NET128	512	256	16	>8	>2	>64/2	1	2	>16	>16	16	32/4	>8/152	>32/4	>8	>16/4	>2	8	2
AB13	Ab306_NET256	1024	128	16	>8	>2	>64/2	1	4	>16	>16	16	16/4	>8/152	>32/4	>8	>16/4	>2	8	4
AB14	Ab306_NET256	2048	512	>16	>8	>2	>64/2	1	>4	>16	>16	32	32/4	>8/152	>32/4	>8	>16/4	>2	32	8
AB15	Ab306_NET256	2048	256	16	>8	>2	>64/2	0.5	4	>16	>16	16	32/4	>8/152	>32/4	>8	>16/4	>2	16	8
AB16	Ab306_NET256	2048	256	16	>8	>2	>64/2	0.5	4	>16	>16	8	32/4	>8/152	>32/4	>8	>16/4	>2	32	8
AB17	Ab306_NET256	4096	512	16	>8	>2	>64/2	0.5	4	>16	>16	16	32/4	>8/152	>32/4	>8	>16/4	>2	16	8
AB18	Ab306_NET128	2048	256	16	>8	>2	>64/2	1	4	>16	>16	16	32/4	>8/152	>32/4	>8	>16/4	>2	32	>8
AB19	Ab306_NET128	2048	256	16	>8	>2	>64/2	0.5	>4	>16	>16	16	32/4	>8/152	>32/4	>8	>16/4	>2	16	8
AB20	Ab306_NET128	4096	256	>16	>8	>2	>64/2	0.5	4	>16	>16	32	32/4	>8/152	>32/4	>8	>16/4	>2	32	8
AB21	Ab306_NET128	2048	256	16	>8	>2	>64/2	0.5	4	>16	>16	32	>32/4	>8/152	>32/4	>8	>16/4	>2	32	8

Examples of mutant strain nomenclature. Ab306_NET8_MV6: derivative mutant from *A. baumannii* 306, isolated at NET 8 mg/L in the presence of 100 mg/L of MV6. Ab306_NET128: derivative mutant from *A. baumannii* 306, isolated at NET 128 mg/L, monotherapy. **Abbreviations**. NET: Netilmicin; MERO: Meropenem; GEN: Gentamicin; CIP: Ciprofloxacin; AUGC: Amoxicillin/clavulanic acid constant 2; COL: Colistin; TGC: Tigecycline; TAZ: Ceftazidime; IMI: Imipenem; AZT: Aztreonam; C/T: Ceftolozane/tazobactam 4; SXT: Trimethoprim/sulfamethoxazole; P/T4: Piperacillin/tazobactam constant 4; FOT: Cefotaxime; CZA: Ceftazidime/Avivactam; ETP: Ertapenem; AMI: Amikacin; TOB: Tobramycin.

Table 4. Variant calling analysis of spontaneous mutants of NET and netilmicin with MV6 (NET/MV6).

NET/MV6						
Mutant	**Gene**	**Mutation**	**Annotation (NCBI)**	**Reference (NCBI)**	**Impact**	**Type**
AB2	ATP-binding protein	SNP 865 G>A/A289T	GBFHJJIP_03385	WP_001207474.1	MOD	MV
	Intergenic TetR/AcrR family	SNP (T>C)	GBFHJJIP_02069_gene-CHR_END	GBFHJJIP_02069	M/L	IR
AB3	ATP-binding protein	SNP 865 G>A/A289T	GBFHJJIP_03385	WP_001207474.1	MOD	MV
AB4	ATP-binding protein	SNP 865 G>A/A289T	GBFHJJIP_03385	WP_001207474.1	MOD	MV
	adeR	DEL 40274 (TCTCCACACTTA>T)	GBFHJJIP_02896_gene/ GBFHJJIP_02897_gene	WP_000459542.1	M/L	IR
AB5	**TetR family**	INS 363 (C>CAT)	GBFHJJIP_01215	GBFHJJIP_01215	**HIGH**	FV
	ATP-binding protein	SNP 865 G>A/A289T	GBFHJJIP_03385	WP_001207474.1	MOD	MV
AB6	**TetR family**	DEL 364 (182 nucleotides>C)	GBFHJJIP_01215	GBFHJJIP_01215	**HIGH**	FV
	HP [domain cpo]	INS 155 (T>TGGACGTGGA)	GBFHJJIP_03384	HMPREF0010_00495	MOD	DII
	ATP-binding protein	SNP 865 G>A/A289T	GBFHJJIP_03385	WP_001207474.1	MOD	MV
AB7	ATP-binding protein	SNP 1069 A>C/T357P SNP 1086 A>C/E362D	GBFHJJIP_03385	WP_001207474.1	MOD	MV
AB8	HP [domain fadD]	SNP 192 T>G/H64Q	GBFHJJIP_03655	D0CB89_ACIB2	MOD	MV
AB9	*adeR*	DEL (TCTCCACACTTA>T)	GBFHJJIP_02896_gene-GBFHJJIP_02897_gene	WP_000459542.1	M/L	IR
	TetR/AcrR family	SNP (T>C)	GBFHJJIP_02069_gene-CHR_END	GBFHJJIP_02069	M/L	IR
AB10	ATP-binding protein	SNP 1069 A>C/T357P SNP 1086 A>C/E362D	GBFHJJIP_03385	WP_001207474.1	MOD	MV
AB11	HP [domain fadD]	SNP 192 T>G/H64Q	GBFHJJIP_03655	D0CB89_ACIB2	MOD	MV
	HP [domain GntR family]	SNP 197 T>C/I66T	GBFHJJIP_03660	HMPREF0010_00945	MOD	MV
NET						
Mutant	**Gene**	**Mutation**	**Annotation (NCBI)**	**Reference (NCBI)**	**Impact**	**Type**
AB12	ATP-binding protein	SNP 865 G>A/A289T	GBFHJJIP_03385	WP_001207474.1	MOD	MV
AB13	ATP-binding protein	SNP 865 G>A/A289T	GBFHJJIP_03385	WP_001207474.1	MOD	MV
AB14	*adeR–adeA*	DEL (TCTCCACACTTA>T)	GBFHJJIP_02896_gene-GBFHJJIP_02897_gene	WP_000459542.1	M/L	IR
	TetR/AcrR family	SNP (A>G)	GBFHJJIP_02069_gene y CHR_END	GBFHJJIP_02069	M/L	IR
AB15	**TetR family**	INS 512 (T>20 nucleotides)	GBFHJJIP_01215	GBFHJJIP_01215	**HIGH**	**FV/S**
	HP [domain fadD]	SNP 192 T>G/H64Q	GBFHJJIP_03655	D0CB89_ACIB2	MOD	MV
	HP [domain GntR family]	SNP 197 T>C/I66T	GBFHJJIP_03660	HMPREF0010_00945	MOD	MV
AB16	**TetR family**	DEL 174 (18 nucleotides>A)	GBFHJJIP_01215	GBFHJJIP_01215	**HIGH**	**FV**
	ATP-binding protein	SNP 865 G>A/A289T	GBFHJJIP_03385	WP_001207474.1	MOD	MV
	ATP-binding protein	SNP 1069 A>C/T357P	GBFHJJIP_03385	WP_001207474.1	MOD	MV
AB17	HP [domain GntR family]	SNP 197 T>C/I66T	GBFHJJIP_03660	HMPREF0010_00945	MOD	MV
AB18	**TetR family**	INS 511 (T>TCTG)	GBFHJJIP_01215	GBFHJJIP_01215	**HIGH**	**DII**
	adeR–adeA	DEL (TCTCCACACTTA>T)	GBFHJJIP_02896_gene-GBFHJJIP_02897_gene	WP_000459542.1	M/L	IR
	TetR/AcrR family	SNP (A>G)	GBFHJJIP_02069	GBFHJJIP_02069	M/L	IR
AB19	ATP-binding protein	SNP 1069 A>C/T357P	GBFHJJIP_03385	WP_001207474.1	MOD	MV
	HP [domain fadD]	SNP 192 T>G/H64Q	GBFHJJIP_03655	D0CB89_ACIB2	MOD	MV
	HP [domain fadD]	SNP 192 T>G/H64Q	GBFHJJIP_03655	D0CB89_ACIB2	MOD	MV
AB20	TetR/AcrR family	SNP (A>G)	GBFHJJIP_02069	GBFHJJIP_02069	M/L	IR
	adeR–adeA	DEL (TCTCCACACTTA>T)	GBFHJJIP_02896_gene-GBFHJJIP_02897_gene	WP_000459542.1	M/L	IR
AB21	HP [domain fadD]	SNP (A>G)	CHR_START/GBFHJJIP_03655	GBFHJJIP_03655	M/L	IR
	TetR/AcrR family	SNP (T>C)	GBFHJJIP_02069	GBFHJJIP_02069	M/L	IR
	HP [domain fadD]	SNP (A>C)	GBFHJJIP_03655	D0CB89_ACIB2	M/L	IR
	PA4642 family protein/tyrosyl-tRNA synthetase	INS (>CAATCAAATCA)	GBFHJJIP_01042_gene/ GBFHJJIP_01043_gene	WP_001218560.1–WP_031969348.1	M/L	IR
	adeR–adeA	DEL (TCTCCACACTTA>)	GBFHJJIP_02896_gene-GBFHJJIP_02897_gene	WP_000459542.1	M/L	IR

Column gene. HP: Hypothetical protein. **Column Mutation.** SNP: Single Nucleotide Polymorphism; DEL: Deletion; INS: Insertion. **Column impact.** MOD: Moderate; M/L: Modifier/Low. **Column Type.** MV: Missense variant; IR: Intergenic Region; FV: Frameshift Variant; S: Stop gained; DII: Disruptive In-frame Insertion. Mutations with a prediction of High Impact are marked in bold.

3. Discussion

Antimicrobial peptides (AMPs) represent a promising alternative for the treatment of *A. baumannii* in the emerging post-antibiotic era. AMPs typically attack bacterial membranes, which form the basis of their anti-*A. baumannii* activity, although some AMPs have been shown to act intracellularly as well. A well-known example is the human cathelicidin

LL-37, a 37-amino acid AMP, whose primary antimicrobial mechanism involves the neutra ization of lipopolysaccharides (LPS) in the bacterial outer membrane [44,45]. Nonetheles an alternative use for AMPs has been explored, involving the enhancement of antibiot efficacy by combining them with short peptides, which may or may not be conjugated the antibiotic [46,47]. The six-amino acid peptide MV6 follows this approach by acting a an adjuvant that enhances the activity of other antimicrobials, specifically aminoglycoside Lacking intrinsic antimicrobial activity, shown by MICs over 2048 mg/L, MV6 reduce selective pressure from antibiotics and minimizes the rate of spontaneous mutations, a demonstrated by MPC determinations. However, the generated mutations appear stab over long periods, with no reversion observed, which may pose a limitation to address i future development stages.

Among all the antimicrobials tested in combination with MV6, only aminoglycoside showed a decrease in MICs, particularly netilmicin, which revealed the higher fold change in MIC reductions. This suggested a potential interaction between MV6 and the mechanism responsible for aminoglycoside resistance, which warranted further exploration. Regre sion analyses indicate a correlation between the expression levels of *adeB* and the MI values of netilmicin, with MICs increasing in proportion to *adeB* overexpression. Althoug the addition of MV6 did not disrupt this correlation, it effectively reduced the MICs i strains with elevated *adeB* expression. A larger sample size will be required to explor this relationship in greater depth. MV6 at 100 mg/L manages to limit the increase i MIC to 32 mg/L, a concentration that could be reduced with a higher MV6 concentratio as demonstrated by the checkerboard assays. To highlight, strain 210 was the only on showing no significant effect of the MIC when combined with MV6 and was the only on lacking the AdeABC efflux pump. This finding reinforces the premise of a potential rela tionship between MV6 activity and *adeABC* expression, which will be further investigate The strong synergistic effect of MV6 on netilmicin, as shown by FICI values considerabl below the synergy threshold of 0.5, along with its low antimicrobial activity provide roon to increase the MV6 dose and achieve a greater effect in enhancing netilmicin's activit achieving concentrations below the susceptibility breakpoint. Future toxicity, cytotoxicit and hemolysis assays will be essential to define the therapeutic window for safe and effe tive use of MV6. Additionally, in vivo efficacy assays will be needed to assess its stabilit and bioavailability once it enters the bloodstream of mice, which may pose a limitation t address in future studies.

The resistance mechanisms of NET and NET/MV6 mutants were determined in strai 306, which was specifically selected for the absence of AMEs and constitutive expression o the *adeABC* efflux pump. Overexpression of the *adeABC* efflux pump is known to confe resistance to a wide range of agents, including aminoglycosides, trimethoprim, fluoro quinolones, chloramphenicol, β-lactams, erythromycin, and tetracyclines [35,48]. Othe efflux pumps, such as AbeM (present in strain 306 according to WGS analysis), AbeD, an ArpAB, have also been associated with aminoglycoside resistance in *A. baumannii* [17,49–51 The resistance profile of the mutants showed a reduction in susceptibility to aminoglyco sides, particularly netilmicin with up to a 64-fold MIC increase, while no changes wer observed with other antibiotics from the aforementioned classes, such as aztreonam o meropenem. This suggests a resistance mechanism more specific to aminoglycosides than to general *adeABC* overexpression.

Mutations identified in these spontaneous mutants included genes related to efflux pumps, specifically TetR-like regulators and ATP-binding proteins. No differences were found between NET mutants and NET/MV6 mutants, indicating that either no specific resistance mechanisms against the MV6 peptide were generated or that any existing mech anisms overlap with those for netilmicin. A hypothesis worth exploring is that MV6 and NET may compete for distinct substrate-specific binding sites within the efflux pump. Thus a mutation causing transporter overexpression could potentially influence susceptibility to both treatments, whether used alone or in combination. Regulators from the TetR family also referred to as TRFs, possess a helix-turn-helix (HTH) DNA-binding domain that typi cally enables them to function as repressors and regulates bacterial AMR [52,53]. More than 2300 nonredundant sequences belonging to this family of regulators have been identified, and it is predicted that *A. baumannii* encodes 42 of them [53–55]. Apart from regulating efflux pumps, such as TetA, which is associated with resistance to tetracycline-like antibiotics, other functions have also been attributed to them [56]. Examples include the gene *adeN*, which belongs to the TRF family and regulates the RND AdeIJK efflux pump [57], and *arpR*, known to regulate another RND efflux pump, ArpAB, which has been related to *A. baumannii*'s opaque/translucent colony phase variation [51]. The present mutated TetR-like regulator could potentially be involved in negatively regulating the expression of a netilmicin-related efflux pump, but its specific activity remains unstudied.

On the other hand, the ATP-binding cassette (ABC) family of transporters relies on the hydrolysis of ATP to ADP to expel substrates across the bacterial membrane [58,59]. The MacB–MacA complex is a representative member of this group, also found in *A. baumannii*, and is known to transport macrolides and gramicidin as substrates [35,60]. Moreover, a potential involvement of this ABC-type transporter in protecting *Serratia marcescens* against aminoglycosides and polymyxins has been reported [61]. Additionally, the ABC transporter MsbA requires an ATP-binding protein to export major lipids, such as LPS and phospholipids, thereby contributing to membrane integrity. The action of MV6 on altering the export of LPS, potentially affecting membrane stability, is a noteworthy area for further investigation. Strains harboring simultaneous mutations in both TetR-like regulators and ATP-binding proteins have been identified, indicating a potential synergistic effect between these alterations. In the Gram-positive *Streptomyces coelicolor* A3, sets of TRFs and adjacent ABC transport systems have been reported, where the TRFs repressed the expression of the ABC transporters within the operon [62], a relation that should be further investigated in *A. baumannii* and aminoglycosides, as it has not been documented yet.

Finally, the identification of mutations in various hypothetical proteins suggests a potential, though uncharacterized, role for these proteins in conferring netilmicin resistance. Combining computational tools such as AlphaFold and Foldseek, which can predict the functions of uncharacterized proteins, may provide valuable insights into their involvement [63]. Transcriptomic analysis and docking simulations may shed light on the specific resistance mechanism to MV6 in combination with netilmicin, strategies that are planned to be explored in the future.

Limitations of This Study

While MV6 has demonstrated strong synergy with netilmicin, a larger and more diverse sample set encompassing various resistance profiles and mechanisms would provide deeper insights into its performance across different scenarios. Additionally, analyzing more strains that overproduce AdeABC would improve our understanding of the relationship between netilmicin resistance and MV6's capacity to counteract the increased activity of this efflux pump. This study has also provided comprehension of the resistance mechanisms against MV6 in combination with netilmicin, which suggest the involvement of altered efflux pump expression. However, the precise mode of action could not be determined with the current experiments, highlighting the need for more specific and targeted assays in future research.

4. Materials and Methods

4.1. Molecules Used in This Study

The MV6-peptide molecules utilized in this study were commercially synthesized by GenicBio Limited (Shanghai, China). The composition and purity of MV6 were confirmed via mass spectrometry and high-performance liquid chromatography (HPLC). In addition, the reference EPI, PAβN (Sigma Aldrich, St. Louis, MO, USA), was selected to compare its activity alongside MV6. The molecular formula of PAβN is $C_{22}H_{25}N_5O$. For experimental use, MV6 and PAβN were dissolved in pure dimethyl-sulfoxide (DMSO).

4.2. Strain Selection and Characterization of Aminoglycoside Modifying Enzymes (AMEs) and Relevant Efflux Pumps

A selective collection of six MDR *A. baumannii* strains with distinct antimicrobial res- tance profiles was used for this study. These strains included *A. baumannii* 80, 81, 210, 30 and the colistin-resistant strain *A. baumannii* CR17 alongside its colistin-sensitive count part *A. baumannii* CS01 [64]. All strains are clinical isolates; strains 80, 81, 210, and 306 we isolated from patients at hospitals in a previous multicenter study [38,65]. To further cha acterize these strains, the presence of AMEs and relevant efflux pump genes was explore by PCR. The core cycling conditions applied were as follows: 95 °C–3 min, [94 °C–1 mi Tm–1 min; 72 °C–X min (1 min per kb)] × 35 cycles, 72 °C–45 s (final extension).

Additionally, the expression of RND-family efflux pumps (*adeABC*, *adeIJK*, and *ad FGH*) was confirmed by reverse-transcriptase real-time PCR (RT-qPCR), targeting the gen encoding the membrane transporter proteins *adeB*, *adeJ*, and *adeG*. The strains were grow overnight in LB broth, diluted 1:100 in fresh medium, and incubated at 37 °C with shakin at 180 rpm until reaching an OD600nm of 0.5. RNA was extracted using the Maxwell (16 LEV simply RNA Blood Kit (Promega, Madison, WI, USA) according to the manufa turer's instructions. To remove any potential DNA contamination, the Ambion DNA-free DNA Removal Kit (Thermo Fisher Scientific, Waltham, MA, USA) was used. Quality contr for the RNA extractions was performed using a Nanodrop ND-1000 (Thermo Fisher), wit acceptable quality parameters being a 260/280 ratio between 1.9 and 2.1 and a 260/230 rati between 1.8 and 2.0.

For reverse transcription and cDNA generation, the PrimeScript RT Reagent K (Takara Bio, Kusatsu, Japan) was used following the manufacturer' s instructions. RT-qPC was performed using the standard protocol from Applied Biosystems™ (Fisher Scientific 95 °C–30 s; (95 °C–15 s; 60 °C–34 s) × 45 cycles, melting curve (95 °C–15 s; 60 °C–15 95 °C–15 s). Primers were designed using the Primer Express™ Software v3.0.1 fro Applied Biosystems™, and different primer concentration combinations were tested t identify the most efficient conditions for the assays. *A. baumannii* ATCC 17978 was used a the reference strain for RT-qPCR quantification. Basal expression levels were controlle with *rpoB* and *gyrB* genes. Biological and technical triplicates were performed for eac strain. Primer sequences and conditions are listed in Table 5.

Table 5. PCR primers used in this study.

Gene	Primer Sequence (5′–3′)	Tm (°C)	Length (bp)	Source
aacC1	1: ATGGGCATCATTCGCACATGTAGG 2: TTAGGTGGCGGTACTTGGGTC	52	456	[66]
aacC2	1. ATTGATTCAGCAGGCCGAAC 2: CTCTTGATGGTGCATGCCTC	59	247	[67]
aacA4	1: TTGCGATGCTCTATGAGTGGCTA 2: CTCGAATGCCTGGCGTGTTT	63	482	[68]
aadA1	1: ATGAGGGAAGCGGTGATCG 2: TTATTTGCCGACTACCTTGGTG	52	792	[66]
aadB	1: ATGGACACAACGCAGGTCGC 2: TTAGGCCGCATATCGCGACC	55	534	[66]
aphA6	1: ATGGAATTGCCCAATATTATTC 2: TCAATTCAATTCATCAAGTTTTA	55	797	[66]
aphA1	1: AAACGTCTTGCTCGAGGC 2: CAAACCGTTATTCATTCGTGA	56	461	[68]

Table 5. *Cont.*

Gene	Primer Sequence (5′–3′)	Tm (°C)	Length (bp)	Source
adeB	1: ATGTCACAATTTTTTATTCGTCGTC 2: TTAGGATGAGATTTTTTTCTTAGAGG	56	3104	[69]
adeJ	1: CTGGCTTATGACACGACTC 2: GGATCCCCATACCACGCTGG	61	988	[69]
adeG	1: GTTGCTCGTGTCGAACTTGC 2: AGGAACGAAACCACCTGGAAC	57	918	[69]
tetA	1: GTAATTCTGAGGACTGTCGC 2: CTGCCTGGACAACATTGCTT	55	950	[16]
tetB	1: TTGGTTAGGGGCAAGTTTTG 2: GTAATGGGCCAATAACACCG	56	659	[16]
adeB (RT-qPCR)	1: CTGCTGTACCGGAGGTATCTGTT 2: GCGCGAATTATCGGGTGTAA	60	~60	This study
adeJ (RT-qPCR)	1: AGGCGAATGGACGTATGGTT 2: AACCGATGACACGCCGTTA	60	~60	This study
adeG (RT-qPCR)	1: CGCGACCGAAATTGTGAAT 2: GATTGTACCCGCTGCAACCT	60	~60	This study
gyrB (RT-qPCR)	1: CTGCAGCAGAAACCCCTTCT 2: ATAATGGCCGCGGTATTCC	60	~60	This study
rpoB (RT-qPCR)	1: TCCATTCCTTGAACACGATGAC 2: CTGCCTGACGTTGCATGTTT	60	~60	This study

Tm: annealing temperature. bp: base pairs.

4.3. Antimicrobial Susceptibility Testing

For minimum inhibitory concentration (MIC) determination, antimicrobial susceptibility testing (AST) was performed following the Clinical and Laboratory Standards Institute (CLSI) guidelines using the microdilution technique in 96-well microtiter plates. The growth medium utilized for AST was the commercial BD Phoenix™ AST Broth (Becton Dickinson, Franklin Lakes, NJ, USA) [70]. A variety of antimicrobials were evaluated, including amikacin, ceftazidime, chloramphenicol, gentamicin, levofloxacin, meropenem, netilmicin, tedizolid, and tobramycin. The MICs of MV6 alone, DMSO, and PAβN were also assessed to exclude potential interactions among these compounds in subsequent AST involving combination treatments. Preliminary studies aimed at identifying an effective concentration of MV6 for combination with antimicrobials and established that a concentration of 100 mg/L was sufficient for combined therapy. Consequently, the MV6–antibiotic combinations were tested using a fixed concentration of 100 mg/L of MV6 along with the serial dilutions of each antibiotic. Three biological replicates were conducted for each MIC determination. Subsequent studies focused on the use of NET. After AST performance, linear regression analysis was performed to assess the relationship between the *adeB* expression (relative quantification) previously mentioned and MIC values of NET alone and combined with MV6.

4.4. Statistical Analysis

To determine whether the expression levels of *adeB*, *adeJ*, and *adeG* differed significantly from those of the control strain *A. baumannii* ATCC 17978, a two-way analysis of variance (ANOVA) was performed. The analysis considered the RQ values as the primary factor, comparing the expression levels of these genes across different clinical strains and the control. The analysis was carried out using IBM SPSS Statistics for Windows, version 23.0 (IBM Corp., Armonk, NY, USA), following its standard procedures for two-way ANOVA. Statistical significance was set at p-value < 0.05. For visualization and statistical analysis of the potential correlation between NET MICs and *ade* genes expression, gene expression values were log transformed ($\log_{10}(x + 1)$) to reduce skewness and variance across sample and AST values were log2 transformed. Data were visualized as a scatter plot with regression line ("lm" model) superimposed onto the plot using ggplot2 package [7 (v. 3.5.1) in R [72] (v. 4.4.0). The data fit to the linear model was assessed by the coefficie of determination (R^2), and its significance was assessed via the *p*-value using the lm functio of stats package (v. 4.4.0) in R. The R^2 value was displayed directly onto the plot for clarit

4.5. Checkerboard Assays

To assess the in vitro interaction between NET and MV6, checkerboard assays wer performed in 96-well microtiter plates using strains 80 and 306. Similar to MIC determin tion, BD Phoenix^{TM} AST Broth (Becton Dickinson, NJ, USA) medium was used for bacteri growth. A column and a row of wells were reserved for controls of each agent individuall to ensure the proper concentration of the agents. Positive and negative controls were als included. Wells in rows contained serial dilutions of MV6 starting from 512 mg/L, whi those in columns varied in concentrations of NET, starting at 128 mg/L and 512 mg/ for strains 306 and 80, respectively. Inoculum at ~5 × 10^5 CFU/mL, NET, MV6, and AS broth were added to a final volume of 200 µL. Incubation followed the CLSI guideline for MIC determination. Three biological replicates were performed. Fractional inhibitor concentration index (FICI) is defined as the summatory of FICs from compound A (MV and compound B (NET). FICIs were calculated and interpreted as follows:

1. FIC_{MV6} = (MIC of MV6 in combination)/(MIC of MV6 alone);
2. FIC_{NET} = (MIC of NET in combination)/(MIC of NET alone);
3. FICI = FIC_{MV6} + FIC_{NET};
4. Synergistic effect if FICI $\leq$ 0.5; additive effect if 0.5 < FICI $\leq$ 1.0; indifferent effect i 1.0 < FICI $\leq$ 2.0; and finally, antagonistic effect if FICI > 2.0.

4.6. Mutant Generation Analysis

Mutant generation was performed following a protocol adapted from Billal et a (2007) [73]. *A. baumannii* 306 strain was selected for this process. After overnight incubatio at 37 °C in Mueller–Hinton Broth (MHB), 1 × 10^8 CFU/mL (0.5 McFarland standard) wa inoculated into 2 mL of MHB supplemented with NET at concentrations ranging from 1× to 8× the MIC. In the case of combination treatments, 100 mg/L of MV6 was also adde The cultures were incubated overnight at 37 °C with shaking at 180 rpm, then plated ont Mueller–Hinton agar (MHA) plates supplemented with the corresponding concentration o NET. Ten mutants from each concentration, if any were recovered, were selected for furthe characterization. The selected mutants underwent susceptibility profiling using DKMGN Sensititre^{TM} custom plates (Thermo Fisher Scientific) for Gram-negative bacteria following the manufacturer's instructions. Two mutants from each treatment were tested for mutatio reversion by incubating them on LB agar plates for 30 days with daily plating of fres inoculum. Every 5 days, the MICs of NET and NET/MV6 were determined following th previously described procedure.

To determine the MPC, defined as the lowest antibiotic concentration that prevent the growth of resistant mutants, one-step mutants were generated [74]. Tubes containing MHB were inoculated with strain 306 and incubated for 24 h at 37 °C. The culture was the diluted to an OD_{600nm} of 0.05 and further incubated until reaching the late-exponentia growth phase. Serial dilutions of *A. baumannii* 306 were plated on MHA containing varying concentrations of NET, both alone and in the presence of MV6 at 100 mg/L. In-plate concentrations of NET ranged from 1 mg/L to 1024 mg/L (Log 2 scale). Control plates without antibiotic were included to monitor the inoculum. After overnight incubation CFUs were counted. The experiments were performed in triplicate, and the mean CFU counts were plotted to analyze mutant generation.

4.7. Whole-Genome Sequencing and Variant Calling

Whole-genome sequencing (WGS) was then performed on the mutants for further analysis. Ten NET/MV6 resistant mutants (one obtained at 8 mg/L, five at 16 mg/L, two at 32 mg/L, and two at 64 mg/L) and ten NET resistant mutants (five obtained at 128 mg/L and five obtained at 256 mg/L) were characterized through WGS. Genomic DNA of the mutants was extracted using the High Pure PCR Template Preparation Kit (Roche Diagnostics, Basel, Switzerland). Indexed paired-end libraries were generated using the Illumina DNA Prep library preparation kit (Illumina Inc., San Diego, CA, USA). The samples were then sequenced in a MiSeq desktop sequencer cartridge (MiSeq Reagent Kit v3, Illumina).

The variant calling analysis was performed using a pipeline developed in NextFlow DSL2. A de novo assembly, mapping, and variant identification were conducted to obtain point mutations: Single nucleotide polymorphisms (SNPs) and insertion/deletions (INDELS). De novo assembly was conducted using SPAdes (v. 3.15.3). After raw reads quality control and filtering, the "–mode novo" was followed, which performs mapping against the assembled reference strain using Bowtie2, (v. 2.2.5) [75] and SAMtools, (v. 1.14) [76]. For variant handling and identification, PicardTools, Genome Analysis Toolkit (GATK), version 4.5.0.0 [77], and FreeBayes (v. 0.9.21.7) were used after genome annotation with Prokka (v. 1.14.6) [78] and Bakta (v. 1.9.4) [79]. From the obtained VCF (variant call format) files, SNPs were listed to meet the following criteria: a quality score > 50, a root mean square (RMS) mapping quality > 25, and a coverage depth > 30. Indels were extracted from the totalpileup files using the following criteria: a quality score > 200, an RMS mapping quality > 25, and a coverage depth > 30. SNPs and INDELs for each isolate were annotated using SnpEff software (v. 4.3) [80].

5. Conclusions

MV6 is a cyclic peptide that lacks direct antimicrobial activity against *A. baumannii* but potentiates the activity of aminoglycosides, such as netilmicin. With a strong synergistic interaction, MV6 can reduce the MIC of netilmicin by several folds, making it a promising candidate for reversing *A. baumannii*'s resistance to this antibiotic. While the exact mode of action remains unclear, this study suggests a potential interaction with aminoglycoside efflux pumps, supported by the mutations observed in resistance mutants. These mutations are often mediated by changes in TetR-like regulators and ATP-binding proteins, both of which are involved in both the expression and activity of bacterial transporters.

6. Future Perspectives

A. baumannii is a critical pathogen, making it urgent to develop new treatment options and thoroughly investigate the resistance mechanisms that enable its extensive multidrug resistance. This study lays the foundation for further exploration of yet unknown TetR regulators and novel efflux pumps involved in aminoglycoside resistance and efflux pump expression in *A. baumannii*, underscoring the extent of what remains to be uncovered about this pathogen's resistance mechanisms. Despite this study's limitations, it highlights the strong relationship between aminoglycoside resistance and altered efflux pump expression. It also emphasizes how the use of adjuvant boosters, such as cyclic peptides, can help restore susceptibility, giving previously ineffective antibiotics a second chance and thereby expanding treatment options.

Author Contributions: Conceptualization, N.R.-C., C.B.-D. and J.V.; methodology, N.R.-C., C.B.-D. and J.V.; validation, N.R.-C., C.B.-D. and J.V.; formal analysis, N.R.-C., J.L., M.A.G.-F. and R.d.P.-J.; investigation, N.R.-C. and M.A.G.-F.; resources, A.O. and J.V.; data curation, N.R.-C., J.L., M.A.G.-F. and R.d.P.-J.; writing—original draft preparation, N.R.-C.; writing—review and editing, N.R.-C., J.L., M.A.G.-F., R.d.P.-J., A.O., C.B.-D. and J.V.; supervision, A.O., C.B.-D. and J.V.; project administration, C.B.-D. and J.V.; funding acquisition, C.B.-D. and J.V. All authors have read and agreed to the published version of the manuscript.

Funding: This work has been supported by Instituto de Salud Carlos III 2020 call for the Strateg Action on Health and co-funded by The European Regional Development Fund (ERDF); PI20/007 titled "Desarrollo de dos nuevas estrategias: Inhibidores de bombas de expulsión activas y AR antisentido, frente a bacterias multiresistentes".

Institutional Review Board Statement: Not applicable.

Informed Consent Statement: Not applicable.

Data Availability Statement: The data presented from the whole-genome sequencing analysis this study are openly available on GitHub "https://github.com/AMRmicrobiology/WGS-Analys VariantCalling (accessed on 14 November 2024)". The rest of the data presented in this study a available upon request from the corresponding author.

Acknowledgments: We acknowledge support from the Spanish Ministry of Science and Innovatio through the "Centro de Excelencia Severo Ochoa 2019-512 2023" Program (CEX2018-000806-S) ar support from the Generalitat de Catalunya through the CERCA Program. We also acknowledg CIBER—Consorcio Centro de Investigación Biomédica en Red (CB 2021), Instituto de Salud Carlos I Ministerio de Ciencia e Innovación, and the European Union–NextGenerationEU for promoting th collaboration presented in this study.

Conflicts of Interest: Authors participating in this study declare that the present research w conducted in the absence of any relationships that could be perceived as potential conflicts of intere The funders had no role in the design of the study; in the collection, analyses, or interpretation data; in the writing of the manuscript; or in the decision to publish the results.

References

1. Peleg, A.Y.; Harald, S.; Paterson, D.L. *Acinetobacter baumannii*: Emergence of a Successful Pathogen. *Clin. Microbiol. Rev.* **2008**, *2* 538–582. [CrossRef]
2. Poirel, L.; Nordmann, P. Carbapenem Resistance in *Acinetobacter baumannii*: Mechanisms and Epidemiology. *Clin. Microbiol. Infe* **2006**, *12*, 826–836. [CrossRef]
3. Vaneechoutte, M.; Young, D.M.; Ornston, L.N.; De Baere, T.; Nemec, A.; Van Der Reijden, T.; Carr, E.; Tjernberg, I.; Dijkshoorn, Naturally Transformable Acinetobacter Sp. Strain ADP1 Belongs to the Newly Described Species Acinetobacter Baylyi. *Ap Environ. Microbiol.* **2006**, *72*, 932–936. [CrossRef] [PubMed]
4. Fournier, P.-E.; Vallenet, D.; Barbe, V.; Audic, S.; Ogata, H.; Poirel, L.; Richet, H.; Robert, C.; Mangenot, S.; Abergel, C.; et a Comparative Genomics of Multidrug Resistance in *Acinetobacter baumannii*. *PLoS Genet.* **2006**, *2*, e7. [CrossRef]
5. WHO. *WHO Bacterial Priority Pathogens List, 2024*; WHO: Geneva, Switzerland, 2024; ISBN 9789240093461.
6. Murray, C.J.; Ikuta, K.S.; Sharara, F.; Swetschinski, L.; Robles Aguilar, G.; Gray, A.; Han, C.; Bisignano, C.; Rao, P.; Wool, E et al. Global Burden of Bacterial Antimicrobial Resistance in 2019: A Systematic Analysis. *Lancet* **2022**, *399*, 629–655. [CrossRe [PubMed]
7. Ma, C.; McClean, S. Mapping Global Prevalence of *Acinetobacter baumannii* and Recent Vaccine Development to Tackle It. *Vaccine* **2021**, *9*, 570. [CrossRef] [PubMed]
8. McConnell, M.J.; Actis, L.; Pachón, J. *Acinetobacter baumannii*: Human Infections, Factors Contributing to Pathogenesis an Animal Models. *FEMS Microbiol. Rev.* **2013**, *37*, 130–155. [CrossRef]
9. Antunes, L.C.S.; Visca, P.; Towner, K.J. *Acinetobacter baumannii*: Evolution of a Global Pathogen. *Pathog. Dis.* **2014**, *71*, 292–30 [CrossRef] [PubMed]
10. Roca, I.; Espinal, P.; Vila-farrés, X.; Vila, J. The *Acinetobacter baumannii* Oxymoron: Commensal Hospital Dweller Turne Pan-Drug-Resistant Menace. *Front. Microbiol.* **2012**, *3*, 148. [CrossRef] [PubMed]
11. Wieczorek, P.; Sacha, P.; Hauschild, T.; Zórawski, M.; Krawczyk, M.; Tryniszewska, E. Multidrug Resistant *Acinetobacte baumannii*—The Role of AdeABC (RND Family) Efflux Pump in Resistance to Antibiotics. *Folia Histochem. Cytobiol.* **2008**, *4* 257–267. [CrossRef]
12. Xu, C.F.; Bilya, S.R.; Xu, W. AdeABC Efflux Gene in *Acinetobacter baumannii*. *New Microbes New Infect.* **2019**, *30*, 100549. [CrossRef
13. Damier-Piolle, L.; Magnet, S.; Brémont, S.; Lambert, T.; Courvalin, P. AdeIJK, a Resistance-Nodulation-Cell Division Pum Effluxing Multiple Antibiotics in *Acinetobacter baumannii*. *Antimicrob. Agents Chemother.* **2008**, *52*, 557–562. [CrossRef]
14. Leus, I.V.; Weeks, J.W.; Bonifay, V.; Smith, L.; Richardson, S.; Zgurskaya, H.I. Substrate Specificities and Efflux Efficiencies of RNI Efflux Pumps of *Acinetobacter baumannii*. *J. Bacteriol.* **2018**, *200*, e00049-18. [CrossRef]
15. Roca, I.; Marti, S.; Espinal, P.; Martínez, P.; Gibert, I.; Vila, J. CraA, a Major Facilitator Superfamily Efflux Pump Associated wit Chloramphenicol Resistance in *Acinetobacter baumannii*. *Antimicrob. Agents Chemother.* **2009**, *53*, 4013–4014. [CrossRef] [PubMed
16. Martí, S.; Fernández-Cuenca, F.; Pascual, Á.; Ribera, A.; Rodríguez-Baño, J.; Bou, G.; Miguel Cisneros, J.; Pachón, J.; Martínez Martínez, L.; Vila, J. Prevalencia de Los Genes TetA y TetB Como Mecanismo de Resistencia a Tetraciclina y Minociclina E Aislamientos Clínicos de *Acinetobacter baumannii*. *Enferm. Infecc. Microbiol. Clin.* **2006**, *24*, 77–80. [CrossRef]

Xian-Zhong, S.; Jing, C.; Tohru, M.; Teruo, K.; Tomofusa, T. AbeM, an H+-Coupled *Acinetobacter baumannii* Multidrug Efflux Pump Belonging to the MATE Family of Transporters. *Antimicrob. Agents Chemother.* **2005**, *49*, 4362–4364. [CrossRef]

Couvé-Deacon, E.; Jové, T.; Afouda, P.; Barraud, O.; Tilloy, V.; Scaon, E.; Hervé, B.; Burucoa, C.; Kempf, M.; Marcos, J.Y.; et al. Class 1 Integrons in *Acinetobacter baumannii*: A Weak Expression of Gene Cassettes to Counterbalance the Lack of LexA-Driven Integrase Repression. *Int. J. Antimicrob. Agents* **2019**, *53*, 491–499. [CrossRef] [PubMed]

Butler, M.S.; Gigante, V.; Sati, H.; Paulin, S.; Al-Sulaiman, L.; Rex, J.H.; Fernandes, P.; Arias, C.A.; Paul, M.; Thwaites, G.E.; et al. Analysis of the Clinical Pipeline of Treatments for Drug-Resistant Bacterial Infections: Despite Progress, More Action Is Needed. *Antimicrob. Agents Chemother.* **2022**, *66*, e0199121. [CrossRef]

Hesterkamp, T. *Antibiotics Clinical Development and Pipeline BT—How to Overcome the Antibiotic Crisis: Facts, Challenges, Technologies and Future Perspectives*; Stadler, M., Dersch, P., Eds.; Springer International Publishing: Cham, Switzerland, 2016; pp. 447–474, ISBN 978-3-319-49284-1.

Provenzani, A.; Hospodar, A.R.; Meyer, A.L.; Leonardi Vinci, D.; Hwang, E.Y.; Butrus, C.M.; Polidori, P. Multidrug-Resistant Gram-Negative Organisms: A Review of Recently Approved Antibiotics and Novel Pipeline Agents. *Int. J. Clin. Pharm.* **2020**, *42*, 1016–1025. [CrossRef] [PubMed]

Gatti, M.; Cosentino, F.; Giannella, M.; Viale, P.; Pea, F. Clinical Efficacy of Cefiderocol-Based Regimens in Patients with Carbapenem-Resistant *Acinetobacter baumannii* Infections: A Systematic Review with Meta-Analysis. *Int. J. Antimicrob. Agents* **2024**, *63*, 107047. [CrossRef] [PubMed]

Scott, C.J.; Zhu, E.; Jayakumar, R.A.; Shan, G.; Viswesh, V. Efficacy of Eravacycline Versus Best Previously Available Therapy for Adults With Pneumonia Due to Difficult-to-Treat Resistant (DTR) *Acinetobacter baumannii. Ann. Pharmacother.* **2022**, *56*, 1299–1307. [CrossRef]

Alosaimy, S.; Morrisette, T.; Lagnf, A.M.; Rojas, L.M.; King, M.A.; Pullinger, B.M.; Hobbs, A.L.V.; Perkins, N.B., 3rd; Veve, M.P.; Bouchard, J.; et al. Clinical Outcomes of Eravacycline in Patients Treated Predominately for Carbapenem-Resistant *Acinetobacter baumannii. Microbiol. Spectr.* **2022**, *10*, e00479-22. [CrossRef]

Landman, D.; Kelly, P.; Bäcker, M.; Babu, E.; Shah, N.; Bratu, S.; Quale, J. Antimicrobial Activity of a Novel Aminoglycoside, ACHN-490, against *Acinetobacter baumannii* and *Pseudomonas aeruginosa* from New York City. *J. Antimicrob. Chemother.* **2011**, *66*, 332–334. [CrossRef] [PubMed]

Keam, S.J. Sulbactam/Durlobactam: First Approval. *Drugs* **2023**, *83*, 1245–1252. [CrossRef] [PubMed]

Nguyen, L.P.; Pinto, N.A.; Vu, T.N.; Lee, H.; Cho, Y.L.; Byun, J.-H.; D'Souza, R.; Yong, D. In Vitro Activity of a Novel Siderophore-Cephalosporin, GT-1 and Serine-Type β-Lactamase Inhibitor, GT-055, against *Escherichia coli, Klebsiella pneumoniae* and *Acinetobacter* spp. Panel Strains. *Antibiotics* **2020**, *9*, 267. [CrossRef] [PubMed]

Saito, H.; Yoshikuni, O.; Megumi, C.; Kazuki, H.; Naomasa, G. Potent In Vitro Antibacterial Activity of DS-8587, a Novel Broad-Spectrum Quinolone, against *Acinetobacter baumannii. Antimicrob. Agents Chemother.* **2013**, *57*, 1978–1981. [CrossRef]

Isler, B.; Doi, Y.; Bonomo, R.A.; Paterson, D.L. New Treatment Options against Carbapenem-Resistant *Acinetobacter baumannii* Infections. *Antimicrob. Agents Chemother.* **2018**, *63*. [CrossRef] [PubMed]

Awan, R.E.; Zainab, S.; Yousuf, F.J.; Mughal, S. AI-Driven Drug Discovery: Exploring Abaucin as a Promising Treatment against Multidrug-Resistant *Acinetobacter baumannii. Health Sci. Rep.* **2024**, *7*, e2150. [CrossRef]

Hua, Y.; Luo, T.; Yang, Y.; Dong, D.; Wang, R.; Wang, Y.; Xu, M.; Guo, X.; Hu, F.; He, P. Phage Therapy as a Promising New Treatment for Lung Infection Caused by Carbapenem-Resistant *Acinetobacter baumannii* in Mice. *Front. Microbiol.* **2018**, *8*, 2659. [CrossRef]

Lood, R.; Winer, B.Y.; Pelzek, A.J.; Diez-Martinez, R.; Thandar, M.; Euler, C.W.; Schuch, R.; Fischetti, V.A. Novel Phage Lysin Capable of Killing the Multidrug-Resistant Gram-Negative Bacterium *Acinetobacter baumannii* in a Mouse Bacteremia Model. *Antimicrob. Agents Chemother.* **2015**, *59*, 1983–1991. [CrossRef] [PubMed]

Nielsen, T.B.; Pantapalangkoor, P.; Luna, B.M.; Bruhn, K.W.; Yan, J.; Dekitani, K.; Hsieh, S.; Yeshoua, B.; Pascual, B.; Vinogradov, E.; et al. Monoclonal Antibody Protects Against *Acinetobacter baumannii* Infection by Enhancing Bacterial Clearance and Evading Sepsis. *J. Infect. Dis.* **2017**, *216*, 489–501. [CrossRef] [PubMed]

McConnell, M.J.; Rumbo, C.; Bou, G.; Pachón, J. Outer Membrane Vesicles as an Acellular Vaccine against *Acinetobacter baumannii. Vaccine* **2011**, *29*, 5705–5710. [CrossRef] [PubMed]

Verma, P.; Tiwari, M.; Tiwari, V. Efflux Pumps in Multidrug-Resistant *Acinetobacter baumannii*: Current Status and Challenges in the Discovery of Efflux Pumps Inhibitors. *Microb. Pathog.* **2021**, *152*, 104766. [CrossRef]

Xing, L.; Barnie, P.A.; Su, Z.; Xu, H. Development of Efflux Pumps and Inhibitors (EPIs) in *A. baumanii. Clin. Microb.* **2014**, *3*, 135. [CrossRef]

Post, V.; Hall, R.M. AbaR5, a Large Multiple-Antibiotic Resistance Region Found in *Acinetobacter baumannii. Antimicrob. Agents Chemother.* **2009**, *53*, 2667–2671. [CrossRef]

Rumbo, C.; Gato, E.; López, M.; Ruiz de Alegría, C.; Fernández-Cuenca, F.; Martínez-Martínez, L.; Vila, J.; Pachón, J.; Cisneros, J.M.; Rodríguez-Baño, J.; et al. Contribution of Efflux Pumps, Porins, and β-Lactamases to Multidrug Resistance in Clinical Isolates of *Acinetobacter baumannii. Antimicrob. Agents Chemother.* **2013**, *57*, 5247–5257. [CrossRef] [PubMed]

Shi, W.F.; Jiang, J.P.; Ml, Z.H. Relationship between Antimicrobial Resistance and Aminoglycoside-Modifying Enzyme Gene Expressions in *Acinetobacter baumannii. Chin. Med. J.* **2024**, *118*, 141–145. [CrossRef]

40. Onohuean, H.; Nwodo, U.U. Polymorphism and Mutational Diversity of Virulence (VcgCPI/VcgCPE) and Resistance Determinants (Aac(3)-IIa, (AacC2, StrA, Sul 1, and 11) among Human Pathogenic Vibrio Species Recovered from Surface Waters in South-Western Districts of Uganda. *J. Genet. Eng. Biotechnol.* **2023**, *21*, 94. [CrossRef] [PubMed]
41. Sacha, P.; Jaworowska, J.; Ojdana, D.; Wieczorek, P.; Czaban, S.; Tryniszewska, E. Occurrence of the *AacA4* Gene among Multidrug Resistant Strains of *Pseudomonas aeruginosa* Isolated from Bronchial Secretions Obtained from the Intensive Therapy Unit at University Hospital in Bialystok, Poland. *Folia Histochem. Cytobiol.* **2012**, *50*, 322–324. [CrossRef] [PubMed]
42. Cabrera, R.; Fernández-Barat, L.; Vázquez, N.; Alcaraz-Serrano, V.; Bueno-Freire, L.; Amaro, R.; López-Aladid, R.; Oscanoa, P.; Muñoz, L.; Vila, J.; et al. Resistance Mechanisms and Molecular Epidemiology of *Pseudomonas aeruginosa* Strains from Patients with Bronchiectasis. *J. Antimicrob. Chemother.* **2022**, *77*, 1600–1610. [CrossRef]
43. Huang, F.C.; Klaus, S.; Herz, S.; Zou, Z.; Koop, H.U.; Golds, T. Efficient Plastid Transformation in Tobacco Using the AphA-6 Gene and Kanamycin Selection. *Mol. Genet. Genom.* **2002**, *268*, 19–27. [CrossRef] [PubMed]
44. Neshani, A.; Sedighian, H.; Mirhosseini, S.A.; Ghazvini, K.; Zare, H.; Jahangiri, A. Antimicrobial Peptides as a Promising Treatment Option against *Acinetobacter baumannii* Infections. *Microb. Pathog.* **2020**, *146*, 104238. [CrossRef]
45. Rangel, K.; Lechuga, G.C.; Provance, D.W.; Morel, C.M.; De Simone, S.G. An Update on the Therapeutic Potential of Antimicrobial Peptides against *Acinetobacter baumannii* Infections. *Pharmaceuticals* **2023**, *16*, 1281. [CrossRef]
46. Panjla, A.; Kaul, G.; Chopra, S.; Titz, A.; Verma, S. Short Peptides and Their Mimetics as Potent Antibacterial Agents and Antibiotic Adjuvants. *ACS Chem. Biol.* **2021**, *16*, 2731–2745. [CrossRef]
47. Chen, C.; Shi, J.; Wang, D.; Kong, P.; Wang, Z.; Liu, Y. Antimicrobial Peptides as Promising Antibiotic Adjuvants to Combat Drug-Resistant Pathogens. *Crit. Rev. Microbiol.* **2024**, *50*, 267–284. [CrossRef] [PubMed]
48. Magnet, S.; Courvalin, P.; Lambert, T. Resistance-Nodulation-Cell Division-Type Efflux Pump Involved in Aminoglycoside Resistance in *Acinetobacter baumannii* Strain BM4454. *Antimicrob. Agents Chemother.* **2001**, *45*, 3375–3380. [CrossRef]
49. Srinivasan, V.B.; Venkataramaiah, M.; Mondal, A.; Rajamohan, G. Functional Characterization of AbeD, an RND-Type Membrane Transporter in Antimicrobial Resistance in *Acinetobacter baumannii*. *PLoS ONE* **2015**, *10*, e0141314. [CrossRef] [PubMed]
50. Tipton, K.A.; Farokhyfar, M.; Rather, P.N. Multiple Roles for a Novel RND Type Efflux System in *Acinetobacter baumannii* AB5075. *Microbiologyopen* **2017**, *6*, e00418. [CrossRef]
51. Kornelsen, V.; Kumar, A. Update on Multidrug Resistance Efflux Pumps in *Acinetobacter* spp. *Antimicrob. Agents Chemother.* **2021**, *65*, e0051421. [CrossRef]
52. Cuthbertson, L.; Nodwell, J.R. The TetR Family of Regulators. *Microbiol. Mol. Biol. Rev.* **2013**, *77*, 440–475. [CrossRef] [PubMed]
53. Ramos, J.L.; Martínez-Bueno, M.; Molina-Henares, A.J.; Terán, W.; Watanabe, K.; Zhang, X.; Gallegos, M.T.; Brennan, R.; Tobes, R. The TetR Family of Transcriptional Repressors. *Microbiol. Mol. Biol. Rev.* **2005**, *69*, 326–356. [CrossRef] [PubMed]
54. Martínez-Bueno, M.; Molina-Henares, A.J.; Pareja, E.; Ramos, J.L.; Tobes, R. BacTregulators: A Database of Transcriptional Regulators in Bacteria and Archaea. *Bioinformatics* **2004**, *20*, 2787–2791. [CrossRef]
55. Casella, L.G.; Weiss, A.; Pérez-Rueda, E.; Antonio Ibarra, J.; Shaw, L.N. Towards the Complete Proteinaceous Regulome of *Acinetobacter baumannii*. *Microb. Genom.* **2017**, *3*, mgen000107. [CrossRef]
56. Beck, C.F.; Mutzel, R.; Barbé, J.; Müller, W. A Multifunctional Gene (TetR) Controls Tn10-Encoded Tetracycline Resistance. *J. Bacteriol.* **1982**, *150*, 633–642. [CrossRef] [PubMed]
57. Rosenfeld, N.; Bouchier, C.; Courvalin, P.; Périchon, B. Expression of the Resistance-Nodulation-Cell Division Pump AdeIJK in *Acinetobacter baumannii* Is Regulated by AdeN, a TetR-Type Regulator. *Antimicrob. Agents Chemother.* **2012**, *56*, 2504–2510. [CrossRef] [PubMed]
58. Abdi, S.N.; Ghotaslou, R.; Ganbarov, K.; Mobed, A.; Tanomand, A.; Yousefi, M.; Asgharzadeh, M.; Kafil, H.S. *Acinetobacter baumannii* Efflux Pumps and Antibiotic Resistance. *Infect. Drug Resist.* **2020**, *13*, 423–434. [CrossRef]
59. De Gaetano, G.V.; Lentini, G.; Famà, A.; Coppolino, F.; Beninati, C. Antimicrobial Resistance: Two-Component Regulatory Systems and Multidrug Efflux Pumps. *Antibiotics* **2023**, *12*, 965. [CrossRef]
60. Okada, U.; Yamashita, E.; Neuberger, A.; Morimoto, M.; van Veen, H.W.; Murakami, S. Crystal Structure of Tripartite-Type ABC Transporter MacB from *Acinetobacter baumannii*. *Nat. Commun.* **2017**, *8*, 1336. [CrossRef] [PubMed]
61. Shirshikova, T.V.; Sierra-Bakhshi, C.G.; Kamaletdinova, L.K.; Matrosova, L.E.; Khabipova, N.N.; Evtugyn, V.G.; Khilyas, I.V.; Danilova, I.V.; Mardanova, A.M.; Sharipova, M.R.; et al. The ABC-Type Efflux Pump MacAB Is Involved in Protection of Serratia Marcescens against Aminoglycoside Antibiotics, Polymyxins, and Oxidative Stress. *mSphere* **2021**, *6*, e00033-21. [CrossRef] [PubMed]
62. Yukun, L.; Shumpei, A.; Takumi, I.; Hiroyasu, O. Regulation of Multidrug Efflux Pumps by TetR Family Transcriptional Repressor Negatively Affects Secondary Metabolism in Streptomyces Coelicolor A3(2). *Appl. Environ. Microbiol.* **2023**, *89*, e01822-22. [CrossRef]
63. Hutson, M. Foldseek Gives AlphaFold Protein Database a Rapid Search Tool. *Nature*, 2023; *Online ahead of print*. [CrossRef]
64. López-Rojas, R.; Jiménez-Mejías, M.E.; Lepe, J.A.; Pachón, J. *Acinetobacter baumannii* Resistant to Colistin Alters Its Antibiotic Resistance Profile: A Case Report From Spain. *J. Infect. Dis.* **2011**, *204*, 1147–1148. [CrossRef]
65. Fernández-Cuenca, F.; Tomás-carmona, M.; Caballero-moyano, F.; Bou, G.; Martínez-martínez, L.; Vila, J.; Pachón, J.; Miguel, J. Actividad de 18 Agentes Antimicrobianos Frente a Aislados Clínicos de *Acinetobacter baumannii*: Segundo Estudio Nacional Multicéntrico (Proyecto GEIH-REIPI-Ab 2010). *Enfermedades Infecc. Microbiol. Clin.* **2013**, *31*, 4–9. [CrossRef]

. Kishk, R.; Soliman, N.; Nemr, N.; Eldesouki, R.; Mahrous, N.; Gobouri, A.; Azab, E.; Anani, M. Prevalence of Aminoglycoside Resistance and Aminoglycoside Modifying Enzymes in *Acinetobacter baumannii* Among Intensive Care Unit Patients, Ismailia, Egypt. *Infect. Drug Resist.* **2021**, *14*, 143–150. [CrossRef]

'. Chegeni, F.B.; Cheraghipour, K.; Shakib, P. Detection of Aacc1 and Aacc2 Genes in Clinical Isolates of Klebsiella Pneumoniae. *Int. J. Med. Investig.* **2020**, *9*, 29–35.

. Rastegar Lari, A.; Valadan Tahbaz, S. Characterization of aminoglycoside resistance mechanisms in *Acinetobacter baumannii* isolates from burn wound colonization mécanismes de résistance aux aminosides *d'Acinetobacter baumannii* isolé de zones brûlées. *Ann. Fires Burn Disaster* **2019**, *32*, 115–121.

. Cosgaya, C.; Ratia, C.; Marí-Almirall, M.; Rubio, L.; Higgins, P.G.; Seifert, H.; Roca, I.; Vila, J. In Vitro and in Vivo Virulence Potential of the Emergent Species of the *Acinetobacter baumannii* (Ab) Group. *Front. Microbiol.* **2019**, *10*, 2429. [CrossRef]

. CLSI. *M100 Performance Standards for Antimicrobial Susceptibility Testing*, 30th ed.; Clinical and Laboratory Standards Institute: Wayne, PA, USA, 2020; Volume 40, ISBN 9781684400669.

. Wickham, H. *Data Transformation BT—Ggplot2: Elegant Graphics for Data Analysis*; Wickham, H., Ed.; Springer International Publishing: Cham, Switzerland, 2016; pp. 203–220, ISBN 978-3-319-24277-4.

2. R Core Team. *R: A Language and Environment for Statistical Computing*; R Core Team: Vienna, Austria, 2024.

3. Billal, D.S.; Fedorko, D.P.; Yan, S.S.; Hotomi, M.; Fujihara, K.; Nelson, N.; Yamanaka, N. In Vitro Induction and Selection of Fluoroquinolone-Resistant Mutants of Streptococcus Pyogenes Strains with Multiple Emm Types. *J. Antimicrob. Chemother.* **2007**, *59*, 28–34. [CrossRef]

4. Drlica, K. The Mutant Selection Window and Antimicrobial Resistance. *J. Antimicrob. Chemother.* **2003**, *52*, 11–17. [CrossRef]

5. Langmead, B.; Salzberg, S.L. Fast Gapped-Read Alignment with Bowtie 2. *Nat. Methods* **2012**, *9*, 357–359. [CrossRef]

6. Danecek, P.; Bonfield, J.K.; Liddle, J.; Marshall, J.; Ohan, V.; Pollard, M.O.; Whitwham, A.; Keane, T.; McCarthy, S.A.; Davies, R.M.; et al. Twelve Years of SAMtools and BCFtools. *Gigascience* **2021**, *10*, giab008. [CrossRef] [PubMed]

7. McKenna, A.; Hanna, M.; Banks, E.; Sivachenko, A.; Cibulskis, K.; Kernytsky, A.; Garimella, K.; Altshuler, D.; Gabriel, S.; Daly, M.; et al. The Genome Analysis Toolkit: A MapReduce Framework for Analyzing next-Generation DNA Sequencing Data. *Genome Res.* **2010**, *20*, 1297–1303. [CrossRef] [PubMed]

8. Seemann, T. Prokka: Rapid Prokaryotic Genome Annotation. *Bioinformatics* **2014**, *30*, 2068–2069. [CrossRef]

9. Schwengers, O.; Jelonek, L.; Dieckmann, M.A.; Beyvers, S.; Blom, J.; Goesmann, A. Bakta: Rapid and Standardized Annotation of Bacterial Genomes via Alignment-Free Sequence Identification. *Microb. Genom.* **2021**, *7*, 685. [CrossRef]

0. Cingolani, P.; Platts, A.; Wang, L.L.; Coon, M.; Nguyen, T.; Wang, L.; Land, S.J.; Lu, X.; Ruden, D.M. A Program for Annotating and Predicting the Effects of Single Nucleotide Polymorphisms, SnpEff: SNPs in the Genome of Drosophila Melanogaster Strain W1118; Iso-2; Iso-3. *Fly* **2012**, *6*, 80–92. [CrossRef]

antibiotics

MDP

Correction

Correction: Roson-Calero et al. Cyclic Peptide MV6, an Aminoglycoside Efficacy Enhancer Against *Acinetobacter baumannii*. *Antibiotics* 2024, *13*, 1147

Natalia Roson-Calero [1,2], Jimmy Lucas [1], María A. Gomis-Font [3,4], Roger de Pedro-Jové [1], Antonio Oliver [3,4], Clara Ballesté-Delpierre [3,5] and Jordi Vila [1,2,3,5,*]

1 Barcelona Institute for Global Health (ISGlobal), 08036 Barcelona, Spain; jimmy.lucas@isglobal.org (J.L.)
2 Department of Basic Clinical Practice, School of Medicine, University of Barcelona, 08036 Barcelona, Spain
3 CIBER de Enfermedades Infecciosas (CIBERINFEC), Instituto Salud Carlos III, 28029 Madrid, Spain
4 Department of Microbiology, Hospital Universitario Son Espases, Health Research Institute of the Balearic Islands (IdISBa), 07120 Palma de Mallorca, Spain
5 Department of Clinical Microbiology, Biomedical Diagnostic Center, Hospital Clinic, 08036 Barcelona, Spa
* Correspondence: jvila@clinic.cat

1. Error in Figure

In the original publication [1], there was a mistake in Figure 1 as published. Th molecule represented in Figure 1 is inaccurate, as the order of the amino acids is reversed, and the amine group of D-Pro(NH_2) must be eliminated. The corrected Figure appears below.

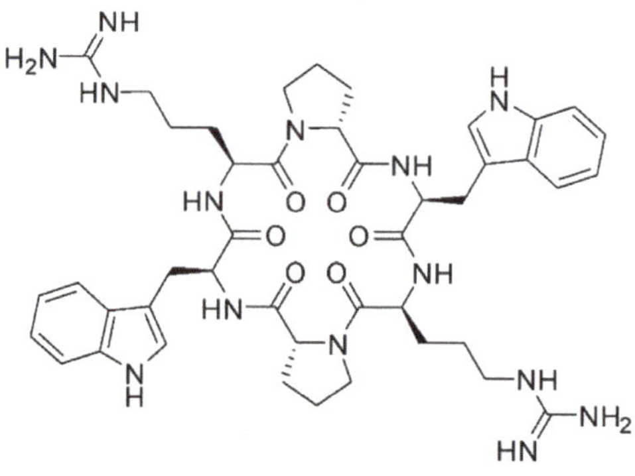

Figure 1. Chemical structure of MV6 cyclic peptide.

2. Text Correction

There was an error in the original publication [1]. The sequence of MV6 specifie in the Section "2.1. MV6 Structure" is incorrect. It includes an additional amino grou in the first D-Proline composing the peptide. Additionally, the correct pattern of MV6 i Arg-D-Pro-Trp, not Trp-D-Pro-Arg.

A correction has been made to Section 2. Results, "2.1. MV6 Structure", first paragraph

The MV6 cyclic peptide was selected for further study against *A. baumannii* from synthetic library of 28 cyclic peptides following a small-scale "shot in the dark" approach in which each peptide was tested in combination with various antibiotics and bacteria species. Its structure consists of six amino acids, two arginine residues (Arg), two D-prolin

Received: 15 January 2025
Accepted: 24 January 2025
Published: 11 February 2025

Citation: Roson-Calero, N.; Lucas, J.; Gomis-Font, M.A.; de Pedro-Jové, R.; Oliver, A.; Ballesté-Delpierre, C.; Vila, J. Correction: Roson-Calero et al. Cyclic Peptide MV6, an Aminoglycoside Efficacy Enhancer Against *Acinetobacter baumannii*. *Antibiotics* 2024, *13*, 1147. *Antibiotics* **2025**, *14*, 174. https://doi.org/10.3390/antibiotics14020174

residues (D-Pro), and two tryptophan residues (Trp), arranged in a cyclic configuration. The final structure is &Arg-D-Pro-Trp-Arg-D-Pro-Trp& (Figure 1).

The authors state that the scientific conclusions are unaffected. This correction was approved by the Academic Editor. The original publication has also been updated.

Reference

Roson-Calero, N.; Lucas, J.; Gomis-Font, M.A.; de Pedro-Jové, R.; Oliver, A.; Ballesté-Delpierre, C.; Vila, J. Cyclic Peptide MV6, an Aminoglycoside Efficacy Enhancer Against *Acinetobacter baumannii*. *Antibiotics* **2024**, *13*, 1147. [CrossRef]

antibiotics

Article

Immunomodulatory Effects of the Tobacco Defensin NaD1

Ekaterina I. Finkina [1,*,†], Ivan V. Bogdanov [1,†], Olga V. Shevchenko [1,2], Serafima I. Fateeva [1,3], Anastasia A. Ignatova [1], Sergey V. Balandin [1] and Tatiana V. Ovchinnikova [1,2,3]

1 M.M. Shemyakin and Yu.A. Ovchinnikov Institute of Bioorganic Chemistry, Russian Academy of Sciences 117997 Moscow, Russia; ovch@ibch.ru (T.V.O.)
2 Moscow Center for Advanced Studies, 123592 Moscow, Russia
3 Department of Bioorganic Chemistry, Lomonosov Moscow State University, 119991 Moscow, Russia
* Correspondence: finkina@mail.ru; Tel.: +7-495-335-0900
† These authors contributed equally to this work.

Abstract: Background/Objectives: Defensins are important components of the innate plant immur system, exhibiting antimicrobial activity against phytopathogens, as well as against fungi pathogen to humans. Along with antifungal activity, plant defensins are also capable of influencing variou immune processes, but not much is known about these effects. In this study, we investigated th immunomodulatory effects of the tobacco defensin NaD1, which possesses a pronounced antifu gal activity. **Methods and Results:** We showed that NaD1 could penetrate the Caco-2 polarize monolayer. Using a multiplex assay with a panel of 48 cytokines, chemokines and growth factor we demonstrated that NaD1 at a concentration of 2 μM had immunomodulatory effects on huma dendritic cells and blood monocytes, mainly inhibiting the production of various immune factor Using the sandwich ELISA method, we demonstrated that NaD1 at the same concentration ha a pronounced immunomodulatory effect on unstimulated THP-1-derived macrophages and thos stimulated by bacterial LPS or fungal zymosan. NaD1 had a dual effect and induced the productio of both pro-inflammatory cytokine IL-1β as well as anti-inflammatory IL-10 on resting and pr inflammatory THP-1-derived macrophages. We also found that the immunomodulatory effects of th tobacco defensin NaD1 and the pea defensin Psd1 differed from each other, indicating nonuniformit in the modes of action of plant defensins. **Conclusions:** Thus, our data demonstrated that the tobacc defensin NaD1 exhibits different immunomodulatory effects on various immune cells. We hypoth sized that influence on human immune system along with antifungal activity, could determine th effectiveness of this peptide under infection in vivo.

Keywords: antimicrobial peptides (AMPs); host defense peptides; plant defensins; tobacco defensi NaD1; pea defensin Psd1; human β-defensin 2 (HBD2); human cathelicidin LL-37; immunomodulator effects; lipopolysaccharide (LPS); zymosan; cytokines

Citation: Finkina, E.I.; Bogdanov, I.V.; Shevchenko, O.V.; Fateeva, S.I.; Ignatova, A.A.; Balandin, S.V.; Ovchinnikova, T.V. Immunomodulatory Effects of the Tobacco Defensin NaD1. *Antibiotics* **2024**, *13*, 1101. https://doi.org/10.3390/antibiotics13111101

Academic Editors: Jean-Marc Sabatier, Marisa Di Pietro and Piyush Baindara

Received: 9 October 2024
Revised: 13 November 2024
Accepted: 14 November 2024
Published: 19 November 2024

1. Introduction

It is well known that host defense antimicrobial peptides (AMPs) such as humar defensins and cathelicidins not only effectively inhibit the growth of pathogenic microor ganisms but also possess immunomodulatory activity, which helps to prevent infection. Ir addition, it has been shown that a number of AMPs from marine organisms [1], insects [2] plants [3] and others also exhibit immunomodulatory properties. Recent data demonstrated that plant defensins were also able to influence the human immune system [4].

Plant defensins are AMPs which protect plants against pathogen and parasite invasion Plant defensins effectively inhibit the growth of human pathogenic fungi of the *Candida Aspergillus* and *Cryptococcus* genera, which, according to phenotypic tests, PCR assays and metagenomic-based studies, are a common cause of life-threatening fungal infections [5] Nowadays, plant defensins are considered as high-potential prototypes of new antifungal drugs [4,6]. These AMPs have similar spatial organizations but are characterized by low amino acid sequence homology, which is probably the reason for the diversity of mechanisms of their antifungal action [4].

Several plant defensins have been demonstrated to have immunomodulatory action on epithelial, immune and endothelial cells. γ-Thionin from *Capsicum chinense* upregulated the expression of TLR2 receptor, cytokines TNF-α, IL-1β and IL-10 and also activated the transcriptional factors of inflammatory response in untreated bovine mammary epithelial cells (bMECs) and those infected by *Staphylococcus aureus* [7,8]. This peptide significantly reduced the internalization of *S. aureus* cells into bMECs, which was not related to its antibacterial activity [7]. The peptide solyC, based on the γ-motif of tomato defensins, exhibited an anti-inflammatory activity, decreasing the production of pro-inflammatory cytokines by LPS-stimulated THP-1 cells [9]. The pea defensin Psd1 up-regulated the expression of human β-defensin 2 (HBD-2) and pro-inflammatory cytokines in epithelial cells Caco-2, decreasing the effects of *Candida albicans* [10]. The pea Psd1 and the lentil Lc-def defensins induced the production of both pro- and anti-inflammatory cytokines/chemokines and growth factors by monocyte-derived dendritic cells and monocytes [10,11]. PaDef from avocado and γ-thionin from *C. chinense* reduced the VEGF-induced proliferation of bovine endothelial cells, as well as other effects of pro-angiogenic factor VEGF [12].

Floral defensin NaD1, a peptide with pronounced antifungal activity, is found in tobacco, which is known to be abundantly rich in biologically active phytochemicals [13]. This peptide is characterized by a complex mechanism of antifungal action, interacts with the fungal cell wall, affects the permeability of the plasma cell membrane by interacting with phosphatidylinositol-4,5-bisphosphate (PI(4,5)P2), penetrates into the cell and causes oxidative cell stress [14,15]. As shown, the development of yeast resistance to NaD1 can take place, but it arises more slowly than to the conventional antimycotic caspofungin [16]. At the same time, to date, nothing is known about the immunomodulatory effects of the tobacco defensin NaD1.

The main goal of this study was to investigate the immunomodulatory action of NaD1. At the first stage, we investigated the ability of NaD1 to penetrate the epithelial barrier by using a monolayer Caco-2 cells model. The effects of this peptide on interleukin production by such immune cells as monocytes and dendritic cells were studied by using the multiplex xMAP assay. Finally, the effects of NaD1 on THP-1-derived macrophages under inflammation caused by such pathogen-associated molecular patterns (PAMPs) as the lipopolysaccharide (LPS) of Gram-negative bacteria and zymosan from yeast cell walls were studied by using the sandwich ELISA method. The host defense antimicrobial peptides HBD2 and LL-37, produced by epithelial and immune cells, as well as the pea defensin Psd1, exhibiting immunomodulatory activity, were used by way of comparison.

2. Results and Discussion

2.1. Cytotoxic Effects of NaD1 Towards Epithelial and Immune Cells

It was shown previously that the tobacco defensin NaD1 exhibited cytotoxic properties. The IC_{50} for umbilical vein endothelial cells (HUVEC), human smooth muscles (CASMC) and human dermal fibroblast cells (AHDF) was approximately 10 μM in the MTT cell viability assay [17]. Therefore, we investigated the cytotoxicity of NaD1 towards epithelial and immune cells by using the resazurin-based method to exclude cytotoxic effects of the peptide on these cells at concentrations of 0.2, 2 and 5 μM applied in subsequent experiments. The membrane-active peptide melittin from the venom of honeybees, exhibiting high hemolytic and nonspecific cytotoxic activity [18], was used for comparation.

Human peripheral blood mononuclear cells (PBMCs) were chosen as a commonly used heterogeneous population of immune cells, consisting of lymphocytes (B cells, T cells and NK cells) and a smaller fraction of monocytic and dendritic cells. NaD1 had a rather low cytotoxic activity on PBMCs. Cell viability of approximately 90% was observed at the peptide concentration of 50 μM (Figure 1A). Even at NaD1 concentration of 180 μM, the viability of PBMCs was more than 50% (Supplementary Materials, Figure S1A). At the same time, melittin induced approximately 50% cell death at a concentration of 2.5 μM (the

calculated cytotoxic concentration of melittin corresponding to 50% viability of PBMC
(CC_{50}) was 2.54 μM) (Figure 1B).

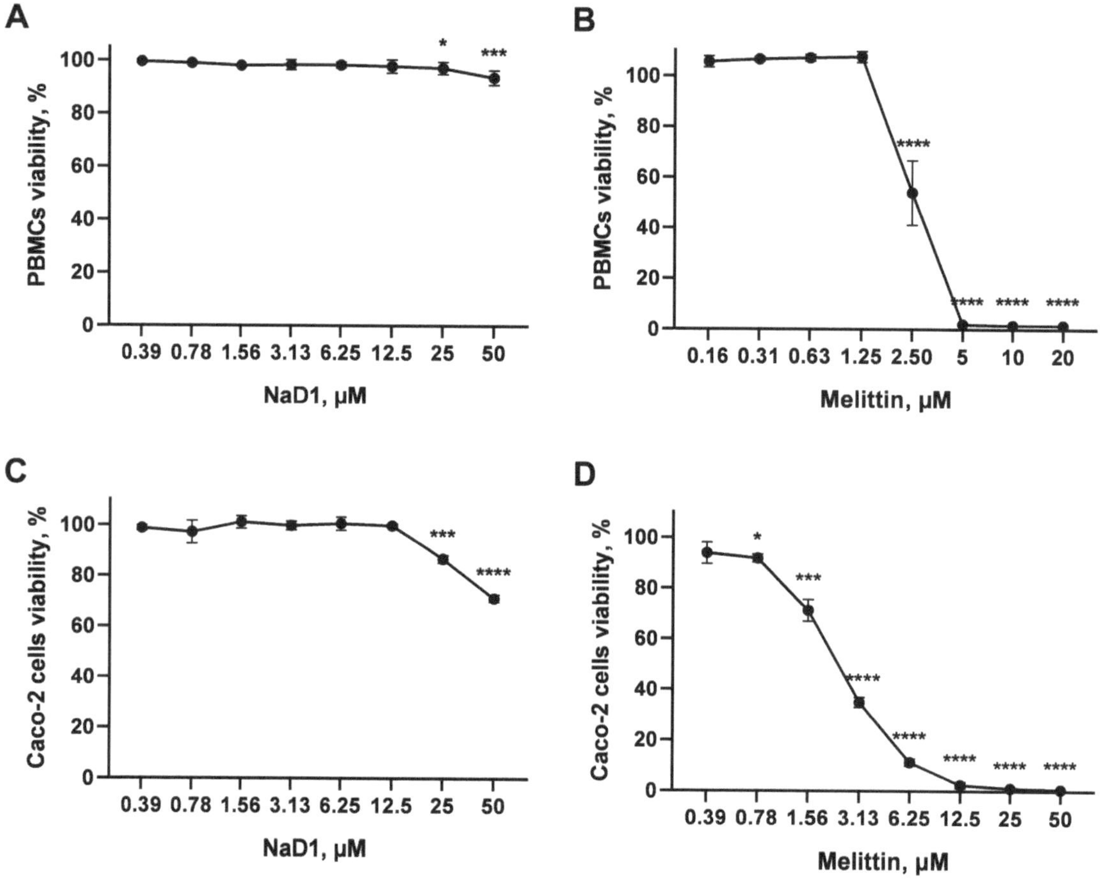

Figure 1. Cytotoxic effects of the tobacco defensin NaD1 towards PBMCs (**A**) and Caco-2 cells i monolayer (**C**). The membrane-active peptide melittin from the venom of honeybees (**B**,**D**) wa used for comparison. Error bars represent a standard deviation (±SD) between two biological an two technical replications. Significance levels are * $p \leq 0.05$, *** $p < 0.001$ and **** $p < 0.0001$. Th significance was calculated by comparing untreated cells (control) with treated by NaD1 or melitti cells. Viability cells in control and experimental samples was compared with un-paired two-sampl *t*-test.

A Caco-2 monolayer mimicking the gastrointestinal epithelial barrier was also use in the cytotoxic assay. The tobacco defensin NaD1 exhibited cytotoxic effects towards th Caco-2 monolayer, but at rather high concentrations; it had no effects on the Caco-2 cells i a monolayer at a concentration of 12.5 μM. Cell viability of approximately 85% and 70% was observed at the peptide concentrations of 25 and 50 μM, respectively (Figure 1C). Fo comparison, a decrease in cell viability was observed even at a concentration of melittin o 0.8 μM. Cell viability was less than 40% at a concentration of this peptide of 3.1 μM (th calculated CC_{50} of melittin was 2.4 μM) (Figure 1D). A similar situation was observed in th case of Caco-2 cells not in a monolayer. Cell viability of approximately 62% was observed a a NaD1 concentration of 53.3 μM. A significant increase in cell death was observed at hig peptide concentrations. Cell viability of only 8% was observed at a NaD1 concentration of 180 μM (the calculated CC_{50} of NaD1 was 61.8 μM) (Supplementary Materials, Figure S1B).

Thus, NaD1 did not exhibit a cytotoxic effect against the tested immune and epithelial cells at a concentration of 12.5 μM. Earlier, it had been shown that the human defensin HBD2 did not increase the level of THP-1 cell death at the concentration of 1.2 μM [19]. The human cathelicidin LL-37 has been previously shown to have a toxic effect on PBMCs at the concentration of 12 μM [20]. The pea defensin Psd1, as we have shown in a previous study, had no cytotoxic activity against PBMCs at the concentration of 50 μM [10].

2.2. Transfer of the Tobacco Defensin NaD1 Across the Caco-2 Polarized Monolayer

To find out whether the tobacco defensin NaD1 can penetrate epithelial barriers we evaluated the permeability of the Caco-2 polarized monolayer to this peptide. Bidirectional transport of NaD1 at the concentration of 5 μM through the polarized Caco-2 monolayer was assessed in two directions: (1) from the apical to basolateral chamber (A→B, absorptive) and (2) from the basolateral to apical chamber (B→A, secretory) (Figure 2).

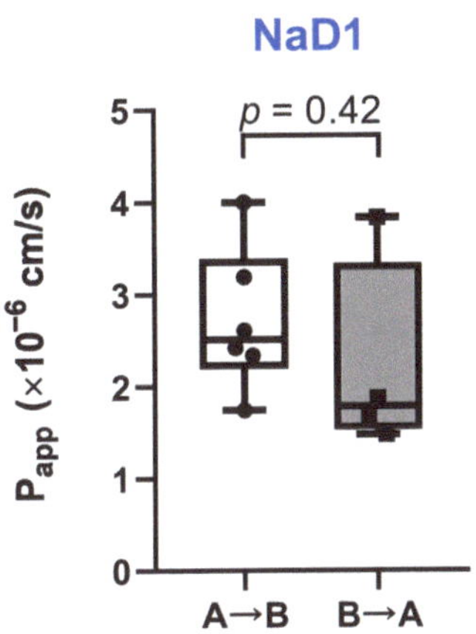

Figure 2. Assessment of bidirectional transport of the tobacco defensin NaD1 through the polarized Caco-2 monolayer. A→B, absorptive transport; B→A, secretory transport; Papp—apparent permeability coefficient. Six and four independent biological replications were used for absorptive and secretory directions, respectively. The normality of Papp coefficient distribution was assessed using Shapiro–Wilk test. Papp coefficients were compared by unpaired two-sample *t*-test.

Mean values of apparent permeability coefficients for absorptive and secretory directions were quite similar: 2.72×10^{-6} and 2.22×10^{-6} cm/s, respectively. Almost equal permeability in both directions suggests the passive transport of NaD1 through the Caco-2 monolayer. As we have shown earlier, the pea defensin Psd1 is probably subjected to active efflux, as apparent permeability coefficients were higher in the secretory direction than in the absorptive one [10]. This may mean that Psd1 is probably a weak substrate of some efflux pumps in Caco-2 cells. At the same time, the apparent permeability coefficients of NaD1 and Psd1 for A→B direction were nearly the same. This difference in the transfer of NaD1 and Psd1 through the Caco-2 polarized monolayer could be explained by discrepancies in their structural organization. It is well known that despite the similarity in spatial structures, plant defensins do not contain conserved regions in their amino acid sequences (Supplementary Materials, Figure S2) [4]. In particular, the primary structure of the tobacco NaD1 is very different from that of the pea Psd1 and has only 26% homology with the latter. In addition, the Psd1 isoelectric point is 7.73, in contrast to that of the more cationic NaD1 (pI 9.08).

Thus, based on the data obtained, we assumed that NaD1 is able to transfer across th human intestinal epithelium and interact with immune cells.

2.3. Immunomodulatory Action of the Tobacco Defensin NaD1 on Human Dendritic Cells and Blood Monocytes

In the next step of this work, we studied the immunomodulatory properties of NaD by using primary monocytes and monocyte-derived immature dendritic cells (DCs). DC represent a type of antigen-presenting cells, playing a key role in adaptive immune respons Their primary functions involve phagocytosis, processing and presentation of capture antigens to T cells. In most tissues, they are present in a state known as "immature" DC unable to activate and stimulate T lymphocytes; their activation and maturation typicall starts when DCs identify danger signals [21]. This is why immature DCs are preser where antigen entrance is expected; for instance, in lower and upper airways, skin and gu epithelial barriers, etc. In the case of immature DCs, NaD1 at the concentration of 2 μM mainly inhibited the production of cytokines and chemokines, including ones with pro and anti-inflammatory action: IL-1RA (from 199 to 133 pg/mL, $p = 0.0311$), CXCL8 (fror 1856 to 1037 pg/mL, $p = 0.0003$), CCL2 (from 514 to 288 pg/mL, $p = 0.0044$), CCL7 (fror 62.22 to 44.12 pg/mL, $p = 0.0237$), CXCL9 (from 76.28 to 16.72 pg/mL, $p = 0.0007$), CCL (from 62.72 to 35.36 pg/mL, $p = 0.0023$), CCL4 (from 204 to 120 pg/mL, $p = 0.0018$) an M-CSF (from 101 to 45.07 pg/mL, $p = 0.0079$). In the case of IL-6, IL-12(p40), TGF-α an GM-CSF, a very slight but statistically significant inhibition of the production was observe (Figure 3). Apparently, NaD1 did not induce the activation and maturation of immatur DCs, according to their cytokine profiles. In our previous studies of the immunomodulator properties of plant defensins, we have shown that the pea defensin Psd1, on the contrar induced an elevation of the production of cytokines IL-1RA, IL-5, IL-6, IL-10, IL-12, IL-1 IL-27 and TNFα and chemokines CXCL8/IL-8, CCL2/MCP-1 and CCL7/MCP-3 by matur DCs [9]. At the same time, the lentil defensin Lc-def, having 48% and 34% homology wit the pea Psd1 and the tobacco NaD1, respectively, did not induce significant changes in th secretion of cytokines and chemokines by mature DCs [11].

Then, primary human monocytes were chosen to study the immunomodulator properties of NaD1. Monocytes circulating in the blood can be recruited and extravasate into tissues during inflammatory processes [22]. Moreover, the recruitment of monocyte to the sites of inflammation is critical for host defense not only when infected, but also i the case of sterile injury [22]. In the case of monocytes, NaD1 at the same concentratio induced a decrease of chemotactic CXCL1/GROα and CXCL8/IL-8 levels and elevatio of the fibroblast growth factor 2 (FGF-2) production; however, these changes were no statistically significant (Figure 3). At the same time, NaD1 induced a slight but statisticall significant ($p = 0.0049$) elevation of the production of pro-inflammatory CCL4/MIP-1β (from 29.57 to 34.19 pg/mL), which is a potent monocyte and lymphocyte chemoattractan recruiting neutrophils, eosinophils, basophils, immature dendritic cells and natural kille cells to the site of inflammation [23]. NaD1 also increased the production of the platele derived growth factor (PDGF-AB/BB) (from 984 to 1457 pg/mL, $p = 0.0232$), which is potent mitogen for cells of mesenchymal origin and plays a significant role in angiogenesi and wound healing [24]. NaD1 also induced an elevation of the PDGF-AA level, but thi effect was not statistically significant (Supplementary Materials, Table S1). Interestingly the lentil defensin Lc-def also increased the production of PDGF-AA, PDGF-AB/BB an CCL3/MIP-1α by monocytes [11], while the pea defensin Psd1 increased the production o both CCL3/MIP-1α and CCL4/MIP-1β by monocytes [10].

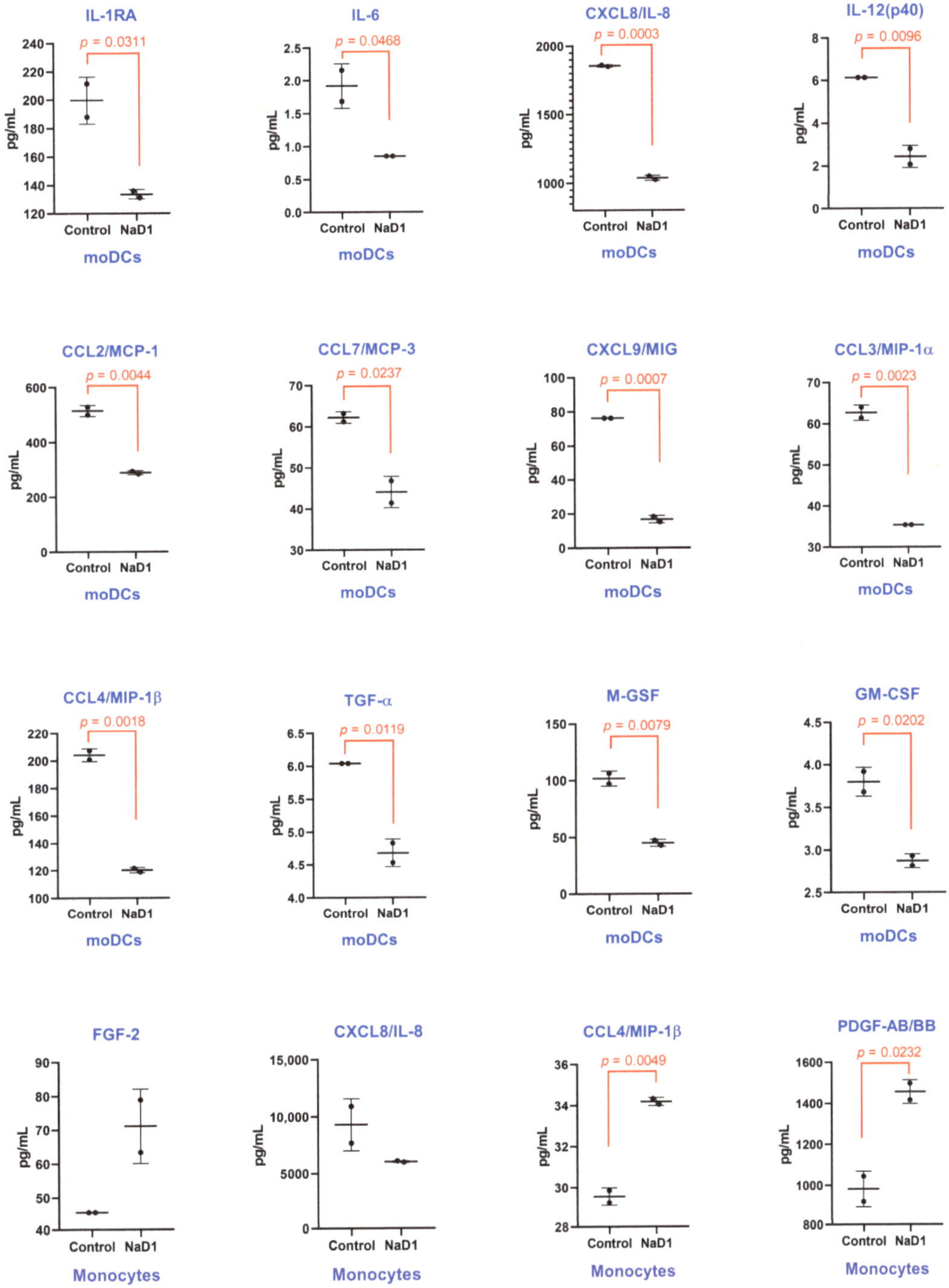

Figure 3. Production of cytokines, chemokines and growth factors upon stimulation of DCs and monocytes by NaD1 at the concentration of 2 μM. Error bars represent a standard deviation (±SD) between two biological replications. The levels in control and experimental wells were compared by unpaired two-sample *t*-test.

According to the cytokine profile, NaD1 appeared to be not capable of activating a inducing the maturation of dendritic cells. Instead, NaD1 inhibited the production of c tokines by immature DCs, and this effect was statistically significant. In primary monocyt NaD1 also induced minor effects. Other plant defensins, such as the pea Psd1 and the len Lc-def, induced much more significant changes in cytokine production by mature DC and monocytes, suggesting that the tobacco NaD1 causes mild immunomodulatory effec on these cell cultures. Taking these results together, we concluded that representativ of plant defensins could induce diverse immunomodulatory effects on various huma immunocompetent cells.

2.4. Immunomodulatory Action of the Tobacco Defensin NaD1 on Unstimulated and Stimulated LPS or Zymosan THP-1-Derived Macrophages

Next, we investigated the effects of NaD1 on THP-1-derived macrophages und inflammation caused by bacterial or fungal PAMPs, such as LPS and zymosan. LPS the most abundant component of the cell wall of Gram-negative bacteria and an agoni of the TLR4 receptor, which stimulates the production of TNF-α, IL-1β, IL-6, IL-8 an other inflammatory cytokines in various cell types in response to pathogens [25]. Zymosa prepared from yeast cell walls mainly consists of polysaccharides including β-gluca and, to a lesser extent, of mannan and activates secretion of inflammatory factors, particular TNF-α and IL-8, by macrophages, monocytes and leukocytes via TLR2 an Dectin-1 receptors [26]. Macrophages maintain tissue homeostasis, resist pathogen invasic and, therefore, are one of the key players in tissue immunity [27]. They are present different tissues and can be activated and polarized depending on their environme into pro-inflammatory (M1) or anti-inflammatory macrophages (M2) [28]. LPS or pr inflammatory cytokines (IFNγ, IL-12) induce the polarization of macrophages with th resting phenotype (M0) to the M1 phenotype while anti-inflammatory cytokines (IL-4, IL-1 and IL-13) can drive their polarization towards M2 [27]. Recently, it has been shown th zymosan induced upregulation of pro-inflammatory genes, intrinsic to M1 macrophage and downregulated the M2 genes [29]. Host defense cationic antimicrobial peptides wit immunomodulatory properties, which are produced by epithelial and immune cells, name the human β-defensin HBD2 and the cathelicidin LL-37, were used for comparison. The pe defensin Psd1 was also used in this experiment. Previously, we have shown a pronounce immunomodulatory effect of Psd1 on DCs or monocytes as well as on a Caco-2/immun cells co-culture upon a fungal infection [10].

To perform this analysis, cytokines IL-1β, IL-6, TNF-α and IL-10 were chosen. I 1β, TNF-α and IL-6 are the major pro-inflammatory cytokines, playing a key role i inflammation, whereas IL-10 is an anti-inflammatory cytokine that suppresses excessiv immune responses and antigen-presentation capacity [30]. THP-1-derived macrophage were pre-treated with antimicrobial peptides at concentrations of 2 or 0.2 μM for 2 followed by stimulation with LPS and zymosan for an additional 24 h. LPS induced th secretion of all cytokines tested, but the most pronounced effect was observed in the cas of IL-6 ($p < 0.0001$) and TNF-α ($p < 0.0001$). Zymosan had the same but less pronounce pro-inflammatory effect, and a statistically significant increase in cytokine production wa observed only for IL-6 ($p = 0.0023$) and TNF-α ($p = 0.0015$) (Figure 4A–D).

At a concentration of 2 μM, AMPs by themselves had the following effects on THP-1 derived macrophages. Only NaD1 significantly increased IL-1β level ($p < 0.0001$). Plan defensins NaD1 ($p = 0.0329$) and Psd1 ($p = 0.001$) induced the production of IL-10. Amon all tested AMPs, only LL-37 slightly increased TNF-α secretion ($p = 0.0454$). All AMPs di not influence IL-6 production (Figure 4A–D).

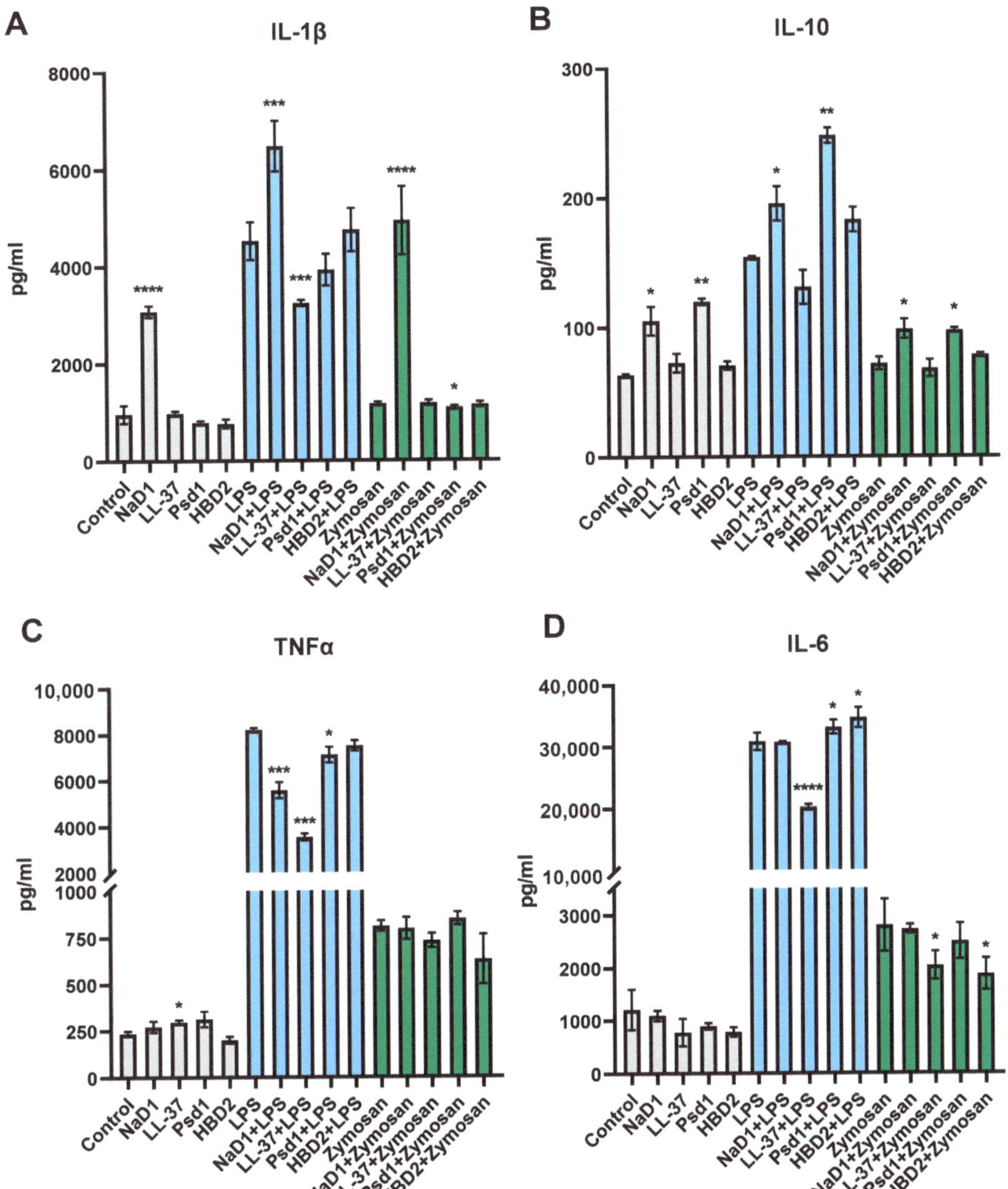

Figure 4. Influence of the tobacco defensin NaD1 and other AMPs at the concentration of 2 μM on production of pro- (**A–D**) and anti-inflammatory (**B**) cytokines either unstimulated or stimulated by LPS or by zymosan THP-1-derived macrophages. Error bars represent a standard deviation (±SD) between two biological and two technical replications. Significance levels are * $p \leq 0.05$, ** $p < 0.01$, *** $p < 0.001$, **** $p < 0.0001$. The significance of difference in cytokine production was calculated by comparing: unstimulated cells (control) with stimulated by AMPs cells (grey bars); stimulated by LPS (blue bars) or zymosan (green bars) cells alone or in the presence of AMPs. Release of the cytokines in control and experimental samples was compared with unpaired two-sample *t*-test.

Along with that, all AMPs in different ways influenced the LPS-induced pro-inflammato response of THP-1-derived macrophages. As expected, LL-37 inhibited the LPS-induce production of all cytokines tested, although for IL-10 this effect was insignificant. Amor defensins, only NaD1 increased the LPS-induced production of IL-1β ($p = 0.0009$). Pla defensins NaD1 and Psd1 increased the LPS-induced production of IL-10, but reduced th LPS-induced secretion of TNF-α. On the contrary, cell pre-incubation with Psd1 ($p = 0.043$ and HBD2 ($p = 0.0121$), but not with NaD1, increased the LPS-stimulated production IL-6 (Figure 4A–D).

AMPs also affected zymosan-induced inflammation. LL-37 slightly decreased th zymosan-induced production of TNF-α and IL-6, although only for IL-6 this effect wa significant ($p = 0.0349$). NaD1 in the presence of zymosan induced the production of IL-1 ($p < 0.0001$) more strongly than without this PAMP. At the same time, Psd1 slightly reduce the production of IL-1β ($p = 0.0121$) in the presence of zymosan. NaD1 ($p = 0.0116$) and Psd ($p = 0.0202$) increased the production of IL-10 in the presence of zymosan, but these effec were less pronounced than in the case of individual AMPs. All of the three defensins did no influence the TNF-α production. HBD2 caused a slight decrease in the zymosan-induce production of IL-6 ($p = 0.0184$) (Figure 4A–D).

At the concentration of 0.2 μM, AMPs by themselves or in the presence of PAMPs ha approximately the same action, but the effects were generally less pronounced (Suppleme tary Materials, Figure S3).

LL-37, also known as hCAP18, has been widely shown to exhibit different effec on various types of human cells. In particular, LL-37 suppresses the LPS-induced ce inflammatory response via the direct binding of LPS [31]. As shown, LL-37 interacts no only with LPS but also with polysaccharides from fungal cell walls, such as mannan, chiti and glucan [32]. In our experiments, we observed similar effects of LL-37 on the LPS- an zymosan-induced production of TNF-α and IL-6, probably due to the ability of this peptid to also bind the fungal PAMP zymosan.

As is known, HBD2 modulated various immune processes. It has been reporte that pre-stimulation of primary human macrophages or THP-1 cells with HBD2 and it subsequent removal from the culture medium resulted in the enhanced production o pro-inflammatory cytokines and chemokines induced following cell stimulation by LPS o zymosan, apparently through activation of the P2X7 receptor [19]. At the same time, co treatment with HBD2 at different concentrations mitigated the release of TNF-α and IL-1 by LPS-stimulated PBMCs [33]. In our case, pre-treatment of THP-1-derived macrophage with HBD2 followed by the addition of LPS or zymosan had no significant effect. At th same time, HBD2 increased or, conversely, decreased PAMP-induced IL-6 production in the case of LPS or zymosan, respectively.

As mentioned above, plant defensins also exhibit immunomodulatory effects. I particular, the peptide solyC, corresponding to the γ-motif of the tomato defensin famil affected THP-1 cells stimulated with LPS but had no effect on unstimulated cells. Co stimulation of THP-1 cells with this peptide and LPS resulted in a decrease of the LPS induced production of the pro-inflammatory cytokines TNF-α and IFN-γ [9]. According to our data, the plant defensins Psd1 and NaD1 had different effects on unstimulated THP-1 derived macrophages and those stimulated by LPS and zymosan. Both peptides induced the synthesis of the anti-inflammatory cytokine IL-10 by THP-1-derived macrophages, bu only NaD1 increased the IL-1β level. Both plant defensins increased IL-10 production and decreased the TNF-α level under LPS-induced inflammation, but NaD1 also significantly increased the secretion of IL-1β and Psd1 slightly increased the IL-6 level. Both plan defensins increased IL-10 production under zymosan-induced inflammation, but only NaD1 also significantly increased the secretion of IL-1β.

As is known, a number of AMPs exhibit immunomodulatory properties and trigge both anti-inflammatory and pro-inflammatory immune responses via distinct mechanisms which depend on biological context [34]. Our data also demonstrated that NaD1 had a pleiotropic action on resting and pro-inflammatory THP-1-derived macrophages, affecting the production of both pro- and anti-inflammatory cytokines. A comparison of the effects of the plant defensins NaD1 and Psd1 revealed that different representatives of plant defensins could have various effects on unstimulated and PAMP-induced pro-inflammatory THP-1-derived macrophages.

Key pro-inflammatory cytokines (IL-1, IL-6 and TNF-α) during infection stimulate the production of acute-phase proteins and attract inflammatory cells. At the same time, an elevated level of IL-6 was found to be associated with the highest risk of death in patients with sepsis [35]. We did not observe any statistically significant changes induced by NaD1 in the production of IL-6. Any immunosuppression or anti-inflammatory signals during the acute-phase response can be responsible for late death in patients with sepsis. For instance, the increase of IL-10 during sepsis was found to be the main predictor of severity and fatal outcome [35]. NaD1 induced an elevation in anti-inflammatory IL-10 production by M0 and M1 macrophages; however, this elevation was mild (from 150 to 180 pg/mL), while the increase in pro-inflammatory IL-1β was much more profound (from 4500 to 6500 pg/mL). Taken together, we demonstrated in this study that the tobacco defensin NaD1 can impact macrophages by inducing changes in the production level of the key cytokines involved in the acute-phase response.

2.5. Possible LPS- and Zymosan-Binding Capacity of the Tobacco Defensin NaD1

As mentioned above, LL-37 (pI 10.61) binds to bacterial LPS and polysaccharides from fungal cell walls, including β-glucan, and electrostatic interactions have been shown to play an important role in peptide–PAMP interaction [31,32]. At the same time, NaD1 is cationic peptide with pI 9.08 that could potentially bind to the negatively charged PAMPs. The ability of the tobacco defensin NaD1 to bind β-glucan and chitin has been shown previously [6,36]. To the best of our knowledge, nothing is known about the ability of NaD1 to interact with LPS. In order to test this and estimate the possible influence of PAMP-binding on immunomodulatory effects, in particular, a decrease in the LPS-induced production of TNF-α, we examined the antimicrobial activity of NaD1 in the presence of LPS or zymosan. LL-37 was used for the comparison in these experiments.

The tobacco defensin NaD1 is known to have a pronounced antifungal activity [6,14–16]. We tested the activity of this peptide against *Candida albicans* ATCC 18804. Taking into consideration a significant decrease in the activity of NaD1 in the presence of sodium chloride at physiological concentrations, a low-salt media such as Sabouraud broth was used. NaD1 effectively inhibited the growth of *C. albicans* at MIC 6.25 μM. LL-37 also exhibited anticandidal activity at MIC 12.5 μM under the test conditions. The next step was to test the influence of LPS and zymosan on the antifungal activity of AMPs.

The antifungal activity of LL-37 was reduced in the presence of both LPS and zymosan (Table 1). Both PAMPs at concentrations of 40 μg/mL and higher doubled the MIC of LL-37 against *C. albicans*. Higher concentrations of LL-37 were also required for the fungicidal effect on yeast-like cells in the presence of LPS or zymosan at concentrations of 100 μg/mL. At the same time, the presence of LPS or zymosan had a slight effect on the anticandidal activity of NaD1. An increase of the MIC of NaD1 was observed only in the presence of LPS at the concentration of 100 μg/mL. The fungicidal effect of NaD1 decreased in the presence of zymosan or LPS. Plating the contents of the wells with NaD1 in MIC concentrations showed a larger number of colonies of *C. albicans* cells in the case of a higher concentration of LPS or zymosan (Supplementary Materials, Figures S4 and S5). A doubling of the MFC, but not MIC, of NaD1 was found in the presence of zymosan at concentrations of 40 μg/mL and higher (Table 1). Thus, we assumed that NaD1 could potentially bind LPS, but to a lesser extent than LL-37. However, unlike LL-37, the immunomodulatory effects of NaD1 most likely did not relate to its ability to bind the PAMPs used.

Table 1. Effects of LPS and zymosan on antifungal activities of the tobacco defensin NaD1 and t human cathelicidin LL-37 towards *Candida albicans* ATCC 18804.

Test Variant	NaD1, μM			LL-37, μM		
	IC_{50}	MIC	MFC	IC_{50}	MIC	MFC
Without PAMP	3.12–6.25	6.25	12.5	6.25–12.5	12.5	25
10 μg/mL LPS	3.12–6.25	6.25	12.5	6.25–12.5	12.5	25
40 μg/mL LPS	3.12–6.25	6.25	12.5	6.25–12.5	**25**	25
100 μg/mL LPS	3.12–6.25	**12.5**	12.5	**12.5–25**	**25**	**>25**
10 μg/mL Zymosan	3.12–6.25	6.25	12.5	6.25–12.5	12.5	25
40 μg/mL Zymosan	3.12–6.25	6.25	**25**	**12.5**	**25**	**>25**
100 μg/mL Zymosan	3.12–6.25	6.25	**25**	**12.5–25**	**25**	**>25**

MIC values of the peptides in the presence of PAMPs exceeding those without them are shown in bold.

2.6. Limitations

In this study, we showed that the tobacco defensin NaD1 exhibits different i munomodulatory effects on various immune cells, including monocyte-derived dendri cells, primary blood monocytes and THP-1-derived macrophages. However, a plenty cells other than the immune cells used are involved in various immune responses, includi in the acute-phase response during infection. This is why the overall immunomodulato action of NaD1 hard to be predicted ex vivo and need to be further investigated in infla matory mouse models. It is also worth noting that the safety of NaD1 should be tested d to the cytotoxic activity of the peptide, which it exhibits in high concentrations.

3. Materials and Methods

3.1. Materials

Candida albicans ATCC 18804 was kept at −70 °C in 10% non-fat milk with 10% glycer Synthetic melittin (>98% pure) was provided by Dr. Sergey V. Sychev in M.M. Shemyak and Yu.A. Ovchinnikov Institute of Bioorganic Chemistry of the Russian Academy Sciences (Moscow, Russia).

3.2. Recombinant Production of Antimicrobial Peptides

The pea defensin Psd1 (UNIPROT P81929) was obtained as described previously [Recombinant tobacco defensin NaD1 (UNIPROT Q8GTM0), human cathelicidin LL- (UNIPROT P49913) and human β-defensin 2 (HBD2, UNIPROT O15263) were obtain by heterologous expression in E. coli cells (Supplementary Materials, Figure S2). DN fragments encoding AMPs were synthesized using PCR with overlapping primers ar inserted into the expression plasmid vector pET-His8-TrxL (Supporting Materials, Figu S6 and Figure S7 and Table S2). Correct plasmid assembly was verified by DNA sequenci performed in two directions. Heterologous expression was carried out in E. coli B 21 (DE3) cells transformed with plasmid constructs pET-His8-TrxL-NaD1/LL-37/HBI using 0.2 mM isopropyl β-D-1-thiogalactopyranoside (IPTG) as an inductor. Recombina peptides were purified from clarified cell lysates using metal chelate chromatograpl cyanogen bromide cleavage of the fusion proteins and two-stage reversed-phase hi performance liquid chromatography (RP-HPLC). Homogeneity and the identity of t recombinant peptide samples were confirmed by MALDI mass spectrometry and C spectroscopy (Supplementary Materials, Figures S8, S9 and Table S3).

3.3. Human Cell Lines and Cultures

Peripheral blood mononuclear cells (PBMCs) from healthy donors were purchas from American Type Culture Collection (ATCC PCS-800-011). Primary monocytes we isolated from PBMCs by adherence to the plastic surface [37]. For that, PBMCs were thaw resuspended in RPMI-1640 culture medium (Corning, Corning, NY, USA) without seru at a concentration of 2×10^6 cells/mL, seeded into the wells of a 24-well plate and plac in a humidified CO_2-intubator (CellXpert C170i, Eppendorf, Hamburg, Germany) for 1 h. After that, non-adherent cells were removed from the wells and attached monocytes were thoroughly washed with PBS and cultured in a CO_2-intubator in complete RPMI-1640 containing 10% human serum (HS Type AB, Capricorn Scientific, Ebsdorfergrund, Germany) and 1×antibiotic-antimycotic (1×AA) solution (Sigma, St. Louis, MO, USA).

Immature monocyte-derived dendritic cells (moDC) were obtained from primary monocytes by a differentiation protocol involving a cytokine cocktail with IL-4 and GM-CSF [38]. For this, the monocytes were cultured in complete RPMI-1640 containing 10% HS, 1×AA solution, 500 U/mL rhIL-4 (Sci-Store, Moscow, Russia) and 800 U/mL rhGM-CSF (Sci-Store) for 3 days. Then, the cells were re-fed with the fresh medium with rhIL-4 and rhGM-CSF and cultured for 4 more days.

THP-1 (ATCC TIB-202) cells were thawed and cultured in complete RPMI-1640, containing 10% FBS (Capricorn Scientific) and 1×AA solution in CO_2-intubator. Monocytic THP-1 cells were differentiated into resting M0 macrophages by stimulation for 24 h with 100 ng/mL phorbol 12-myristate 13-acetate (PMA) in complete RPMI-1640. Then, the attached macrophage-like cells were washed out with culture medium without serum and incubated for another 48 h in PMA-free complete culture medium with 10% HS and 1×AA solution.

Caco-2 (ATCC HTB-37) cells were thawed and cultured in complete DMEM/F-12 culture medium (Corning), supplemented with 10% FBS in a humidified CO_2-intubator. After the culture was subcultivated three times, the cells were seeded on cell inserts (PET, 0.4 μm, 0.6 cm^2 surface area, SPL Life Sciences, Pocheon-si, Republic of Korea) at a density of 2.5×10^5 cells/cm^2. Cells were grown for 21 days with re-feeding with the fresh medium every 2–3 days. To evaluate integrity of Caco-2 monolayer, a transepithelial electrical resistance (TEER) was measured by Millicell ERS-2 Voltohmmeter (Merck-Millipore, Burlington, MA, USA). Only inserts with TEER > 400 Ω cm^2 were used in the transport assay.

C. albicans ATCC 18804 cells in stock were inoculated onto modified YPD (yeast extract 5 g/L, peptone 10 g/L, glucose 10 g/L) agar plates and incubated for 24 h at 37 °C.

3.4. Cytotoxicity Assay

The cytotoxic effects of the tobacco defensin NaD1 towards PBMCs or Caco-2 cells were investigated in 96-well plates using the resazurin method as previously described [10]. In brief, 4×10^5 per well Caco-2 cells not in a monolayer as well as Caco-2 cells in a monolayer in DMEM/F12 (1:1) medium with 10% FBS or 2×10^6 PBMCs per well in RPMI-1640 also with 10% FBS were incubated with serial dilutions of NaD1 at final concentrations from 0.39 to 50 μM or from 2.1 to 180 μM for 24 h. After that, resazurin (Sigma, St. Louis, MO, USA) was added at a final concentration of 70 μM and the plates were incubated overnight (16–18 h). Untreated cells and cells treated by non-ionogenic detergent Triton X-100 were used as negative and positive controls, respectively. The membrane-active peptide melittin from honeybee venom was used for comparison. The cell viability was estimated by resorufin fluorescence registered at 595 HM via the following equation: cell viability (%) = $(F_{sample}/F_{control}) \times 100\%$. Experiment was carried out twice in duplicate.

3.5. Labeling of the Tobacco Defensin NaD1 with FITC

The labeling of NaD1 with FITC was performed as previously reported in [21].

3.6. Caco-2 Permeability Assay

The permeability assay was conducted in transfer buffer (HBSS solution, containing 1 mM $CaCl_2$, 1 mM $MgCl_2$, 10 mM D(+)glucose, pH 7.4). For the measuring transport of NaD1 in the absorptive direction (from the apical to basolateral chamber, A→B), 0.7 mL of the transport buffer was placed in the basolateral chamber and 0.4 mL of 5 μM FITC-labeled NaD1 in the transport buffer was placed in the apical chamber. For the measuring transport of NaD1 in the secretory direction (from the basolateral to apical chamber, B→A), 0.7 mL of 5 μM FITC-labeled NaD1 was placed in the basolateral chamber and 0.4 mL of the transport buffer was placed in the apical chamber. The permeability assay through th Caco-2 polarized monolayer was conducted for 90 min in 6 independent inserts for th absorptive direction and 4 independent inserts for the secretory direction. The Caco- permeability assay was performed twice on two different days.

The apparent permeability coefficients (Papp) were calculated for each insert accordin to the following equation: Papp = dQ/dt × (1/(A × C0)), where dQ/dt is an amou of product present in the basolateral or apical chamber as a function of time (nM/s), is an area of the insert (in cm^2) and C0 is an initial concentration of NaD1 in the apic or basolateral chamber (nM/mL). In order to verify the cell monolayer, the appare permeability of a paracellular marker, Lucifer Yellow (Sigma), was estimated.

3.7. Stimulation of Human Cell Cultures with NaD1

For the study of the immunomodulatory properties of NaD1, monocytes and immatu dendritic cells (moDC) were seeded into the wells of a 24-well plate at densities of 4 × 1 and 1.3 × 10^5 cells per well, respectively, in complete RPMI-1640 with 10% HS and A solution. Then, 24 h later, the culture medium was replaced by a fresh one either with 2 μ NaD1 for stimulation of the cells or without if for the control wells. The cultures were ke in a humidified CO_2-intubator for 24 h and then the samples of the culture media we taken and frozen.

THP-1-derived macrophages were seeded into the wells of a 24-well plate at a densit of 2.7 × 10^5 cells/well in complete RPMI-1640 with 10% HS and 1×AA solution. The 24 h later, the culture medium was replaced by the fresh complete RPMI-1640 containin lipopolysaccharide from *Escherichia coli* 0111:B4 (LPS, Sigma, St. Louis, MO, USA) o zymosan from the cell wall of *Saccharomyces cerevisiae* (Serva, Heidelberg, Germany) a concentrations of 1 or 10 μg/mL, respectively; antimicrobial peptides NaD1, Psd1, LL-37 o HBD2 at concentrations of 2 or 0.2 μM; and combinations of antimicrobial peptides wit LPS or zymosan at the same concentrations. Insoluble zymosan was resuspended in steril water and sonicated twice for 30 s; after that, the stock solution was heated to 80 °C fo 20 min. Peptides were added 2 h before stimulation with LPS and zymosan. After tha PAMPs solutions were added and the cells were cultured for 24 h. Complete RPMI-164 only was used in the control wells. Culture supernatants were collected and stored a −70 °C prior to the cytokines' assessment.

3.8. Multiplex Assessment of Cytokine Production by Monocytes and Immature Dendritic Cells upon Stimulation with NaD1

The absolute levels of 48 cytokines, chemokines and growth factors were determined a a protein level by multiplex xMAP technology (Luminex, Austin, TX, USA). For this, a MIL LIPLEX Human Panel A kit (HCYTA-60K-PX48, Merck, Darmstadt, Germany) was use and the following 48 analytes were evaluated in two biological replications: sCD40L, EG CCL11/Eotaxin-1, FGF-basic/FGF-2, Flt-3 ligand, CX3CL1/Fractalkine, G-CSF, GM-CS GROα, IFNα2, IFN-γ, IL-1α, IL-1β, IL-1RA, IL-2, IL-3, IL-4, IL-5, IL-6, IL-7, CXCL8/IL 8, IL-9, IL-10, IL-12(p40), IL-12(p70), IL-13, IL-15, IL-17A/CTLA8, IL-17E/IL-25, IL-17 IL-18, IL-22, IL-27, CXCL10/IP-10, CCL2/MCP-1, CCL7/MCP-3, M-CSF, CCL22/MDC CXCL9/MIG, CCL3/MIP-1α, CCL4/MIP-1β, PDGF-AA, PDGF-AB/BB, CCL5/RANTES TGF-α, TNF-α, TNF-β and VEGF-A. The fluorescent data were obtained on a MAGPIX sys tem (Merck) operated with xPONENT 4.2 software (Merck). Final analysis was performe in MILLIPLEX Analyst v5.1 software (Merck).

3.9. ELISA Assay of Cytokine Production by THP-1-Derived Macrophages

The production of pro- and anti-inflammatory cytokines IL-1β, IL-6, TNF-α and IL-1 by stimulated and non-stimulated THP-1-derived macrophages were estimated by using ELISA kits (Vector-Best, Koltsovo, Russia). Briefly, 96-well plates with adsorbed mono clonal antibodies to the corresponding cytokine were used. Undiluted or diluted 5, 20 o 80 times in the case of IL-10, TNF-α and IL-1β or IL-6, respectively, culture supernatants from the wells of 24-well plates were taken for analysis. Biotinylated polyclonal antibodies to cytokines, streptavidin conjugated to HRP and TMB substrate were further used for the detection of immune complexes. A panel of calibration solutions containing different concentrations of cytokines was used in each experiment. Experiments were performed using two biological and two technical repeats. The release of the cytokines in the control and experimental samples was compared with an unpaired two-sample *t*-test using GraphPad Prism v.8.0.1 (GraphPad Software, Inc., San Diego, CA, USA). The p values ≤ 0.05 were considered significant.

3.10. Antifungal Activity of NaD1 and LL-37

The antifungal assay was performed by the microdilution method using 96-well microplates as described [11]. Briefly, *C. albicans* ATTC 18804 cells in stock were inoculated onto Sabouraud agar plates with 2% glucose and incubated for 24 h at 37 °C. After replating, cells were inoculated in Sabouraud broth and cultured at 37 °C for 2 h. The cell concentration was determined using a LUNA-II cell counter (Logos Biosystems, Anyang-si, Republic of Korea). Yeast cells in Sabouraud broth diluted to concentration of 4×10^4 cells/mL were mixed with equal volumes of serial two-fold dilutions of the peptides in water and the plates were incubated at 30 °C for 24 h. The final peptide concentrations in the wells were 25, 12.5, 6.25, 3.13, 1.56, 0.78 and 0.39 µM. Controls without peptides were also tested. The wells of the microplate were previously blocked with 0.1% BSA. Yeast growth was assessed using an inverted light microscope and also by measuring absorbance at 630 nm. The minimum inhibitory concentrations IC_{50} and MIC were defined as the lowest peptide concentrations inhibiting fungal growth by at least 50 and 100%, respectively. To assess the minimum fungicidal concentrations (MFCs), the entire contents of the plate wells with peptides at MIC concentrations and higher were seeded on Sabouraud agar and incubated at 37 °C for 24 h. The MFCs corresponded to the minimal peptide concentration, in which no colony growth was observed.

To estimate the effects of LPS and zymosan on the antifungal activity of NaD1 and LL-37, these PAMPs were added in culture media at concentrations of 10, 40 and 100 µg/mL [39,40]. All of the experiments were performed twice in triplicate.

4. Conclusions

In this work, for the first time, we investigated immunomodulatory properties of the plant defensin NaD1 from tobacco flowers, which has a pronounced antimicrobial activity against pathogenic fungi. Using a monolayer of Caco-2 cells as a model system, we showed that NaD1 has the ability to permeate epithelial barriers. Using the multiplex xMAP assay, we revealed that NaD1 at the concentration of 2 µM exhibited an immunomodulatory effect on immune cells, such as primary monocytes and immature dendritic cells. In contrast to other plant defensins, such as the pea Psd1 and the lentil Lc-def, the action of NaD1 is mainly inhibitory, since the production of a wide range of cytokines/chemokines/growth factors by these cells is reduced in the presence of NaD1. Moreover, using the sandwich ELISA method, we demonstrated that NaD1, also at the concentration of 2 µM, had a pronounced immunomodulatory effect on unstimulated THP-1-derived macrophages and those stimulated by bacterial LPS or fungal zymosan. We showed that NaD1 had a pleiotropic action on resting and LPS- or zymosan-stimulated pro-inflammatory THP-1-derived macrophages, affecting the production of both pro- and anti-inflammatory cytokines. We also noted that the immunomodulatory effects of NaD1 on THP-1-derived macrophages under inflammation in vitro were somewhat different from those of the pea Psd1 and it was unlikely that these effects were due to the ability of tobacco defensin to bind such PAMPs as LPS or zymosan. These data demonstrated a lack of uniformity in the immunomodulatory action of plant defensins and suggested that not only differences in antimicrobial activities, but also various effects on the human immune system could influence the effectiveness of these peptides under infection in vivo. Our results indicate the need for further in vitro and in vivo study of the immunomodulatory

effects of these plant AMPs, including a mice model of acute inflammation induced by L
or other TLR-stimulating agents.

Supplementary Materials: The following supporting information can be downloaded at: htt //www.mdpi.com/article/10.3390/antibiotics13111101/s1, Figure S1: Cytotoxic effects of the tobac defensin NaD1 at high concentrations towards PBMCs (A) and Caco-2 cells not reached monolayer (Figure S2: Comparison of the amino acid sequences of antimicrobial peptides used; Figure Influence of the tobacco defensin NaD1 and other AMPs at concentration of 0.2 μM on producti of pro- and anti-inflammatory cytokines by unstimulated or stimulated by LPS or zymosan THP derived macrophages; Figure S4: Effect of LPS on candidacidal action of the tobacco defensin Na and human cathelicidin LL-37; Figure S5: Effect of zymosan on candidacidal action of the tobac defensin NaD1 and human cathelicidin LL-37; Figure S6: Agarose electrophoresis of amplico encoding LL-37 (A), NaD1 (C) and HBD2 (E) and assembled plasmid constructs pET-His8-TrxL-L 37 (B), pET-His8-TrxL-NaD1 (D) and pET-His8-TrxL-HBD2 (F); Figure S7: Schematic representati of the plasmid vectors pET-His8-TrxL-NaD1/LL-37/HBD2; Figure S8: MALDI mass spectra of t recombinant antimicrobial peptides; Figure S9: Circular dichroism spectra of antimicrobial peptid (0.3 mM) in an aqueous solution and in micellar detergents (30 mM); Table S1. Absolute levels 48 cytokines, chemokines and growth factors assessed by multiplex xMAP technology; Table S2: L of overlapping primers; Table S3: Antimicrobial peptide secondary structure estimation (%) predict from far-UV CD spectra.

Author Contributions: Conceptualization, E.I.F. and I.V.B.; funding acquisition, T.V.O.; investigati E.I.F., I.V.B., O.V.S., S.I.F., S.V.B. and A.A.I.; analysis and validation of data—E.I.F., I.V.B. and T.V. visualization—E.I.F., I.V.B., S.V.B. and O.V.S.; writing—original draft, E.I.F. and I.V.B.; writing—revie and editing, T.V.O.; supervision, T.V.O. All authors have read and agreed to the published version the manuscript.

Funding: The study was supported by the Russian Science Foundation grant No. 22-14-003 https://rscf.ru/en/project/22-14-00380/ (accessed on 13 November 2024).

Institutional Review Board Statement: Not applicable.

Informed Consent Statement: Not applicable.

Data Availability Statement: Data are contained within the article and Supplementary Materials.

Conflicts of Interest: The authors declare no conflicts of interest.

References

1. Kang, H.K.; Lee, H.H.; Seo, C.H.; Park, Y. Antimicrobial and Immunomodulatory Properties and Applications of Marine-Derive Proteins and Peptides. *Mar. Drugs* **2019**, *17*, 350. [CrossRef] [PubMed]
2. Li, Z.; Mao, R.; Teng, D.; Hao, Y.; Chen, H.; Wang, X.; Wang, X.; Yang, N.; Wang, J. Antibacterial and immunomodulato activities of insect defensins-DLP2 and DLP4 against multidrug-resistant *Staphylococcus aureus*. *Sci. Rep.* **2017**, *7*, 12124. [CrossRe [PubMed]
3. Báez-Magaña, M.; Díaz-Murillo, V.; López-Meza, J.E.; Ochoa-Zarzosa, A. Immunomodulatory effects of thionin Thi2.1 fro Arabidopsis thaliana on bovine mammary epithelial cells. *Int. Immunopharmacol.* **2018**, *57*, 47–54. [CrossRef] [PubMed]
4. Finkina, E.I.; Shevchenko, O.V.; Fateeva, S.I.; Tagaev, A.A.; Ovchinnikova, T.V. Antifungal Plant Defensins as an Alternative To to Combat Candidiasis. *Plants* **2024**, *13*, 1499. [CrossRef] [PubMed]
5. Pham, D.; Sivalingam, V.; Tang, H.M.; Montgomery, J.M.; Chen, S.C.; Halliday, C.L. Molecular Diagnostics for Invasive Fung Diseases: Current and Future Approaches. *J. Fungi* **2024**, *10*, 447. [CrossRef]
6. Parisi, K.; McKenna, J.A.; Lowe, R.; Harris, K.S.; Shafee, T.; Guarino, R.; Lee, E.; van der Weerden, N.L.; Bleackley, M.R.; Anderso M.A. Hyperpolarisation of Mitochondrial Membranes Is a Critical Component of the Antifungal Mechanism of the Plant Defensi Ppdef1. *J. Fungi* **2024**, *10*, 54. [CrossRef]
7. Díaz-Murillo, V.; Medina-Estrada, I.; López-Meza, J.E.; Ochoa-Zarzosa, A. Defensin γ-thionin from Capsicum chinense ha immunomodulatory effects on bovine mammary epithelial cells during *Staphylococcus aureus* internalization. *Peptides* **2016**, *7* 109–118. [CrossRef]
8. Báez-Magaña, M.; Alva-Murillo, N.; Medina-Estrada, I.; Arceo-Martínez, M.T.; López-Meza, J.E.; Ochoa-Zarzosa, A. Pla Defensin γ-Thionin Induces MAPKs and Activates Histone Deacetylases in Bovine Mammary Epithelial Cells Infected Wit *Staphylococcus aureus*. *Front. Vet. Sci.* **2020**, *7*, 390. [CrossRef]

Rigano, M.M.; Romanelli, A.; Fulgione, A.; Nocerino, N.; D'Agostino, N.; Avitabile, C.; Frusciante, L.; Barone, A.; Capuano, F.; Capparelli, R. A novel synthetic peptide from a tomato defensin exhibits antibacterial activities against *Helicobacter pylori*. *J. Pept. Sci.* **2012**, *18*, 755–762. [CrossRef]

10. Bogdanov, I.V.; Fateeva, S.I.; Voropaev, A.D.; Ovchinnikova, T.V.; Finkina, E.I. Immunomodulatory Effects of the Pea Defensin Psd1 in the Caco-2/Immune Cells Co-Culture upon *Candida albicans* Infection. *Int. J. Mol. Sci.* **2023**, *24*, 7712. [CrossRef]
11. Finkina, E.I.; Bogdanov, I.V.; Ignatova, A.A.; Kanushkina, M.D.; Egorova, E.A.; Voropaev, A.D.; Stukacheva, E.A.; Ovchinnikova, T.V. Antifungal Activity, Structural Stability, and Immunomodulatory Effects on Human Immune Cells of Defensin from the Lentil Lens culinaris. *Membranes* **2022**, *12*, 855. [CrossRef] [PubMed]
12. Falcón-Ruiz, E.A.; López-Meza, J.E.; Ochoa-Zarzosa, A. The plant defensins PaDef and γ-thionin inhibit the endothelial cell response to VEGF. *Peptides* **2023**, *165*, 171008. [CrossRef] [PubMed]
13. Selwal, N.; Tabassum, Z.; Rahayu, F.; Dwi Yulia, N.; Sugiono, S.; Endarto, O.; Riajaya, P.D.; Djajadi, D.; Khamidah, A.; Wani, A.K. Therapeutic potential and phytoremediation capabilities of the tobacco plant: Advancements through genetic engineering and cultivation techniques. *Biocatal. Agric. Biotechnol.* **2023**, *52*, 102845. [CrossRef]
14. Payne, J.A.; Bleackley, M.R.; Lee, T.H.; Shafee, T.M.; Poon, I.K.; Hulett, M.D.; Aguilar, M.I.; van der Weerden, N.L.; Anderson, M.A. The plant defensin NaD1 introduces membrane disorder through a specific interaction with the lipid, phosphatidylinositol 4,5 bisphosphate. *Biochim. Biophys. Acta* **2016**, *1858*, 1099–1109. [CrossRef] [PubMed]
15. Hayes, B.M.; Bleackley, M.R.; Anderson, M.A.; van der Weerden, N.L. The plant defensin NaD1 enters the cytoplasm of *Candida albicans* via endocytosis. *J. Fungi* **2018**, *4*, 20. [CrossRef]
16. McColl, A.I.; Bleackley, M.R.; Anderson, M.A.; Lowe, R.G.T. Resistance to the Plant Defensin NaD1 Features Modifications to the Cell Wall and Osmo-Regulation Pathways of Yeast. *Front. Microbiol.* **2018**, *9*, 1648. [CrossRef]
17. Baxter, A.A.; Poon, I.K.H.; Hulett, M.D. The plant defensin NaD1 induces tumor cell death via a non-apoptotic, membranolytic process. *Cell Death Discov.* **2017**, *3*, 16102. [CrossRef]
18. Zhang, H.Q.; Sun, C.; Xu, N.; Liu, W. The current landscape of the antimicrobial peptide melittin and its therapeutic potential. *Front. Immunol.* **2024**, *15*, 1326033. [CrossRef]
19. Wanke, D.; Mauch-Mücke, K.; Holler, E.; Hehlgans, T. Human beta-defensin-2 and -3 enhance pro-inflammatory cytokine expression induced by TLR ligands via ATP-release in a P2X7R dependent manner. *Immunobiology* **2016**, *221*, 1259–1265. [CrossRef]
20. Zharkova, M.S.; Orlov, D.S.; Golubeva, O.Y.; Chakchir, O.B.; Eliseev, I.E.; Grinchuk, T.M.; Shamova, O.V. Application of Antimicrobial Peptides of the Innate Immune System in Combination With Conventional Antibiotics -A Novel Way to Combat Antibiotic Resistance? *Front. Cell Infect. Microbiol.* **2019**, *9*, 128. [CrossRef]
21. Banchereau, J.; Steinman, R.M. Dendritic cells and the control of immunity. *Nature* **1998**, *392*, 245–252. [CrossRef] [PubMed]
22. Kratofil, R.M.; Kubes, P.; Deniset, J.F. Monocyte Conversion During Inflammation and Injury. *Arterioscler. Thromb. Vasc. Biol.* **2017**, *37*, 35–42. [CrossRef] [PubMed]
23. Menten, P.; Wuyts, A.; Van Damme, J. Macrophage inflammatory protein-1. *Cytokine Growth Factor Rev.* **2002**, *13*, 455–481. [CrossRef] [PubMed]
24. Antoniades, H.N. PDGF: A multifunctional growth factor. *Baillieres Clin. Endocrinol. Metab.* **1991**, *5*, 595–613. [CrossRef] [PubMed]
25. Ngkelo, A.; Meja, K.; Yeadon, M.I.; Kirkham, P.A. LPS induced inflammatory responses in human peripheral blood mononuclear cells is mediated through NOX4 and Giα dependent PI-3kinase signalling. *J. Inflamm.* **2012**, *9*, 1. [CrossRef]
26. Sato, M.; Sano, H.; Iwaki, D.; Kudo, K.; Konishi, M.; Takahashi, H.; Takahashi, T.; Imaizumi, H.; Asai, Y.; Kuroki, Y. Direct binding of Toll-like receptor 2 to zymosan, and zymosan-induced NF-kappa B activation and TNF-alpha secretion are down-regulated by lung collectin surfactant protein A. *J. Immunol.* **2003**, *171*, 417–425. [CrossRef]
27. Mohammadi, A.; Blesso, C.N.; Barreto, G.E.; Banach, M.; Majeed, M.; Sahebkar, A. Macrophage plasticity, polarization and function in response to curcumin, a diet-derived polyphenol, as an immunomodulatory agent. *J. Nutr. Biochem.* **2019**, *66*, 1–16. [CrossRef]
28. Yunna, C.; Mengru, H.; Lei, W.; Weidong, C. Macrophage M1/M2 polarization. *Eur. J. Pharmacol.* **2020**, *877*, 173090. [CrossRef]
29. Venkatachalam, G.; Giri, J.; Mallik, S.; Arumugam, G.S.; Arulmani, M.; Dewangan, V.K.; Doble, M.; Zhao, Z. Immunomodulatory zymosan/ι-carrageenan/ agarose hydrogel for targeting M2 to M1 macrophages (antitumoral). *RSC Adv.* **2024**, *14*, 11694–11705. [CrossRef]
30. Rasquinha, M.T.; Sur, M.; Lasrado, N.; Reddy, J. IL-10 as a Th2 Cytokine: Differences Between Mice and Humans. *J. Immunol.* **2021**, *207*, 2205–2215. [CrossRef]
31. Scheenstra, M.R.; van Harten, R.M.; Veldhuizen, E.J.A.; Haagsman, H.P.; Coorens, M. Cathelicidins Modulate TLR-Activation and Inflammation. *Front. Immunol.* **2020**, *11*, 1137. [CrossRef] [PubMed]
32. Tsai, P.W.; Yang, C.Y.; Chang, H.T.; Lan, C.Y. Human antimicrobial peptide LL-37 inhibits adhesion of *Candida albicans* by interacting with yeast cell-wall carbohydrates. *PLoS ONE* **2011**, *6*, e17755. [CrossRef] [PubMed]
33. Koeninger, L.; Armbruster, N.S.; Brinch, K.S.; Kjaerulf, S.; Andersen, B.; Langnau, C.; Autenrieth, S.E.; Schneidawind, D.; Stange, E.F.; Malek, N.P.; et al. Human β-Defensin 2 Mediated Immune Modulation as Treatment for Experimental Colitis. *Front. Immunol.* **2020**, *11*, 93. [CrossRef] [PubMed]
34. Lee, E.Y.; Lee, M.W.; Wong, G.C.L. Modulation of toll-like receptor signaling by antimicrobial peptides. *Semin. Cell Dev. Biol.* **2019**, *88*, 173–184. [CrossRef]

35. Chaudhry, H.; Zhou, J.; Zhong, Y.; Ali, M.M.; McGuire, F.; Nagarkatti, P.S.; Nagarkatti, M. Role of cytokines as a double-edge sword in sepsis. *In Vivo* **2013**, *27*, 669–684.
36. Bleackley, M.R.; Dawson, C.S.; Payne, J.A.; Harvey, P.J.; Rosengren, K.J.; Quimbar, P.; Garcia-Ceron, D.; Lowe, R.; Bulone, V.; va der Weerden, N.L.; et al. The interaction with fungal cell wall polysaccharides determines the salt tolerance of antifungal pla defensins. *Cell Surf.* **2019**, *5*, 100026. [CrossRef]
37. Wahl, L.M.; Smith, P.D. Isolation of Monocyte/Macrophage Populations. *Curr. Protoc. Immunol.* **1995**, *16*, 7.6.1–7.6.8. [CrossRe
38. Nair, S.; Archer, G.E.; Tedder, T.F. Isolation and generation of human dendritic cells. *Curr. Protoc. Immunol.* **2012**, *7*, 7.32.1–7.32.2 [CrossRef]
39. Chen, Z.; Wang, L.; He, D.; Liu, Q.; Han, Q.; Zhang, J.; Zhang, A.M.; Song, Y. Exploration of the Antibacterial and An Inflammatory Activity of a Novel Antimicrobial Peptide Brevinin-1BW. *Molecules* **2024**, *29*, 1534. [CrossRef]
40. Han, J.; Jyoti, M.A.; Song, H.Y.; Jang, W.S. Antifungal Activity and Action Mechanism of Histatin 5-Halocidin Hybrid Peptid against *Candida* ssp. *PLoS ONE* **2016**, *11*, e0150196. [CrossRef]

antibiotics

ˑticle

;omputational Prediction and Structural Analysis of ι-Hairpinins, a Ubiquitous Family of Antimicrobial Peptides, Jsing the Cysmotif Searcher Pipeline

ına A. Slavokhotova [1,2,*], Andrey A. Shelenkov [1] and Eugene A. Rogozhin [2,3]

1 Central Research Institute of Epidemiology, Novogireevskaya Str., 3a, 111123 Moscow, Russia
2 Shemyakin and Ovchinnikov Institute of Bioorganic Chemistry RAS, Miklukho-Maklaya Str., 16/10, 117437 Moscow, Russia; rea21@list.ru
3 All-Russian Institute for Plant Protection, Podbelskogo Str., 196608 Saint-Petersburg-Pushkin, Russia
* Correspondence: slavokhotova@cmd.su

Abstract: Background: α-Hairpinins are a family of antimicrobial peptides, promising antimicrobial agents, which includes only 12 currently revealed members with proven activity, although their real number is supposed to be much higher. α-Hairpinins are short peptides containing four cysteine residues arranged in a specific Cys-motif. These antimicrobial peptides (AMPs) have a characteristic helix—loop—helix structure with two disulfide bonds. Isolation of α-hairpinins by biochemical methods is cost- and labor-consuming, thus requiring reliable preliminary in silico prediction. Methods: In this study, we developed a special algorithm for the prediction of putative α-hairpinins on the basis of characteristic motifs with four (4C) and six (6C) cysteines deduced from translated plant transcriptome sequences. We integrated this algorithm into the Cysmotif searcher pipeline and then analyzed all transcriptomes available from the One Thousand Plant Transcriptomes project. Results: We predicted more than 2000 putative α-hairpinins belonging to various plant sources including algae, mosses, ferns, and true flowering plants. These data make α-hairpinins one of the ubiquitous antimicrobial peptides, being widespread among various plants. The largest numbers of α-hairpinins were revealed in the *Papaveraceae* family and in *Papaver somniferum* in particular. Conclusions: By analyzing the primary structure of α-hairpinins, we concluded that more predicted peptides with the 6C motif are likely to have potent antimicrobial activity in comparison to the ones possessing 4C motifs. In addition, we found 30 α-hairpinin precursors containing from two to eight Cys-rich modules. A striking similarity between some α-hairpinin modules belonging to diverse plants was revealed. These data allowed us to assume that the evolution of α-hairpinin precursors possibly involved changing the number of Cys-rich modules, leading to some missing middle and C-terminal modules, in particular.

Keywords: antimicrobial peptide; plant defense; α-hairpinin; plant transcriptome

ation: Slavokhotova, A.A.; elenkov, A.A.; Rogozhin, E.A. mputational Prediction and ˑuctural Analysis of α-Hairpinins, Jbiquitous Family of Antimicrobial ptides, Using the Cysmotif archer Pipeline. *Antibiotics* **2024**, *13*, 19. https://doi.org/10.3390/ tibiotics13111019

ademic Editors: Piyush Baindara d Marisa Di Pietro

ceived: 20 August 2024
vised: 25 October 2024
cepted: 26 October 2024
blished: 30 October 2024

1. Introduction

During the last decade, humanity faced the problem of antibiotic resistance of pathogenic and opportunistic bacteria and the reduction in the effect of using antibiotics to overcome the infections caused by these bacterial agents. According to the study published in *The Lancet*, in the year 2019, nearly 1.3 million people died from diseases caused by antimicrobial-resistant pathogens, while the total number of fatal cases associated partly with antimicrobial resistance (AMR) was estimated to be about 5 million [1]. Consequently, searching for novel antimicrobial agents with various types of action became one of the major challenges in human healthcare.

Antimicrobial peptides (AMPs) have recently been a point of a special interest as naturally occurring molecules with a broad spectrum of antifungal, antibacterial, and antiviral activities [2,3]. These small proteins have been found in almost all living organisms from bacteria to eukaryotes [4] and have been considered as the innate immunity componer serving as the first line of defense [5]. Being widely diverse, AMPs share some comm features, including small size, positive surface charge, and amphiphilic structure [6]. N merous studies showed that many AMPs interact with membranes of pathogens and indu membrane permeabilization, thus killing the microorganisms [7]. This mechanism of acti suggests small risks of AMR developing, making AMPs a prospective source of natu antibiotics [8].

Plant AMPs were first discovered in the early 1970s [9], and currently, their numb exceeds 800 [10]. The majority of plant AMPs have a compact spatial structure due to t presence of several cysteine residues forming the so-called cysteine motif (Cys-motif) [1 The structure of many AMPs was previously shown to be stabilized by disulfide bon and cysteines play an essential role in such bond formation [12,13]. According to Cys-mo and some other characteristics, AMPs are divided into several families, including a sm family of α-hairpinins [14]. The latter are small peptides with a very specific Cys-motif th can be denoted as $C^1X_3C^2X_nC^3X_3C^4$, where C^i shows cysteine residues (i = 1. . .4), X c independently be any amino acid residue except cysteine, and subscripts 3 and n indica the number of such non-cysteine residues. This cysteine arrangement facilitates a comm helix—loop—helix spatial structure comprising two α-helices oriented antiparallel a joined by a loop (Figure 1) [14]. Motif sequences can be highly heterogeneous, even with the same plant species, with cysteines sometimes being the only conservative residues fact that significantly complicates the possibility of detecting a new α-hairpinin using homology search only.

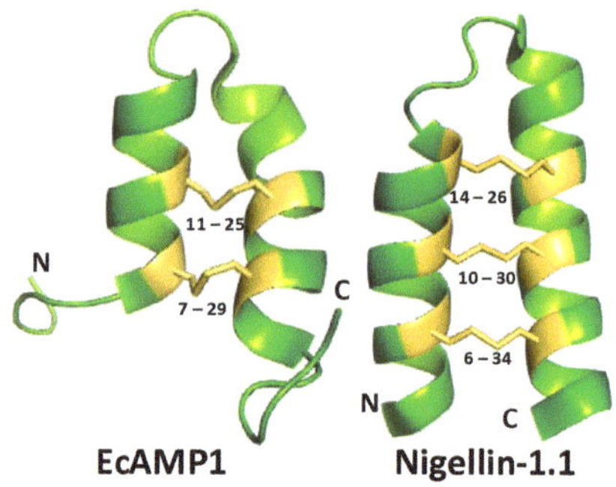

Figure 1. Examples of α-hairpinin spatial structures. On the left, Ec-Amp1 (PDB code 2L2R) with fo cysteines is displayed, and on the right, Nigellin-1.1 (PDB code 2NB2) with six cysteines is show Disulfide bridges are shown as yellow sticks, cysteine residues are highlighted in yellow, and th positions are specified.

It should be noted that until recently, α-hairpinins were considered to contain on four cysteines. However, according to the primary Cys-motif of the precursors, the fif or sixth cysteines located one to three amino acids apart from the main ones can also b presented. Moreover, novel antibacterial α-hairpinin with six cysteines has been recent discovered in a ghost pepper [15]. In addition, α-hairpinin with six cysteines and th characteristic helix—loop—helix structure was isolated from *Nigella sativa* (Figure 1, PD ID: 2NB2, https://www.rcsb.org/structure/2nb2, accessed on 29 September 2024). Thu we can conclude that α-hairpinins are peptides with the specific 4–6 cysteine residue mo sharing common helix—loop—helix structure (Figure 1).

Representatives of this family exhibit a broad spectrum of activity [14]. For examp several α-hairpinins can bind a trypsin used by insects for protein digestion and inhibit i proteinase activity [16–18]. Some α-hairpinins display antiviral activity; in particular, th can inhibit protein synthesis of invaders and possess strong ribosome-inactivating activity [1 20]. Finally, the majority of α-hairpinins show strong antifungal activity against a wide rang of fungi [14]. Some of these peptides were also active against bacterial pathogens [21]. Th mechanism of α-hairpinin action was studied on the antifungal peptide EcAMP1, which exhibited activity via initial binding with components of a fungal cell wall, glycoproteins, proteins–amyloids, and glycans, in particular. After binding, EcAmp1 was distributed evenly over the cell surface and interacted with the plasma membrane, which provoked the peptide internalization into the cell surface and presumably induced apoptosis [22].

Biochemical isolation of plant AMPs is always time- and labor-consuming since acidic extraction followed by multistage stepwise chromatography is usually used [23]. On the other hand, a wide range of transcriptomic and genomic data of diverse plant species are currently available in public databases. Therefore, it is highly preferable to use these data for AMP prediction. There are many platforms for a novel AMP search (e.g., ACEP [24], amPEPpy [25], AmPEP [26], sAMPpred-GAT [27]), most of which are based on machine-learning methods. However, they usually lack specificity in their predictions and do not allow the revealing of the peptides belonging to a particular class. A recent study highlighted an existing bias in AMP predictors when dealing with disordered regions within AMP sequences and considered this as a specificity-limiting factor [28]. To the best of our knowledge, no existing algorithm is able to correctly predict α-hairpinins in wide-scale analysis.

Recently, we developed the Cysmotif searcher computational pipeline for AMP predictions in silico [29]. Using this software, several plant transcriptomes were analyzed to reveal their specific AMP profiles [30,31]. In the current study, we developed a novel algorithm for the prediction of α-hairpinin peptides. This algorithm was later integrated into Cysmotif searcher software and is now available for users in Github (https://github.com/fallandar/cysmotifsearcher, accessed 20 October 2024). We analyzed more than 1200 transcriptomes available in the One Thousand Plant Transcriptomes project (1KP) [32] that resulted in the prediction of more than 2000 putative α-hairpinins. Besides classification of putative α-hairpinins and their distribution among plant families, the diversity of α-hairpinin precursors was also analyzed. Below, we will refer to all predicted α-hairpinin-like peptides as 'α-hairpinins' for the sake of brevity, although not all of them might exhibit antimicrobial activity in vitro and in vivo.

We believe that the data obtained will facilitate the isolation and activity confirmation for this AMP family, which can ultimately lead to the discovery of potent antimicrobial agents against various plant and, possibly, human pathogens.

2. Results

2.1. Prediction of α-Hairpinin Peptides Using the 4C Cys-Motif

A general α-hairpinin motif, $C^1X_3C^2X_{4\text{-}20}C^3X_3C^4$, was deduced from the primary structure of α-hairpinins with proven antimicrobial activity and named 4C by us. It includes three non-cysteine amino acid residues between the first and the second cysteines, as well as between the third and fourth ones. The number of residues between the second and the third cysteines in the motif can vary from 4 to 20. For example, both sequences $\mathbf{C^1}$AVR$\mathbf{C^2}$KLTM$\mathbf{C^3}$VRD$\mathbf{C^4}$ (Seq. 1) and $\mathbf{C^1}$WMP$\mathbf{C^2}$SLQPD$\mathbf{C^3}$LTW$\mathbf{C^4}$ (Seq. 2) possess the motif generalized above, although their similarity is low, and they have different numbers of residues (four and five, respectively) between the second and the third cysteines. At the same time, a sequence $\mathbf{C^1}$AVRC^{2}KLT$\mathbf{C^3}$VRD$\mathbf{C^4}$ (Seq. 3) does not contain the motif, although it is highly similar to Seq. 1, since it includes three residues between the second and the third cysteines, and this number is not within the range of 4–20.

The motif provided above was used for an in silico AMP search in transcriptomes available from the 1KP [32]. As a result, 1269 predicted α-hairpinins were found (Figure 2A, Table S1). Among them, 869 contained four cysteine residues arranged in the 4C motif that varied by different numbers of amino acids between the second and the third cysteines shown as 'X_n'. The top five n values were 6, 4, 13, 5, and 7 occurred in 106, 92, 81, 78, and 68 predicted α-hairpinins, respectively (Figure 2C).

Considerably less predicted AMPs (pAMPs), namely, 289, had five cysteines in their primary structure. Among them, 84 detected peptides possessed a 4C+1 motif, which meant that one more cysteine located further than four amino acids ahead of the C1 was prese in addition to the 'classic' 4C motif. The n values for this motif are shown in Figure 2A. total of 205 α-hairpinins had a motif named 5C ($C^5X_{1–3}C^1X_3C^2X_{4–20}C^3X_3C^4$). This mot was separated from the 4C+1 by us since the fifth cysteine located so close to the first o might be involved in the formation of an elongated α-helix. The most frequent were 5 putative peptides with eight amino acids between the second and the third cysteines, an their number was 93; other n values are indicated in Figure 2D and Table S1.

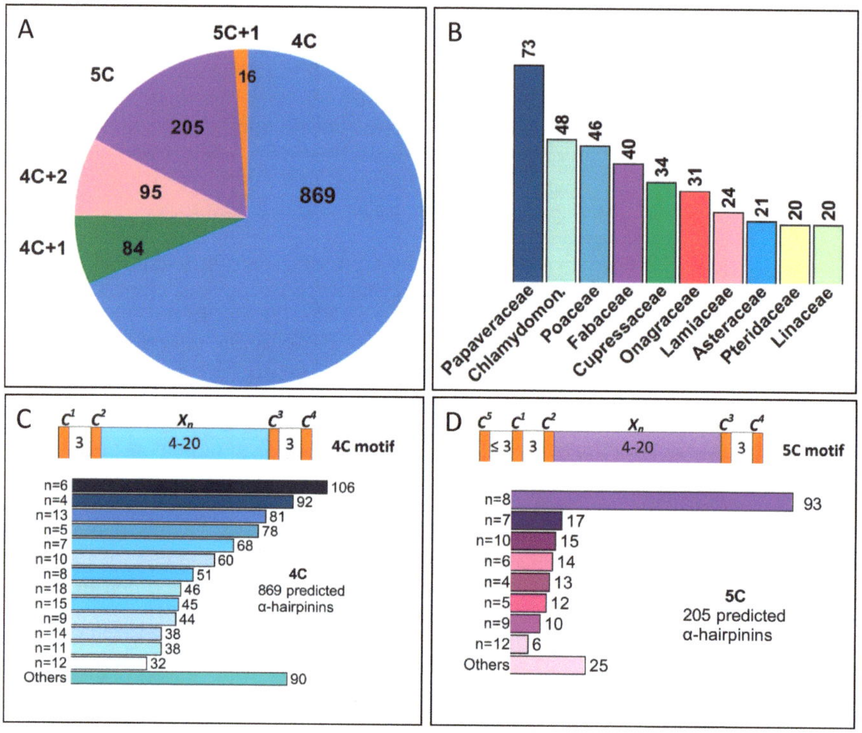

Figure 2. Diversity of the first class of putative α-hairpinins predicted using the 4C motif. (**A**) Classificatio of α-hairpinins depending on the Cys-motif. Outside of a pie chart, the designations of the motifs are show (**4C**, $C^1X_3C^2X_{4\text{-}20}C^3X_3C^4$; **4C + 1,** $C^5X_{\geq 4}C^1X_3C^2X_{4\text{-}20}C^3X_3C^4$; **4C + 2,** $C^6XC^5X_{\geq 4}C^1X_3C^2X_{4\text{-}20}C^3X_3C^4$; **5C** $C^5X_{\leq 3}C^1X_3C^2X_{4\text{-}20}C^3X_3C^4$; **5C + 1,** $C^6X\,C^5X_{\leq 3}C^1X_3C^2X_{4\text{-}20}C^3X_3C^4$); inside the pie chart, the number display the amount of α-hairpinins with the corresponding Cys-motif. (**B**) The top 10 plant familie including putative α-hairpinins with the 4C motif and its derivatives. (**C**,**D**) Schemes and bar charts o two prevailing Cys-motifs. Cysteine residues are displayed as orange boxes and signified as $C^{1\text{-}5}$, wher the superscripts designate a serial number in the motif. Numbers between cysteines independentl denote any amino acids except cysteine; **Xn**, any amino acids except cysteine between the secon and the third cysteines, independently. Bar charts display the number of predicted α-hairpinins wit different n values.

Finally, 111 α-hairpinins predicted in silico had six cysteines, including 95 pAMPs wit 4C+2 motifs. Putative 4C+2 peptides possessed two additional cysteines located further tha three amino acids apart from C1 besides the basic 4C motif (Figure S1B, Table S1). Among thes predicted α-hairpinins, the most common motif was $C^6X_2C^5X_{30}C^1X_3C^2X_7C^3X_3C^4$, which occurred in 68 peptides. The other 16 predicted peptides were arranged to a group sharing the 5C + 1 motif that contained one additional cysteine to the main motif 5C (Table S1).

The distribution of 1269 α-hairpinins among plant families and groups, as well as their expression in various plant organs, are reported in Figure 2B. These pAMPs were mostly found in plants belonging to the *Papaveraceae* family (73 predicted peptides); the second abundant family was, surprisingly, *Chlamydomonadaceae* (48 pAMPs), followed by *Poaceae* (46 pAMPs) and *Fabaceae* (40 pAMPs). *Cupressaceae*, *Onagraceae*, *Lamiaceae*, *Asteraceae*, *Pteridaceae*, and *Linaceae* had from 34 to 20 predicted α-hairpinins, while other families had less than 20 predicted peptides. The predicted α-hairpinins belonged mainly to dicots (607 pAMPs), algae (267 pAMPs), ferns (168 pAMPs), and monocots (118 pAMPs). Significantly smaller numbers of pAMPs were revealed in mosses (43), horn- and liverworts (43), lycophytes (22), and eusporangiate monilophytes (16), while less than 10 α-hairpinins were found in Euglenozoa, basalmost angiosperms, *Dinophyceae*, *Gnetales*, and *Cycadales* families. Predicted α-hairpinins were mainly expressed in leaves (548 pAMPs), algae cells (195 pAMPs), flowers (137 pAMPs), shoots (128 pAMPs), stems (50 pAMPs), and roots (21 pAMPs) (Table S2).

The sequence logo for the most widespread 4C motif, $C^1X_3C^2X_6C^3X_3C^4$, is shown in Figure 3. There were no dominating residues in non-cysteine positions, except for slightly prevalent serine (S) in the eighth position.

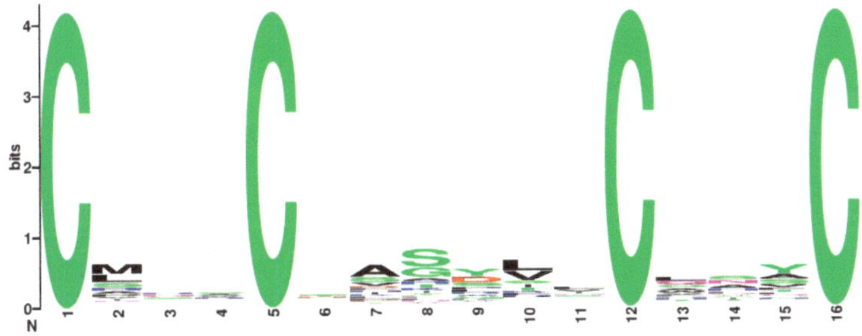

Figure 3. A sequence logo of the most prevalent 4C motif, $C^1X_3C^2X_6C^3X_3C^4$.

Furthermore, we analyzed the sequences of the predicted α-hairpinins in order to check whether they can possess antimicrobial activity with antifungal and antibacterial activity prediction web-servers. The results of this analysis are presented in Table S2. In total, 83% of the peptides found were predicted to possess antifungal or antibacterial activity, or both. However, these results are greatly affected by changing the possible proteolysis site, which is rather hard to reliably detect in silico but can alter the length of a resulting peptide and, in turn, its predicted activity. Thus, the real number of 4C peptides with antifungal activity can be even higher.

2.2. Prediction of α-Hairpinins Using the 6C Cys-Motif

In addition to the basic 4C α-hairpinin motif, a novel one with six cysteines was recently discovered. This motif, denoted 6C ($C^1X_3C^2X_3C^3X_{4\text{-}20}C^4X_3C^5X_3C^6$), was used for in silico search within 1KP plant transcriptomes. Surprisingly, the 6C motif was widely distributed, with 1069 pAMPs assigned to this second class of putative α-hairpinins. In particular, 1057 of them had six cysteines, while the remaining 12 included one or two cysteines in addition. Among the 1057 peptides, the most frequent were those with 10, 9, 12, 11, 6, and 8 amino acids between the third and the fourth cysteines observed in 558, 134, 110, 64, 54, and 51 pAMPs, respectively (Figure 4A, Table S1). Considerably less predicted peptides contained 13, 4, and 7 amino acids in these positions (29, 18, and 13 pAMPs, respectively), and less than 10 pAMPs were contained in each group of peptides with 5 or 14–18 amino acids between the third and the fourth cysteines. Concerning the predicted peptides with additional cysteine residues, eight of them had the seventh cysteine located 42 amino acids apart from the 6C motif, while two pAMPs had a more compact motif and contained eight amino acids between the seventh cysteine and the main motif (Table S1).

Moreover, two predicted α-hairpinins had eight cysteines in their Cys-motif, which we located close to each other ($C^8X_4C^7X_2C^1X_3C^2X_3C^3X_8C^4X_3C^5X_3C^6$).

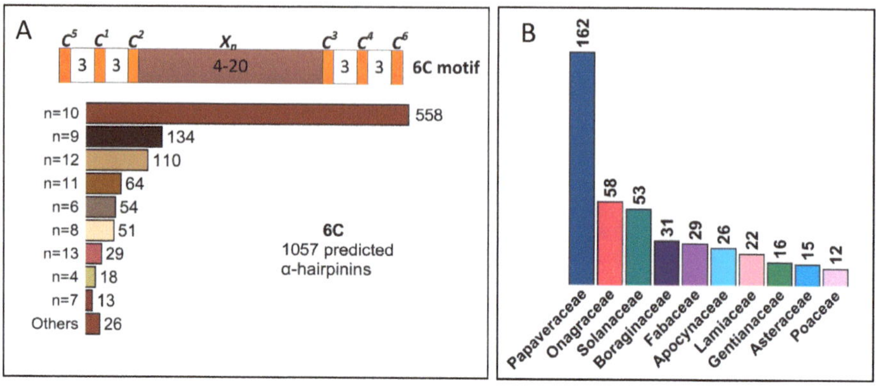

Figure 4. Variety of α-hairpinin-like peptides with the 6C motif. (**A**) A bar chart displaying a distributi of different n values (number of non-cysteine amino acids between C2 and C3) found in the predicte 6C motif-possessing peptides. (**B**) Top 10 plant families abundant with 6C putative α-hairpinins.

We also studied the distribution of pAMPs with 6C motifs among plant taxons togeth with their expression in different parts of the plant. Similarly to 4C α-hairpinins, the 6 pAMPs were abundant in the members of the *Papaveraceae* family that contained 16 predicted peptides (Figure 4B, Table S2). A total of 58, 53, 30, 29, 26, and 22 pAMP were detected in *Onagraceae*, *Solanaceae*, *Boraginaceae*, *Fabaceae*, *Apocynaceae*, and *Lamiacea* respectively, which was significantly less than in *Papaveraceae*. Interestingly, besides th listed families, as many as 143 families contained from 2 to 15 α-hairpinins with the 6 motif (Table S2). The described motif was found predominantly in dicots and was muc less common, e.g., in monocots (about 60 pAMPs) and mosses (22 pAMPs) (Table S2). Th predicted 6C α-hairpinins were detected in different plant organs; the most abundant we leaves (more than 500 pAMPs in total), flowers and fruits (almost 300 pAMPs), and shoo (118 pAMPs), while roots contained considerably less AMPs (Table S2).

In total, 95% of the revealed 6C peptides were predicted to possess antimicrobi activity, mostly antifungal, with third-party prediction servers. The activity predictio results are shown in Table S2.

A sequence logo for the most widespread 6C motif, $C^1X_3C^2X_3C^3X_{10}C^4X_3C^5X_3C^6$, i shown in Figure 5. In this case, the prevalence of particular non-cysteine residues was mor prominent than for the 4C motif. For example, charged residues aspartic acid (D), arginine (R and lysine (K) dominated in the 3rd, 13th, 15th, 19th, 22nd, 23rd, and 27th positions, whi polar threonine (T) and hydrophobic valine (V) held 7th and 10th positions, respectively.

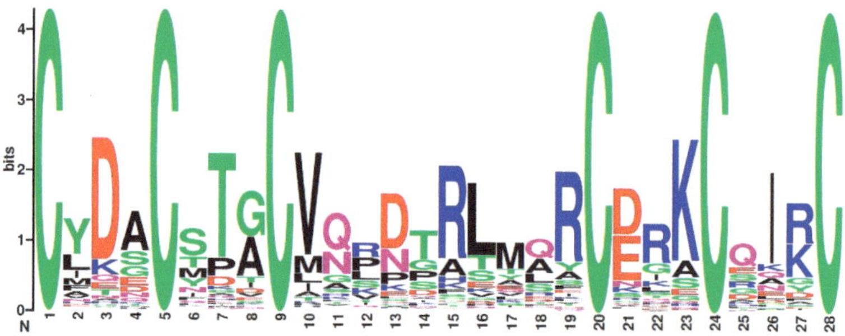

Figure 5. A sequence logo of the most prevalent 6C motif, $C^1X_3C^2X_3C^3X_{10}C^4X_3C^5X_3C^6$.

2.3. Prediction of Precursor Proteins with Several α-Hairpinin Domains

A special algorithm predicting α-hairpinin precursor proteins with more than one α-hairpinin module was developed. As a result, 30 precursors were identified with a number of α-hairpinin modules varying from 2 to 8. A total of 16 precursors had two AMP modules, including five relatively short putative precursors in which α-hairpinin sequences were separated by spacers and had short C-terminal regions at their ends (group 2M α-hairpinins; Table S1, Figure S2). Another 10 predicted precursors possessed a structure typical for vicilin seed storage proteins: variable N-terminal part with two α-hairpinin modules followed by N- and C-terminal cupin 7S vicilin-like domains (group 2M vicilin-like α-hairpinins; Table S1, Figure S2). One more predicted precursor had two α-hairpinin domains on the opposite termini separated by a large middle part exhibiting similarity only to hypothetical proteins with unknown functions (group 2M termini α-hairpinin; Table S1, Figure S2).

Besides two-modular α-hairpinin precursors, eight precursors with three domains, four with four domains, one with seven domains, and one with eight domains were detected in this work. Among them, four three-modular precursors contained also two cupin 7S vicilin-like domains (Table S1, Figure S2), while another four three-modular precursors, as well as precursors with four, seven, and eight α-hairpinin modules, did not include these vicilin-like domains and were not terminated by the short C-region (Table S1, Figure S2).

A total of 19 out of 30 predicted prepropeptides exhibited similarity to vicilin proteins; however, 6 of them did not contain cupin 7S vicilin-like domain (Table S1). These six putative peptides, as well as some other α-hairpinin precursors available in GenBank, were incorrectly annotated as vicilin seed storage proteins, when in fact they had a typical modular structure. Moreover, two detected precursors exhibited similarity to antimicrobial peptides from various Poacea plants, while another two precursors had high homology with MBP-1 α-hairpinin [21] previously isolated from corn (*Zea mays*; Table S1). Seven predicted proteins had considerable similarity only with uncharacterized proteins, including a precursor from *Anemone hupenhensis* with two closely located Cys-rich domains having 6C and 5C motifs simultaneously.

3. Discussion

3.1. In Silico Prediction of More Than 2000 α-Hairpinins

Until recently, it was thought that the family of α-hairpinins was a small group of AMPs including only 12 representatives with proven antimicrobial activity [14,15,33]. All these peptides were isolated from different plants by basic procedures including extraction and stepwise purifications. In this work, we applied a completely different approach and developed a special algorithm for in silico α-hairpinin prediction. Two Cys-motifs, namely, 4C and 6C, were used for the detection of α-hairpinins in the transcriptomes available in the 1KP project [32]. Surprisingly, more than 2000 putative α-hairpinins were found. Our pipeline relies on a conserved pattern search followed by several filtration steps, and it does not involve 3D structure prediction or antimicrobial activity verification. Nevertheless, most of the α-hairpinins revealed were found to possess antimicrobial activity by third-party prediction software (83% for 4C and 95% for 6C putative peptides, respectively). Thus, Cysmotif searcher should be used carefully, taking into account the described limitations, as with any other in silico prediction software.

It is worth noting that our pipeline has been recently used to reveal a novel 6C α-hairpinin peptide from ghost pepper [15]. The authors used the motifs developed by us as a starting point for revealing novel CC-AMP1-like α-hairpinin and confirmed its activity against bacterial pathogens from the ESKAPE group in vitro. In another publication, the authors used Cysmotif searcher for α-hairpinin mining in the genome of lima bean and revealed the PlHrp1 peptide, which was shown to possess antifungal and antibacterial activity using in silico analysis [34]. These recent discoveries confirm the possibility of Cysmotif searcher application for α-hairpinin mining in plant transcriptomes and genomes.

3.2. α-Hairpinins Are Ubiquitous Components of the Plant Defense System

In this study, more than 1200 4C and 1000 6C α-hairpinins were predicted, which co pletely changed our notions regarding this family. First, α-hairpinins seem to be ubiquitou since they were observed in more than 200 plant families and occurred in various pla taxons from primitive plants and algae to true flowering plants such as *Asterids* and *Rosi* Interestingly, the family most abundant with α-hairpinins was *Papaveraceae*, which includ numerous latex-bearing plants rich in different bioactive compounds such as alkaloid carotenoids, phenols, and terpenoids [35]. Secondly, our previous study detected abou 4000 putative defensins and 3000 putative lipid-transfer proteins [36] within 1KP, bein known as one of the major components of plant immunity [37,38], and their number wa comparable to the one of α-hairpinins found in this work. These data together with variou types of α-hairpinin antimicrobial activity may point to the unique role of α-hairpinin in plant defense. Thirdly, one particular plant (*Papaver somniferum*) contained more tha 40 various α-hairpinins with different 4C and 6C motifs (Table S2). It should be mentione that *Papaver somniferum* is an underestimated source of plant AMPs; although it is abunda with α-hairpinins (shown in this study) and thionins [39], we failed to find any informatio regarding peptides with proven antimicrobial activity isolated from this plant. In order t facilitate the investigations in this field, our study revealed a collection of α-hairpinins wit 4C and 6C motifs, so that particular peptides can be selected and then expressed in vitro fo the further investigation of their antimicrobial activities, spatial structures, or mechanism of action.

The chi-squared test allowed us to reveal that the distribution of α-hairpinins amon groups (e.g., core eudicots, conifers), families, and organs of plants was not as expecte according to the frequencies of the corresponding cohorts in the initial 1KP dataset wit α-values lower than 10^{-10}. The numbers for particular species were too low to perform th test. The most significant differences of expected and observed values were found for basa eudicots (85 motifs vs. 56 expected), core eudicots (76 vs. 114), and green algae (205 vs. 166 For the families and organs, the most significant differences were revealed for *Asteraceae* (2 vs. 41 expected), *Chlamydomonadaceae* (48 vs. 22), and *Papaveraceae* (73 vs. 33), as well as fo leaves (sum for all categories including 'leaves' or 'leaf': 398 vs. 501 expected). Howeve the only reliable conclusion found in these data could be a confirmation of *Papaveraceae* as source of α-hairpinins.

We believe that the number and variety of predicted α-hairpinins in plants als indicates a significant role of these peptides in plant immunity. Furthermore, our results ar in good coincidence with other studies devoted to AMP repertoire prediction. For exampl a transcriptome of healthy *Stellaria media* seedlings contained 18 predicted α-hairpini transcripts, while 12 such sequences were found in the transcriptome of shoots infecte with *Fusarium oxysporum* [31]. Other examples were transcriptomic studies of *Peltophoru dubium* and *Leymus arenarius* seedlings that contained 10 and 16 α-hairpinin sequence respectively [30,40]. At the same time, α-hairpinins were not found in some transcriptome although the goal of a study was to reveal the maximum numbers of defense peptides [41 It is also worth noting that α-hairpinins were found in various plant organs: putativ peptides with 4C motifs mostly occurred in leaves, algae cells, and flowers, while th majority of 6C putative peptides were observed in flowers and fruits, leaves, and shoots.

3.3. The Diversity of Detected Cys-Motifs

Another focus of this study was an analysis of Cys-motif diversity presenting amon the predicted α-hairpinins. All the putative peptides were divided into two large classe possessing either the 4C or 6C motif. The first finding was that putative 6C peptides wer more uniform than 4C peptides. In particular, 1057 out of 1069 6C α-hairpinins shared th same 6C motif and contained only six cysteines. In comparison, 4C α-hairpinins could b divided into five groups including the largest one with putative peptides sharing the 4C motif, as well as considerably smaller groups of predicted α-hairpinins with five and si cysteines. We speculate that additional cysteine residues may be involved in the formatio of coils in α-helices, shorten unstructured tails, and provide the peptide with enhanced stability. Similar peptides with one or two additional cysteines were detected earlier in some modules of precursors of Sm-Amp-X, Tk-Amp-X [14], and MiAMP2c [42], and the presence of mature peptides with elongated Cys-motifs was confirmed by mass spectrometry and biochemical isolation [42]. The next assumption is that the 6C α-hairpinin class most likely contains more active AMPs than the first one. The prediction of antimicrobial activity in silico proved this assumption, since the fraction of 6C peptides possessing antifungal and/or antibacterial activity was higher than the one for 4C peptides (95% versus 83%, respectively). This conclusion also coincides with 'n' values and their frequencies among putative peptides. Previously known α-hairpinins with proven antimicrobial activity had the distance between the second and the third cysteines equal to 11–13 amino acids [14], but we detected a number of 4C peptides with 4–7 residues in this segment. At the same time, almost 900 putative 6C peptides had n values equal to 9–13 amino acids. In comparison, the number of 4C peptides with the same distance in the loop was only 250. Therefore, it seems plausible that considerably more peptides might be functional among 6C predicted α-hairpinins; they could occur as mature peptides in a plant and exhibit antimicrobial activity.

3.4. α-Hairpinin Precursors Could Evolve by Changing the Number of Cys-Rich Modules

A total of 30 α-hairpinin precursors containing from two to eight Cys-rich modules were predicted in this work using a specially developed algorithm. It is known that α-hairpinin precursors are divided into two types depending on the presence of the vicilin-like region [14]. Indeed, both of the types were found in the current study. The first type of precursors, termed vicilin-like, consists of a signal peptide, several Cys-rich α-hairpinin modules, and a vicilin-like C-terminal region [14,42,43]. Due to proteolysis, Cys-rich α-hairpinin modules are released from prepropeptide, become mature, and acquire their antimicrobial activity [14,44]. The hydrophobic C-terminal part serves as a seed storage protein and contains two cupin-7S vicilin domains that are peculiarly widespread within legumes, playing a role in sucrose binding and oxidative stress response, and they are also widely known as major food allergens [45]. We detected 13 vicilin-like precursors with two cupin-7S vicilin domains and two or three α-hairpinin modules.

The second type of precursor, named modular, usually exhibits no similarity to vicilins and consists of a signal peptide followed by a cassette with up to 12 α-hairpinin modules, and it ends with a short C-terminal domain [14]. Again, mature α-hairpinins start to function as antimicrobial peptides upon proteolysis of the precursor. Remarkably, this modular type of precursor is also typical for various antimicrobial peptides from animals including spiders [46,47] and scorpions [48,49]. We detected 17 plant modular precursors containing from two to eight α-hairpinin motifs.

We analyzed putative α-hairpinin precursors using multiple alignments of the whole precursors or their separated Cys-rich modules. As a result, two main conclusions were drawn. First, different plants belonging to distinct families can contain highly similar α-hairpinin modules or even whole precursors, and possibly have one common AMP ancestor. Second, the diversity of α-hairpinin precursors found in this work or those deposited to GenBank is based on changing the number of α-hairpinin modules (Figure 6). Specifically, N-terminal Cys-rich modules are mainly similar, while some middle and C-terminal modules can be removed.

These conclusions are common for both vicilin-like and modular α-hairpinin precursors. In particular, two highly similar Cys-rich modules of vicilin-like precursors isolated from plants belonging to primitive fern (*Danaea nodosa*), true flowering *Caryophyllaceae* (*Silene latifolia*), and *Acteraceae* (*Matricaria matricarioides*) families were observed (Figure S3A,B). Then, N-terminal α-hairpinin modules of a succulent (*Delosperma echinatum*), a parasitic plant (*Orobanche fasciculate*), and plants from *Caryophyllaceae* (*Saponaria officinalis*) and *Lamiaceae* (*Lavandula angustifolia*) families were very similar as well (Figure S3C). Moreover, these precursors exhibited rather high similarity throughout the whole protein length, which supports the idea of one possible common precursor for some vicilin-like prepropeptid In modular α-hairpinin precursors, three out of five Cys-rich motifs of *Setaria italica* (Ge Bank ID XP_022685044; Figure S4A) were absolutely identical and the other two differ by single amino acid substitutions. Highly similar to them were modules from Australi cereals *Thyridolepis multiculmis* and *Neurachne minor* predicted in this study (Figure S4 When their whole precursors were aligned, the high similarity was observed between t first two domains together with 13 C-terminal amino acids, while C-terminal Cys-ri modules were missed in precursors from *T. multiculmis* and *N. minor* (Figure 6 and Figu S4B). This finding confirms the idea of developing α-hairpinin variety by removing son C-terminal Cys-rich modules. One more discovery supporting this assumption was t revelation of precursors from *Pycnanthemum tenuifolium* and *Microstegium vimineum* wi three and seven Cys-rich motifs, respectively, which were similar to MBP-1 containing eig modular prepropeptides. Precursors from *P. tenuifolium* had only four mismatches amor 166 amino acids (Figure S5), while precursors from *M. vimineum* contained considerab more substitutions. However, it exhibited high similarity along the whole precursor leng and contained 11 identical C-terminal amino acids, while lacking the last eighth doma (Figure 6).

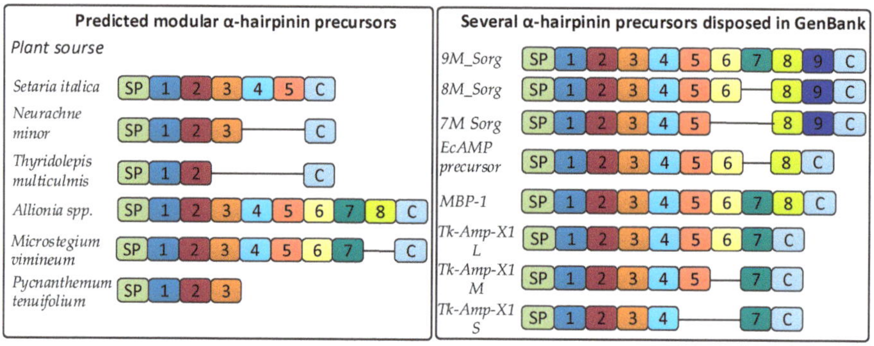

Figure 6. Modular α-hairpinin precursors differ by the amount of C-terminal Cys-rich motifs. SP, si nal peptide; 1–8, α-hairpinin modules; C, C-terminal region. Precursor from *S. italica* has GenBank I XP_022685044. Precursors from *N. minor*, *T. multiculmis*, *Allionia* spp., *M. vimineum*, and *P. tenuifoliu* have the following designations in Table S1: 3M_BXAY, 2M_WCOR, 8M_EGOS, 7M_YPIC, ar 3M_DYFF. 9M_Sorg, 8M_Sorg, and 7M_Sorg are precursors with GenBank IDs XP_002449751. XP_021316824.1, and XP_021316825.1, respectively. MBP-1 is a precursor of antimicrobial peptic MBP-1 from *Z. mays* (NP_001142639). EcAMP precursor was deduced from partial sequences wi GenBank IDs KF478770 and KF478781. Tk-Amp-X1 is proteolysed from L, M, and S precursors wi Uniprot accession numbers HF562352, HF562351, and HF562347, respectively.

We also analyzed some cereal AMP precursors available in GenBank or described i the papers, and again found that within the cassette, some modules close to C-terminu can be removed, while in N-terminal regions, the last Cys-rich motif and C-termin region remained intact. For example, *Sorghum bicolor* contains three different α-hairpini precursors with 7–9 Cys-rich modules. The sixth and the seventh modules could be omitte when forming middle and short forms of precursors, while the last two Cys-rich moti are presented in all isoforms (Figure 6 and Figure S6). Tk-Amp-X precursors isolate from *Triticum kiharae* [14] had a similar organization; in particular, the long form containe seven α-hairpinin modules, while short and middle precursors had reduced cassettes, i which the fifth and/or the sixth C-terminal modules were excluded (Figure 6). Moreove precursor prepropeptide from *Echinochloa crus galli* [14], which was highly similar to eigh modular MBP-1, included a truncated cassette without the seventh domain (Figure 6 The same was also found in some vicilin-like α-hairpinin precursors. In particular, th precursors from *D. echinatum*, *O. fasciculate*, *S. officinalis*, and *L. angustifolia* mentioned above contained a quite similar N-terminal signal peptide, two first α-hairpinin modules, and two cupin-like domains, while the third Cys-rich motif was present only in a precursor from *L. angustifolia* (Figure S3C).

Remarkably, during BLAST annotation, we many times observed the proteins similar to our predicted precursors, which possessed additional middle or next-to-last modules; however, we did not find any α-hairpinin precursors with a modified order of modules. We may speculate that elongation of α-hairpinin precursors by the addition of some Cys-rich modules led to increasing antimicrobial activity since more mature AMPs will be released from elongated precursor upon proteolysis, which, in turn, will lead to increasing the number of active AMPs and amplification of the antimicrobial activity as a whole, helping the plant to resist pathogen invasion. Arguably, proteolysis of different modules can be dependent on a stage of plant development reaching maximum in immature seeds and during seedling development, which are the most vulnerable stages.

3.5. The Limitations of the Current Study

The current study represents an in silico investigation of α-hairpinin features that can be used for the prediction of putative peptides from this family in plant transcriptomic sequences. Although the motifs used for such a prediction underwent careful curation using rather stringent criteria, it should be mentioned that unambiguous and specific in silico identification of α-hairpinins is a challenging task. Thus, the putative α-hairpinins predicted by Cysmotif searcher should be subjected to an additional selection process according to specific investigation goals before they will be ready for experimental verification. The investigators with a significant background in the field of plant AMP isolation and verification can readily propose additional filtration steps for the predicted peptide set to achieve better specificity or applicability in a particular case.

Another limitation is that we used the plant transcriptome dataset from the 1KP project for our prediction, and more recent and/or more specific datasets can provide slightly better results. In addition, our goal was to capture the general patterns for the α-hairpinin family and not to predict as many potentially active peptides as possible. For this reason, we excluded rarely occurring motifs from the downstream analysis, but they can potentially represent more specific cases and thus be of interest in some future investigations. The list of these single motifs can be found in Table S3.

To conclude, α-hairpinins predicted by Cysmotif searcher provide a good starting point for future in vitro and in vivo activity investigations, but the limitations described above should always be taken into account when performing such analyses.

4. Materials and Methods

4.1. Development of the Criteria for α-Hairpinin Prediction

Most part of AMPs represent cysteine-rich peptides possessing a special type of signature sequences called cysteine motifs. The structure of such motifs is rather flexible, and sequence similarity between the motifs of a particular AMP family could be rather low, so that sometimes only cysteine residues have conserved positions in them. This makes their searching and classification a complex task when using homology-based methods only. Previously, we developed the computational pipeline 'Cysmotif searcher' [29] for revealing such motifs and classifying the peptides containing them into AMP families. The motifs used in the pipeline were deduced from literature data and AMP databases and manually curated by us to achieve the required sensitivity and specificity. The AMP families that could be revealed included defensins, thionins, cyclotides, snakins, hevein-like peptides, and lipid-transfer proteins. We revealed more than 10,000 potential AMPs [36] in 1267 plant transcriptomes from the 1KP project [32].

However, an important AMP class of α-hairpinins [14] was not covered by the motifs developed at that time due to insufficient data available. In order to reveal α-hairpinin sequences, we deduced several motifs using the verified α-hairpinin data currently availab in public databases and the literature. The list of motifs is shown in Table S2.

The difficulty of predicting α-hairpinins lies in the fact that their Cys-motifs a widespread and could be a part of Cys-motifs belonging to other AMPs such as defensi lipid-transfer proteins, and thionins. For this reason, all the predicted peptides we additionally manually curated and annotated by BLAST search, and only the sequenc that have passed all filtration steps were characterized as putative α-hairpinins. Th curation criteria for inclusion and exclusion of the peptides from α-hairpinin class a shown in Table 1.

Table 1. Inclusion and exclusion criteria for the manual curation of potential α-hairpinin peptides

Inclusion Criteria	Exclusion Criteria
1. The presence of the 4C motif ($CX_3CX_{4\text{–}20}CX_3C$)	1. The absence of the 4C/6C core motif
2. The presence of the 6C motif ($CX_3CX_3CX_{4\text{–}20}CX_3CX_3C$)	2. The presence of 'CC' before or after the 4C/6C core motif
3. The presence of one or two additional cysteines located before or after the 4C/6C core motif	3. The presence of three or more cysteines, which are not arranged ir the 4C/6C core motif, before or after the core motif itself
	4. The presence of X_1CX_3C after the motif
	5. BLAST annotation of a peptide as belonging to some other proteir family, except vicilins and unknown proteins

Inclusion criteria were the presence of either the 4C or 6C motifs. The possibility a presence of one or two additional cysteines before or after the motif was based on th reports that some precursors of α-hairpinins contained additional cysteines, which can l located either quite close to the source 4C/6C motif or at a significant distance from it.

Exclusion criteria were the absence of the 4C/6C motifs. Furthermore, the presen of additional cysteine residues beyond the α-hairpinin motif was also assessed. If doub cysteines before or after the motif were present, this sequence was excluded, since th feature was a characteristic of other plant AMPs (thionins, hevein-like peptides, lipi transfer proteins etc.). Moreover, if there were three or more cysteines outside the co motifs, they had to be arranged in an additional α-hairpinin motif or its part, otherwise th sequence was excluded. The reason for such exclusion is that there are some plant AM families that have 4C/6C core motifs as a part of more broad Cys-motifs, but other cystein in such motifs have a different arrangement. Another exclusion criterion was the presen of X_1CX_3C immediately after the motif, since this is the characteristic C-tail sequence defensins. Finally, if BLAST annotation characterized a sequence as representing no hairpinin protein (e.g., thionin) except vicilins and unknown proteins, it was also exclude from further study.

We used the same set of 1267 plant transcriptomes as in our previous investigations [3 in order to make the results of two studies comparable with each other and exclude no hairpinin motifs as described above.

4.2. Description of the Prediction Pipeline

Searching for motifs in translated transcriptomic sequences from the 1KP project wa performed by Cysmotif searcher. The version used to obtain the results in this manuscri was 3.3.3 (source code is available at https://github.com/fallandar/cysmotifsearch under GNU GENERAL PUBLIC LICENSE Version 3, accessed on 29 September 202 Novel motifs are included in the file 'motifs_hairpinin.txt' available from the reposito given above.

The flowchart of the pipeline is shown in Figure 7.

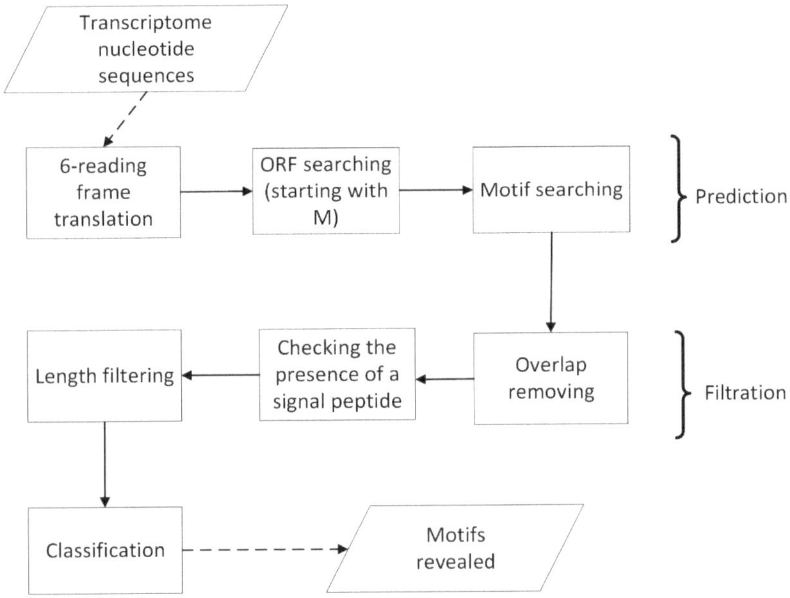

Figure 7. A flowchart of the Cysmotif searcher prediction pipeline operation.

In general, the pipeline includes three stages, namely, prediction of the cysteine motifs, their filtration, and classification of the results. Cysmotif searcher accepts nucleotide or translated transcriptomic sequences and motif list as input, performs 6-reading frame translation when necessary, and then reveals open reading frames (ORFs) starting from methionine residues in resulting amino acid sequences. The presence of methionine provides additional evidence that these sequences are unlikely to be the artifacts. The transcripts are not subjected to homology- or similarity-based filtration to avoid the algorithm overfitting. In the next step, checking for the presence of the motifs from the provided list in the revealed ORFs is performed. If two motifs overlap each other, then the one containing greater number of cysteine residues is selected. In addition, no cysteines are allowed to appear after the motif revealed in the same ORF, except for the case of motif extension or searching for modular motifs, as described below. These steps are intended to select the most reliable motifs from the transcriptomes. Next, the amino acid sequences containing motifs are filtered based on the presence of a signal peptide in them (using SignalP 5.0 [50]), and the sequences not containing a signal peptide are excluded from further processing. Finally, the sequences are filtered based on the length of a mature peptide (<=600 aa), again to exclude possible artifacts. The peptides that have passed all the filters are classified based on the motifs revealed in them. In the case of α-hairpinin searching, the classification includes 4C and 6C classes.

An additional set of dedicated software (scripts) developed by us was used to collect the descriptive statistics on plant families possessing each type of motif. The script cysmotif_stat.sh can be found at Github ((https://github.com/fallandar/cysmotifsearcher, accessed on 29 September 2024). This dedicated script collects the motif distribution among plant groups, families, species, and parts. It takes the output file of the main pipeline and annotation file from 1KP project as input and distributes the sequences revealed into cohorts based on the annotation provided. The first level includes high-order groups (e.g., core eudicots), the second includes families, the third level is for species, and the fourth one is for tissues. The output is provided as text files in tabular form.

Upon obtaining a filtered and curated set of potential α-hairpinin sequences, we excluded the motifs revealed in only one sequence in order to increase the reliability of the analysis. Our main goal was to provide future activity determination experimen with more reliable and generalized data, not to capture the complete possible diversi of potential α-hairpinins. Single sequences can represent rare peptides, but they can al arise from sequencing errors or artifacts introduced by transcriptome assembly softwa However, such sequences can be of interest for some researchers, so we added them a separate Table S3, but we did not include them in the downstream analysis since th were not supposed to reflect the general properties of the plant families or groups to whi they belonged.

In order to check whether the peptides revealed could in fact possess antimicrobi activity against some fungi or bacteria, we submitted all the sequences obtained to pred tion servers. Antibacterial activity was assessed by AntiBP3 [51] (https://webs.iiitd.ed in/raghava/antibp3, accessed on 24 September 2024, threshold value = 0.5), and antifung activity—by Antifp [52] (http://webs.iiitd.edu.in/raghava/antifp, accessed on 24 Septe ber 2024, threshold value = −0.3) and AfpTransferPred [53] (https://selectFigureht.or afptransferpred/, accessed on 24 September 2024, threshold value = 0.5). In the latter ca a peptide was considered antifungal if both servers predicted this.

It is essential to verify the results of all in silico prediction procedures for transcriptom sequences using some random shuffling algorithm to exclude possible artifacts [54]. For t purpose of additional verification of the prediction consistency and exclusion of possib false positives, the motif searching was additionally conducted in the amino acid sequenc built with a random number generator. We took ORFs, in which the motifs were foun performed their random shuffling using the Fisher–Yates shuffle procedure, and subjecte the obtained amino acid sequences to motif searching. Only two motifs correspondir to 4C peptides and no motifs corresponding to 6C peptides were revealed, which mak less than 0.2% of the corresponding numbers for the real transcriptomic sequences. Th procedure confirms that the probability of revealing the exact motif structure by chance very low.

4.3. *Prediction of Modular α-Hairpinin Precursors*

Another important addition to the pipeline was the procedure of revealing the s quences possessing 'modular' motifs, which included several repeats (up to 6) of a partic lar motif in the same ORF. These repeats could have similar, but not identical, sequence This improvement was achieved by a semi-automated algorithm, the implementation which is a script cysmotif_modular.sh available in Github (see link above). The algorith includes running cysmotif_searcher.pl several times with the same set of motifs to searc but the output of the previous run becomes an input to the next run. The motifs reveale are marked in lower case in the corresponding amino acid sequence of a peptide, an cysmotif_searcher.pl is instructed with the –u option to skip conversion of input sequenc to upper case. Such a trick, together with a processing logic set to report only the first on from two non-overlapping motifs within a single sequence, allows the program to igno the previously revealed motifs and search for additional ones within the same sequence order to reveal multiple domains in a peptide. The exemplary sequential pipeline calls a as follows:

cysmotif_searcher.pl -i AAAA_translated.fasta.bz2 -m motifs_hairpinin.txt -t -b -n 5 -l 600 –c;

cysmotif_searcher.pl -i AAAA_translated_motifs_orfonly_withM.fasta -m motifs_ha pinin.txt -t -b -n 55 -l 600 -c –u.

The description of all options can be found in Github (see above).

An example of a modular structure is provided in Figure S2. Previously, a sequenc including additional cysteines beyond the motif was classified as belonging to the artifici 'cysteine-rich peptide' class and excluded from further analysis due to uncertainty regardin its tertiary structure [36].

4.4. Statistics Collection and Result Presentation

Sequence logos for 4C and 6C motifs were generated using the Weblogo resource (https://weblogo.berkeley.edu/logo.cgi, accessed on 20 October 2024).

Then, we collected the descriptive statistics on the motif distribution among plant families and parts of the plants using the cysmotif_stat.sh script described above. The full list of the peptide sequences revealed is available in Table S1 (multidomain), Table S2 (4C and 6C motifs), and Table S3 (4C and 6C sequences with motifs revealed only once, which were excluded from the downstream analysis).

We also performed statistical analysis using the chi-square test to check whether the distribution of the revealed α-hairpinins among various transcriptome groups (plant families, organs, etc.) simply reflects the frequency of such groups in the total number of transcriptomes or not. For example, our null hypothesis was that the distribution of α-hairpinins is independent of the plant family. We used the *p*-value of 0.01 as a threshold value in the test. To ensure the significance of the results and criterion applicability, we excluded the groups containing less than six sequences from the analysis.

5. Conclusions

In this study, we developed a novel algorithm for reliably detecting α-hairpinins in plant transcriptomic sequences and expanded it to reveal peptides possessing a complex modular structure, which cannot be predicted by a simple homology search. Several important findings were revealed in this study based on the application of this novel algorithm to the analysis of transcriptomic sequences from the 1KP project:

i. being widely widespread among various plant families, α-hairpinins are especially prevalent in *Papaveraceae*, in particular, *Papaver somniferum*, which contains 43 various α-hairpinins with different Cys-motifs belonging to two large classes 4C and 6C;
ii. more than 2000 putative α-hairpinins were predicted, among which the peptides with the 6C motif are more likely to possess strong antimicrobial activity and are more important to isolate by biochemical methods from the corresponding plants, at least according to in silico activity prediction;
iii. the diversity of α-hairpinin precursors was possibly developed by changing the number of α-hairpinin modules; specifically, some middle and C-terminal Cys-rich modules can be removed without losing the peptide function;
iv. the unambiguous identification of α-hairpinins represents a very difficult and challenging task, and thus a careful curation of the sequences obtained using the provided motifs is needed since their prediction was based solely on the transcriptomic sequences; researchers with a significant background in the field can propose their own curation and filtration steps, which will supplement the steps included in the Cysmotif searcher pipeline.

The data and software provided will facilitate the transcriptome-based mining and possible isolation of peptides with high antimicrobial activity from different plants, which can ultimately contribute to developing better antimicrobial drugs.

Supplementary Materials: The following supporting information can be downloaded at https://www.mdpi.com/article/10.3390/antibiotics13111019/s1. Figure S1: A bar chart displaying a distribution of different X values found in predicted peptides with 4C + 1 and 4C + 2 motifs. Figure S2: Structure of predicted modular α-hairpinin precursors. Figure S3: Multiple alignments of α-hairpinin-modules referring to the first class of precursors. Figure S4: Multiple alignments of α-hairpinin-modules referring to the second class of precursors. Figure S5: Multiple alignments of selected modular α-hairpinin precursors. Figure S6: Multiple alignments of modular α-hairpinin precursors found in *Sorghum bicolor*. Table S1: Variety of α-hairpinin Cys motifs found using Cysmotif searcher. Table S2: Distribution of predicted α-hairpinins among plant taxons. Table S3: The list of motifs appearing only once in the dataset, which were excluded from the downstream analysis.

Author Contributions: Conceptualization, A.A.S. (Anna A. Slavokhotova) and A.A.S. (Andrey A. Shelenkov); methodology, A.A.S. (Andrey A. Shelenkov); software, A.A.S. (Andrey A. Shelenkov);

investigation, A.A.S. (Anna A. Slavokhotova); resources, A.A.S. (Andrey A. Shelenkov); data curatic A.A.S. (Andrey A. Shelenkov); writing—original draft preparation, A.A.S. (Anna A. Slavokhotov writing—review and editing, A.A.S. (Anna A. Slavokhotova), A.A.S. (Andrey A. Shelenkov), ar E.A.R.; visualization, A.A.S. (Anna A. Slavokhotova); supervision, A.A.S. (Andrey A. Shelenko and E.A.R.; funding acquisition, E.A.R. All authors have read and agreed to the published version the manuscript.

Funding: This work was supported by the Russian Science Foundation (grant 19-76-30005-P).

Data Availability Statement: The supporting data are available in Tables S1–S3 and Figures S1–S Source code of the software used and user instructions are available on Github (https://github.con fallandar/cysmotifsearcher, accessed 24 October 2024).

Conflicts of Interest: The authors declare no conflicts of interest.

References

1. Antimicrobial Resistance Collaborators. Global burden of bacterial antimicrobial resistance in 2019: A systematic analysis. *Lanc* **2022**, *399*, 629–655. [CrossRef] [PubMed]
2. Huan, Y.; Kong, Q.; Mou, H.; Yi, H. Antimicrobial Peptides: Classification, Design, Application and Research Progress in Multip Fields. *Front. Microbiol.* **2020**, *11*, 582779. [CrossRef]
3. Moretta, A.; Scieuzo, C.; Petrone, A.M.; Salvia, R.; Manniello, M.D.; Franco, A.; Lucchetti, D.; Vassallo, A.; Vogel, H.; Sgambato, A et al. Antimicrobial Peptides: A New Hope in Biomedical and Pharmaceutical Fields. *Front. Cell Infect. Microbiol.* **2021**, *11*, 66863 [CrossRef]
4. Maroti, G.; Kereszt, A.; Kondorosi, E.; Mergaert, P. Natural roles of antimicrobial peptides in microbes, plants and animals. *Re Microbiol.* **2011**, *162*, 363–374. [CrossRef] [PubMed]
5. Pasupuleti, M.; Schmidtchen, A.; Malmsten, M. Antimicrobial peptides: Key components of the innate immune system. *Crit. Re Biotechnol.* **2012**, *32*, 143–171. [CrossRef] [PubMed]
6. Zasloff, M. Antimicrobial peptides of multicellular organisms. *Nature* **2002**, *415*, 389–395. [CrossRef]
7. Diamond, G.; Beckloff, N.; Weinberg, A.; Kisich, K.O. The roles of antimicrobial peptides in innate host defense. *Curr. Pharm. De* **2009**, *15*, 2377–2392. [CrossRef]
8. Rima, M.; Rima, M.; Fajloun, Z.; Sabatier, J.M.; Bechinger, B.; Naas, T. Antimicrobial Peptides: A Potent Alternative to Antibiotic *Antibiotics* **2021**, *10*, 1095. [CrossRef]
9. Fernandez de Caleya, R.; Gonzalez-Pascual, B.; Garcia-Olmedo, F.; Carbonero, P. Susceptibility of phytopathogenic bacteria t wheat purothionins in vitro. *Appl. Microbiol.* **1972**, *23*, 998–1000. [CrossRef]
10. Shi, G.; Kang, X.; Dong, F.; Liu, Y.; Zhu, N.; Hu, Y.; Xu, H.; Lao, X.; Zheng, H. DRAMP 3.0: An enhanced comprehensive dat repository of antimicrobial peptides. *Nucleic Acids Res.* **2022**, *50*, D488–D496. [CrossRef]
11. Campos, M.L.; Liao, L.M.; Alves, E.S.F.; Migliolo, L.; Dias, S.C.; Franco, O.L. A structural perspective of plant antimicrobia peptides. *Biochem. J.* **2018**, *475*, 3359–3375. [CrossRef] [PubMed]
12. Decker, A.P.; Mechesso, A.F.; Wang, G. Expanding the Landscape of Amino Acid-Rich Antimicrobial Peptides: Definitio Deployment in Nature, Implications for Peptide Design and Therapeutic Potential. *Int. J. Mol. Sci.* **2022**, *23*, 12874. [CrossRe [PubMed]
13. Dias Rde, O.; Franco, O.L. Cysteine-stabilized alphabeta defensins: From a common fold to antibacterial activity. *Peptides* **2015**, *7* 64–72. [CrossRef]
14. Slavokhotova, A.A.; Rogozhin, E.A. Defense Peptides From the alpha-Hairpinin Family Are Components of Plant Innat Immunity. *Front. Plant Sci.* **2020**, *11*, 465. [CrossRef]
15. Culver, K.D.; Sadecki, P.W.; Jackson, J.K.; Brown, Z.A.; Hnilica, M.E.; Wu, J.; Shaw, L.N.; Wommack, A.J.; Hicks, L.M. Identificatio and Characterization of CC-AMP1-like and CC-AMP2-like Peptides in *Capsicum* spp. *J. Proteome Res.* **2024**, *23*, 2948–296 [CrossRef]
16. Park, S.S.; Abe, K.; Kimura, M.; Urisu, A.; Yamasaki, N. Primary structure and allergenic activity of trypsin inhibitors from th seeds of buckwheat (*Fagopyrum esculentum* Moench). *FEBS Lett.* **1997**, *400*, 103–107. [CrossRef] [PubMed]
17. Oparin, P.B.; Mineev, K.S.; Dunaevsky, Y.E.; Arseniev, A.S.; Belozersky, M.A.; Grishin, E.V.; Egorov, T.A.; Vassilevski, A.A Buckwheat trypsin inhibitor with helical hairpin structure belongs to a new family of plant defence peptides. *Biochem. J.* **201** *446*, 69–77. [CrossRef]
18. Conners, R.; Konarev, A.V.; Forsyth, J.; Lovegrove, A.; Marsh, J.; Joseph-Horne, T.; Shewry, P.; Brady, R.L. An unusual helix-turr helix protease inhibitory motif in a novel trypsin inhibitor from seeds of Veronica (*Veronica hederifolia* L.). *J. Biol. Chem.* **2007**, *28* 27760–27768. [CrossRef]
19. Ng, Y.M.; Yang, Y.; Sze, K.H.; Zhang, X.; Zheng, Y.T.; Shaw, P.C. Structural characterization and anti-HIV-1 activities o arginine/glutamate-rich polypeptide Luffin P1 from the seeds of sponge gourd (*Luffa cylindrica*). *J. Struct. Biol.* **2011**, *174*, 164–17 [CrossRef]

0. Li, F.; Yang, X.X.; Xia, H.C.; Zeng, R.; Hu, W.G.; Li, Z.; Zhang, Z.C. Purification and characterization of Luffin P1, a ribosome-inactivating peptide from the seeds of Luffa cylindrica. *Peptides* **2003**, *24*, 799–805. [CrossRef]
1. Duvick, J.P.; Rood, T.; Rao, A.G.; Marshak, D.R. Purification and characterization of a novel antimicrobial peptide from maize (*Zea mays* L.) kernels. *J. Biol. Chem.* **1992**, *267*, 18814–18820. [CrossRef] [PubMed]
2. Vasilchenko, A.S.; Yuryev, M.; Ryazantsev, D.Y.; Zavriev, S.K.; Feofanov, A.V.; Grishin, E.V.; Rogozhin, E.A. Studying of cellular interaction of hairpin-like peptide EcAMP1 from barnyard grass (*Echinochloa crusgalli* L.) seeds with plant pathogenic fungus *Fusarium solani* using microscopy techniques. *Scanning* **2016**, *38*, 591–598. [CrossRef] [PubMed]
3. Sharma, P.; Kaur, J.; Sharma, G.; Kashyap, P. Plant derived antimicrobial peptides: Mechanism of target, isolation techniques, sources and pharmaceutical applications. *J. Food Biochem.* **2022**, *46*, e14348. [CrossRef] [PubMed]
4. Fu, H.; Cao, Z.; Li, M.; Wang, S. ACEP: Improving antimicrobial peptides recognition through automatic feature fusion and amino acid embedding. *BMC Genom.* **2020**, *21*, 597. [CrossRef]
5. Lawrence, T.J.; Carper, D.L.; Spangler, M.K.; Carrell, A.A.; Rush, T.A.; Minter, S.J.; Weston, D.J.; Labbe, J.L. amPEPpy 1.0: A portable and accurate antimicrobial peptide prediction tool. *Bioinformatics* **2021**, *37*, 2058–2060. [CrossRef] [PubMed]
6. Bhadra, P.; Yan, J.; Li, J.; Fong, S.; Siu, S.W.I. AmPEP: Sequence-based prediction of antimicrobial peptides using distribution patterns of amino acid properties and random forest. *Sci. Rep.* **2018**, *8*, 1697. [CrossRef]
7. Yan, K.; Lv, H.; Guo, Y.; Peng, W.; Liu, B. sAMPpred-GAT: Prediction of antimicrobial peptide by graph attention network and predicted peptide structure. *Bioinformatics* **2023**, *39*, btac715. [CrossRef]
8. Bello-Madruga, R.; Torrent Burgas, M. The limits of prediction: Why intrinsically disordered regions challenge our understanding of antimicrobial peptides. *Comput. Struct. Biotechnol. J.* **2024**, *23*, 972–981. [CrossRef]
9. Shelenkov, A.A.; Slavokhotova, A.A.; Odintsova, T.I. Cysmotif Searcher Pipeline for Antimicrobial Peptide Identification in Plant Transcriptomes. *Biochem. Biokhimiia* **2018**, *83*, 1424–1432. [CrossRef]
0. Slavokhotova, A.A.; Shelenkov, A.A.; Odintsova, T.I. Prediction of *Leymus arenarius* (L.) antimicrobial peptides based on de novo transcriptome assembly. *Plant Mol. Biol.* **2015**, *89*, 203–214. [CrossRef]
1. Slavokhotova, A.A.; Shelenkov, A.A.; Korostyleva, T.V.; Rogozhin, E.A.; Melnikova, N.V.; Kudryavtseva, A.V.; Odintsova, T.I. Defense peptide repertoire of Stellaria media predicted by high throughput next generation sequencing. *Biochimie* **2017**, *135*, 15–27. [CrossRef]
2. Matasci, N.; Hung, L.H.; Yan, Z.; Carpenter, E.J.; Wickett, N.J.; Mirarab, S.; Nguyen, N.; Warnow, T.; Ayyampalayam, S.; Barker, M.; et al. Data access for the 1000 Plants (1KP) project. *Gigascience* **2014**, *3*, 17. [CrossRef] [PubMed]
3. Slezina, M.P.; Odintsova, T.I. Plant Antimicrobial Peptides: Insights into Structure-Function Relationships for Practical Applications. *Curr. Issues Mol. Biol.* **2023**, *45*, 3674–3704. [CrossRef] [PubMed]
4. Silva Alves, M.C.; Cabral da Silva, R.C.; Quaresma, S. Characterization of α-hairpinin in the genome of lima bean (*Phaseolus lunatus*). *Int. J. Mol. Biol. Open Access* **2024**, *7*, 112–116. [CrossRef]
5. Kielich, N.; Mazur, O.; Musidlak, O.; Gracz-Bernaciak, J.; Nawrot, R. Herbgenomics meets Papaveraceae: A promising -omics perspective on medicinal plant research. *Brief. Funct. Genomics* **2023**, *23*, 579–594. [CrossRef]
6. Shelenkov, A.; Slavokhotova, A.; Odintsova, T. Predicting Antimicrobial and Other Cysteine-Rich Peptides in 1267 Plant Transcriptomes. *Antibiotics* **2020**, *9*, 60. [CrossRef] [PubMed]
7. Nawrot, R.; Barylski, J.; Nowicki, G.; Broniarczyk, J.; Buchwald, W.; Gozdzicka-Jozefiak, A. Plant antimicrobial peptides. *Folia Microbiol.* **2014**, *59*, 181–196. [CrossRef]
8. Lay, F.T.; Anderson, M.A. Defensins--components of the innate immune system in plants. *Curr. Protein Pept. Sci.* **2005**, *6*, 85–101. [CrossRef]
9. Hong, K.; Austerlitz, T.; Bohlmann, T.; Bohlmann, H. The thionin family of antimicrobial peptides. *PLoS ONE* **2021**, *16*, e0254549. [CrossRef]
0. Rodriguez-Decuadro, S.; da Rosa, G.; Radio, S.; Barraco-Vega, M.; Benko-Iseppon, A.M.; Dans, P.D.; Smircich, P.; Cecchetto, G. Antimicrobial peptides in the seedling transcriptome of the tree legume *Peltophorum dubium*. *Biochimie* **2021**, *180*, 229–242. [CrossRef]
1. Istomina, E.A.; Korostyleva, T.V.; Kovtun, A.S.; Slezina, M.P.; Odintsova, T.I. Transcriptome-Wide Identification and Expression Analysis of Genes Encoding Defense-Related Peptides of *Filipendula ulmaria* in Response to *Bipolaris sorokiniana* Infection. *J. Fungi* **2024**, *10*, 258. [CrossRef] [PubMed]
2. Marcus, J.P.; Green, J.L.; Goulter, K.C.; Manners, J.M. A family of antimicrobial peptides is produced by processing of a 7S globulin protein in Macadamia integrifolia kernels. *Plant J. For. Cell Mol. Biol.* **1999**, *19*, 699–710. [CrossRef] [PubMed]
3. Yamada, K.; Shimada, T.; Kondo, M.; Nishimura, M.; Hara-Nishimura, I. Multiple functional proteins are produced by cleaving Asn-Gln bonds of a single precursor by vacuolar processing enzyme. *J. Biol. Chem.* **1999**, *274*, 2563–2570. [CrossRef] [PubMed]
4. Zhang, J.; Payne, C.D.; Pouvreau, B.; Schaefer, H.; Fisher, M.F.; Taylor, N.L.; Berkowitz, O.; Whelan, J.; Rosengren, K.J.; Mylne, J.S. An Ancient Peptide Family Buried within Vicilin Precursors. *ACS Chem. Biol.* **2019**, *14*, 979–993. [CrossRef]
5. Zhang, Y.; Che, H.; Li, C.; Jin, T. Food Allergens of Plant Origin. *Foods* **2023**, *12*, 2232. [CrossRef]
6. Sachkova, M.Y.; Slavokhotova, A.A.; Grishin, E.V.; Vassilevski, A.A. Structure of the yellow sac spider *Cheiracanthium punctorium* genes provides clues to evolution of insecticidal two-domain knottin toxins. *Insect Mol. Biol.* **2014**, *23*, 527–538. [CrossRef]

47. Vassilevski, A.A.; Sachkova, M.Y.; Ignatova, A.A.; Kozlov, S.A.; Feofanov, A.V.; Grishin, E.V. Spider toxins comprising disulfic rich and linear amphipathic domains: A new class of molecules identified in the lynx spider *Oxyopes takobius*. *FEBS J.* **2013**, 2 6247–6261. [CrossRef] [PubMed]
48. Juichi, H.; Ando, R.; Ishido, T.; Miyashita, M.; Nakagawa, Y.; Miyagawa, H. Chemical synthesis of a two-domain scorpion tox LaIT2 and its single-domain analogs to elucidate structural factors important for insecticidal and antimicrobial activities. *J. Pe Sci.* **2018**, *24*, e3133. [CrossRef]
49. Zhu, S.; Gao, B.; Aumelas, A.; del Carmen Rodriguez, M.; Lanz-Mendoza, H.; Peigneur, S.; Diego-Garcia, E.; Martin-Eauclai M.F.; Tytgat, J.; Possani, L.D. MeuTXKbeta1, a scorpion venom-derived two-domain potassium channel toxin-like peptide wi cytolytic activity. *Biochim. Biophys. Acta* **2010**, *1804*, 872–883. [CrossRef]
50. Almagro Armenteros, J.J.; Tsirigos, K.D.; Sonderby, C.K.; Petersen, T.N.; Winther, O.; Brunak, S.; von Heijne, G.; Nielsen, SignalP 5.0 improves signal peptide predictions using deep neural networks. *Nat. Biotechnol.* **2019**, *37*, 420–423. [CrossRef]
51. Bajiya, N.; Choudhury, S.; Dhall, A.; Raghava, G.P.S. AntiBP3: A Method for Predicting Antibacterial Peptides against Gra Positive/Negative/Variable Bacteria. *Antibiotics* **2024**, *13*, 168. [CrossRef] [PubMed]
52. Agrawal, P.; Bhalla, S.; Chaudhary, K.; Kumar, R.; Sharma, M.; Raghava, G.P.S. In Silico Approach for Prediction of Antifung Peptides. *Front. Microbiol.* **2018**, *9*, 323. [CrossRef] [PubMed]
53. Lobo, F.; Gonzalez, M.S.; Boto, A.; Perez de la Lastra, J.M. Prediction of Antifungal Activity of Antimicrobial Peptides by Transf Learning from Protein Pretrained Models. *Int. J. Mol. Sci.* **2023**, *24*, 270. [CrossRef] [PubMed]
54. Slavokhotova, A.; Korostyleva, T.; Shelenkov, A.; Pukhalskiy, V.; Korottseva, I.; Slezina, M.; Istomina, E.; Odintsova, T. Transcri tomic Analysis of Genes Involved in Plant Defense Response to the Cucumber Green Mottle Mosaic Virus Infection. *Life* **202** *11*, 1064. [CrossRef]

antibiotics

rticle

Jnveiling a New Antimicrobial Peptide with Efficacy against ?. *aeruginosa* and *K. pneumoniae* from Mangrove-Derived ?*aenibacillus thiaminolyticus* NNS5-6 and Genomic Analysis

;amfa Sermkaew [1,2], Apichart Atipairin [1,2], Sucheewin Krobthong [3], Chanat Aonbangkhen [3,4], odying Yingchutrakul [5], Jumpei Uchiyama [6] and Nuttapon Songnaka [1,2,*]

1 School of Pharmacy, Walailak University, Thasala, Nakhon Si Thammarat 80160, Thailand; namfa.se@wu.ac.th (N.S.); apichart.at@wu.ac.th (A.A.)
2 Drug and Cosmetics Excellence Center, Walailak University, Thasala, Nakhon Si Thammarat 80160, Thailand
3 Center of Excellence in Natural Products Chemistry (CENP), Department of Chemistry, Faculty of Science, Chulalongkorn University, Bangkok 10330, Thailand; sucheewin.k@chula.ac.th (S.K.); chanat.a@chula.ac.th (C.A.)
4 Center of Excellence on Petrochemical and Materials Technology, Chulalongkorn University, Bangkok 10330, Thailand
5 National Center for Genetic Engineering and Biotechnology, National Science and Technology Development Agency, Pathum Thani 12120, Thailand; yodying.yin@biotec.or.th
6 Department of Bacteriology, Graduate School of Medicine, Dentistry and Pharmaceutical Sciences, Okayama University, Okayama 700-8558, Japan; uchiyama@okayama-u.ac.jp
* Correspondence: nuttapon.so@wu.ac.th

Abstract: This study focused on the discovery of the antimicrobial peptide (AMP) derived from mangrove bacteria. The most promising isolate, NNS5-6, showed the closest taxonomic relation to *Paenibacillus thiaminolyticus*, with the highest similarity of 74.9%. The AMP produced by *Paenibacillus thiaminolyticus* NNS5-6 exhibited antibacterial activity against various Gram-negative pathogens, especially *Pseudomonas aeruginosa* and *Klebsiella pneumoniae*. The peptide sequence consisted of 13 amino acids and was elucidated as Val-Lys-Gly-Asp-Gly-Gly-Pro-Gly-Thr-Val-Tyr-Thr-Met. The AMP mainly exhibited random coil and antiparallel beta-sheet structures. The stability study indicated that this AMP was tolerant of various conditions, including proteolytic enzymes, pH (1.2–14), surfactants, and temperatures up to 40 °C for 12 h. The AMP demonstrated 4 μg/mL of MIC and 4–8 μg/mL of MBC against both pathogens. Time-kill kinetics showed that the AMP acted in a time- and concentration-dependent manner. A cell permeability assay and scanning electron microscopy revealed that the AMP exerted the mode of action by disrupting bacterial membranes. Additionally, nineteen biosynthetic gene clusters of secondary metabolites were identified in the genome. NNS5-6 was susceptible to various commonly used antibiotics supporting the primary safety requirement. The findings of this research could pave the way for new therapeutic approaches in combating antibiotic-resistant pathogens.

Keywords: antimicrobial peptide; antimicrobial resistance; bacterial genome; biosynthetic gene cluster; *Klebsiella pneumoniae*; Mangrove; mass spectrometry; NNS5-6; *Paenibacillus thiaminolyticus*; *Pseudomonas aeruginosa*

itation: Sermkaew, N.; Atipairin, A.; robthong, S.; Aonbangkhen, C.; ingchutrakul, Y.; Uchiyama, J.; ongnaka, N. Unveiling a New ntimicrobial Peptide with Efficacy gainst *P. aeruginosa* and . *pneumoniae* from 1angrove-Derived *Paenibacillus* *iiaminolyticus* NNS5-6 and Genomic nalysis. *Antibiotics* **2024**, *13*, 846. ttps://doi.org/10.3390/ ntibiotics13090846

cademic Editors: Piyush Baindara nd Marisa Di Pietro

eceived: 7 August 2024
evised: 26 August 2024
ccepted: 3 September 2024
ublished: 5 September 2024

1. Introduction

Antimicrobial resistance (AMR) poses a significant challenge to global public health. Its severity is exacerbated by the overuse and inappropriate application of antimicrobial agents in humans, animals, and plants. This contributes to an increase in new resistant microorganisms in the environment. Presently, AMR results in over 700,000 deaths annually worldwide, creating substantial economic and health impacts [1]. The "ESKAPE" pathogens a group that includes *Enterococcus faecium*, *Staphylococcus aureus*, *Klebsiella pne-moniae*, *Acinetobacter baumannii*, *Pseudomonas aeruginosa*, and *Enterobacter* species [2]. Th group is categorized as "critical" on the World Health Organization's (WHO) priori list of bacterial pathogens, underscoring the urgent need for novel antibiotics to addre drug resistance. *Pseudomonas aeruginosa*, a ubiquitous Gram-negative bacterium, is wide recognized as an opportunistic pathogen notorious for its antibiotic resistance. It is leading cause of nosocomial infections and ventilator-associated pneumonia, presentin a substantial challenge in clinical settings because of its ability to form biofilms, whic complicates treatment [3,4]. *P. aeruginosa* contributes to more than 5% of infectious exace bations in patients with chronic obstructive pulmonary disease (COPD) and is associate with elevated mortality rates among these individuals [3]. Similarly, *Klebsiella pneumonia* has emerged as a virulent pathogen, attributable to the rising number of severe infection Given the evolutionary diversity among clinical strains, it plays a significant role in variou infection models, including pneumonia, liver abscesses, and gastrointestinal tract coloniz tion [5]. Exploring new antimicrobial agents is critical in addressing the persistent threat c infectious diseases caused by *P. aeruginosa* and *K. pneumoniae*.

Microorganisms, especially bacteria, are well-suited for the large-scale synthesis c bioactive compounds. Since the discovery of penicillin, microbial secondary metabolite have been a primary source of novel antimicrobial agents. Marine bacteria have emerged a promising sources of therapeutic compounds, including antibacterial, antiviral, antifunga and anticancer agents [6]. The mangrove ecosystem, characterized by harsh environmenta conditions such as high tides, hypersaline waters, and large temperature fluctuations, pr vides a unique habitat that fosters the production of bioactive compounds [7,8]. Genomi studies suggest that bacteria from harsh environments, with their often-large genome have the potential to produce a diverse array of secondary metabolites, surpassing previou estimates. This highlights the importance of exploring these unique habitats for novel a timicrobial compounds [9]. Attention has been focused on antimicrobial peptides (AMPs which have been investigated for the treatment of infections [10]. AMPs are an essentia component of the innate immune system present in all living organisms, playing a crucia role as the frontline defense against pathogens. They vary in the length of their amino aci residues and are composed of charged and hydrophobic amino acids. In general, AMP are unstructured and potentially form amphipathic alpha-helical or beta-sheet structure in bacterial cell membranes. These peptides disrupt membranes without target–recepto specificity and have high-affinity binding interactions; therefore, they are less likely t induce resistance in pathogens [11].

The objective of this study was to isolate mangrove bacteria capable of producin an AMP effective against *P. aeruginosa* and *K. pneumoniae*. The amino acid sequence an properties of the AMP were characterized. The genomic information of the promisin bacterial isolate was explored, including the biosynthetic gene clusters for secondar metabolites. This research seeks to identify new antimicrobial resources and provid valuable insights to develop alternative agents for facing the emergence of antimicrobia resistant infections.

2. Results

2.1. Antimicrobial Investigation of Bacterial Isolates from Mangrove Sediments

Mangrove sediments were collected from five different locations at Banlaem Mar grove, Thasala District, Nakhon Si Thammarat Province, Thailand. The pH of the sediment ranged from 6.26 to 8.00, and the salinity was between 9.00 and 10.00 ppt. Colonies grow on Mueller Hinton (MH) agar, Zobell Marine (ZM) agar, and Starch Casein (SC) aga were screened for antibacterial activities against *P. aeruginosa* TISTR 357 using the sof agar overlay technique. The antibacterial activity of the isolates was expanded to includ activity against various bacterial pathogens. Only one isolate, NNS5-6, cultured on MH agar, exhibited antibacterial activity against *P. aeruginosa* TISTR 357. The cell-free super natant (CFS) collected from a 24-h preculture of NNS5-6 in MH broth showed an expande antibacterial spectrum against Gram-negative bacteria, including *P. aeruginosa* TISTR 357, *K. pneumoniae* TISTR 1383, *Escherichia coli (E. coli)* TISTR 887, *Salmonella typhimurium (S. typhimurium)* TISTR 1469, and *Vibrio parahaemolyticus (V. parahaemolyticus)* TISTR 1596, with inhibition zones ranging from 12.70 ± 0.25 to 14.84 ± 0.15 mm. In contrast, there was no activity against Gram-positive bacteria, such as *Staphylococcus aureus (S. aureus)* TISTR 517 and methicillin-resistant *Staphylococcus aureus* (MRSA) strain 2468 (Table 1). The standard antibiotics vancomycin and colistin were used as positive controls to compare the degree of antibacterial activity of the active isolate. Therefore, the NNS5-6 isolate was selected for further antibacterial studies.

Table 1. Antibacterial activities of the CFS of NNS5-6 against bacterial pathogens. The antibacterial spectrum was studied using the agar well diffusion method and compared with standard antibiotics.

Isolate	Zone of Inhibition (mm ± SD; n = 3)						
	S. aureus TISTR 517	MRSA Strain 2468	*E. coli* TISTR 887	*K. pneumoniae* TISTR 1383	*P. aeruginosa* TISTR 357	*S. typhimurium* TISTR 1469	*V. parahaemolyticus* TISTR 1596
NNS5-6	0.00 ± 0.00	0.00 ± 0.00	14.24 ± 0.15	13.34 ± 0.53	14.84 ± 0.15	12.70 ± 0.25	13.09 ± 0.64
Vancomycin (30 μg)	21.59 ± 0.51	21.76 ± 0.78	0.00 ± 0.00	0.00 ± 0.00	0.00 ± 0.00	0.00 ± 0.00	0.00 ± 0.00
Colistin (1 μg)	0.00 ± 0.00	0.00 ± 0.00	18.13 ± 0.88	18.81 ± 0.39	18.05 ± 0.67	20.4 ± 0.89	18.8 ± 0.76

2.2. Production Kinetics of Antibacterial Compounds of NNS5-6

This study investigated the production of antibacterial compounds by NNS5-6 over the incubation period with culture growth monitored through cell suspension turbidity. The antibacterial activity of the collected CFS at different incubation times was determined by measuring inhibition zones against various bacterial pathogens. The antibacterial activity was observed during the early stationary phase of the growth curve. The results revealed that the initial and maximum activity of the antibacterial compounds production by NNS5-6 was found at 12 h and 20 h of incubation, respectively. The maximum inhibition zone of NNS5-6 CFS at 20 h was in the range of 15.50 ± 0.20 to 16.89 ± 0.50 mm. The antibacterial activity declined after 20 h of incubation until it diminished at 24 h during the stationary phase of growth, except in E. coli TISTR 887, where the antibacterial activity of NNS5-6 CFS remained until 96 h of incubation. In contrast, the Gram-positive bacteria, S. aureus TISTR 517 and MRSA strain 2468, were not inhibited by NNS5-6 CFS at any incubation period. NNS5-6 demonstrated the ability to produce antibacterial compounds against Gram-negative bacteria within 20 h of incubation (Figure 1).

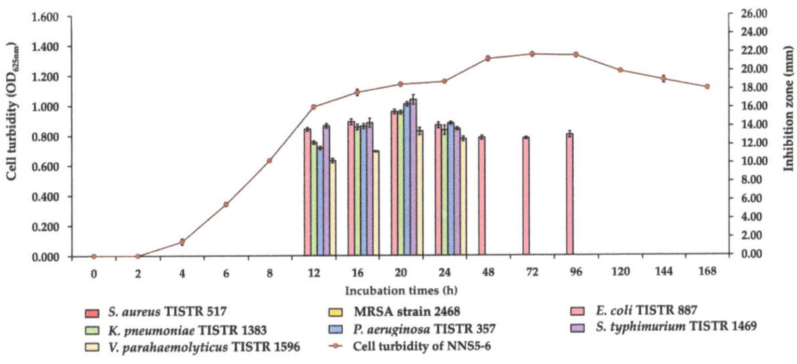

Figure 1. The growth profile of NNS5-6 and its production kinetics of antibacterial components were determined during the incubation periods. The growth curve was reported by the turbidity of the cell suspension (primary *y*-axis). The antibacterial activity against bacterial pathogens is presented by the inhibition zones (secondary *y*-axis).

2.3. Purification of Antibacterial Compounds of NNS5-6

The antibacterial components produced by a twenty-hour-old NNS5-6 culture were p rified through sequential steps, including ammonium sulfate precipitation, cation-exchan chromatography, and size-exclusion chromatography. The fractions obtained at each st were tested for antibacterial activity against *P. aeruginosa* TISTR 357 using the agar w diffusion assay. A bioassay-guided approach was used to identify the fraction with a tibacterial activity (Figure S1a). The active fraction was collected to determine the presen of the active peptide band, and the purity of the purified peptide was confirmed using 15% gel of SDS-PAGE. The SDS-PAGE analysis revealed a single stained protein band Lane 1 of the half-excised gel, with a molecular weight below approximately 5 kDa. T same location of the inhibition zone was observed in the other half-excised gel overla with *P. aeruginosa* TISTR 357, demonstrating the effectiveness of the purification procedur (Figure S1b). The efficiency of these procedures was assessed using a purification balan sheet. The specific activity increased with each purification step, indicating the effectivene of the procedures in purifying the active peptide. The final purification step resulted in 13.17-fold increase in purification power and yielded 10.55% of the active peptide compar with the initial crude product (Table 2).

Table 2. The purification balance sheet of the antibacterial components of NNS5-6. The four purifi tion steps were used to obtain the purified compound.

Purification Procedure	Volume (mL)	Total Dried Weight (mg)	Activity (AU/mL)	Total Activity (AU)	Specific Activity (AU/mg)	Purification Factor	%Yield
Crude product	976.50	453.30	20.00	19,530.00	43.08	1.00	100.00
Salt precipitation	72.78	102.40	80.00	5822.40	56.86	1.32	29.81
Cation-exchange chromatography	42.67	37.92	80.00	3413.60	90.02	2.09	17.48
Size-exclusion chromatography	12.88	3.63	160.00	2060.80	567.40	13.17	10.55

2.4. De Novo Amino Acid Sequence of the Purified AMP and Determination of Its Secondary Structure

The purified AMP of NNS5-6 was subjected to amino acid sequencing using ta dem mass spectrometry. The molecular weight of the peptide was determined by ma spectrometry. The peptide fragmentation detected the parent molecule having a mass 1297.61 Da in the positive ion mode. The de novo algorithm was used to predict the amir acid sequence from b-ion and y-ion fragmentations. The AMP, composed of 13 amino ac residues, was sequenced as Val-Lys-Gly-Asp-Gly-Gly-Pro-Gly-Thr-Val-Tyr-Thr-Met, wi an average local confidence (ALC) score of 75% (Figure 2a). The results of de novo sequer ing indicated that the final amino acid, methionine, was oxidized with an incorporate oxygen atom. Therefore, the molecular weight of the peptide was 1280.6121 Da. The phys ochemical properties of the peptide were predicted using ProtParam on the Expasy serv (https://web.expasy.org/protparam/, accessed on 10 July 2024). The analysis reveale a theoretical pI of 5.81. The peptide contained one positively charged amino acid, lysin and one negatively charged residue, aspartic acid. The predicted net charge was 0 at p 7.4. The grand average of hydropathy (GRAVY) value of the peptide was calculated to b −0.231, predicted by hydrophobic amino acids valine, proline, and methionine, as well a hydrophilic amino acids lysine, glycine, aspartic acid, threonine, and tyrosine.

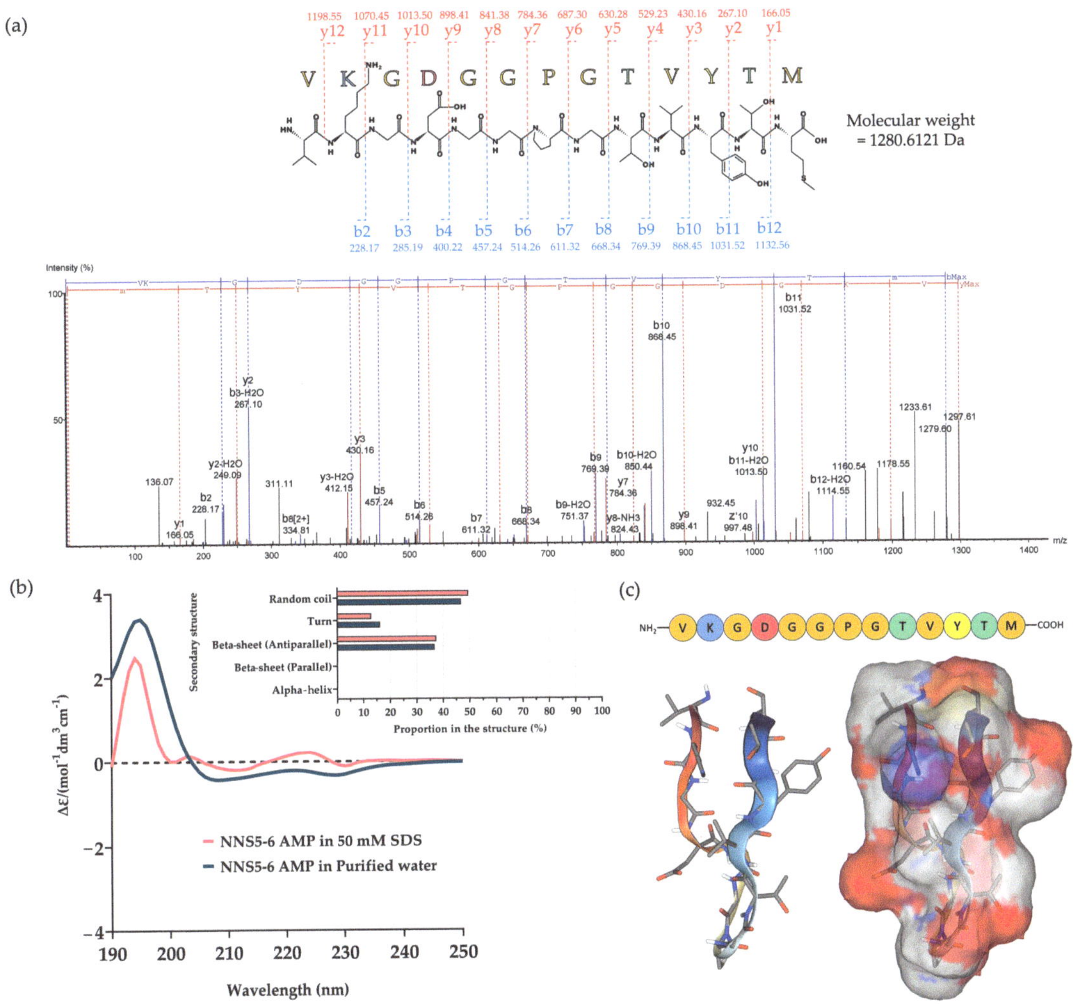

Figure 2. The amino acid sequence of the purified AMP was elucidated by de novo amino acid sequencing using a molecular fragmentation technique (dashed blue line and dashed red line indicate b-ion and y-ion, respectively) from mass spectrometry (**a**). The secondary structure of NNS5-6 AMP was determined by CD spectroscopy. The experiment was designed to compare the structural changes in the dissolved AMP in purified water and 50 mM SDS. The proportion of secondary structures was analyzed by BeStSel and compared in different solvents (**b**). The 3D molecular model of NNS5-6 AMP was predicted using PEP-FOLD4. The beta-sheet structure of the AMP is shown in a cartoon style. The molecular surface reveals the hydrophobic region represented by grey and the hydrophilic regions with negative and positive electrostatic potentials represented by red and blue, respectively (**c**).

Circular dichroism (CD) spectroscopy was used to determine the secondary structure of NNS5-6 AMP in different solvents. Purified water was used to study its native conformation, while 50 mM SDS, above the critical micelle concentration, was employed to simulate the negatively charged environment of bacterial cell membranes for studying AMP interactions. The CD spectra of NNS5-6 AMP showed similar delta epsilon in both purified water and 50 mM SDS solution. The CD spectra scanned in the range of 190–240 nm revealed a positive band with a magnitude at 195 nm and a negative band with a magnitude at 215 nm in both solvent systems. The CD spectra indicated a beta-she conformation of the peptide in both solvents. In addition, the negative band at 200–210 n suggested the presence of a random coil structure. The secondary structure componen of the NNS5-6 AMP were calculated using CD spectra via the BeStSel web-based servic In purified water, the NNS5-6 AMP showed structural components of 46.9% random co conformation, 36.8% antiparallel beta-sheet, and 16.3% turn conformation. A slight diffe ence in secondary structure components was observed when NNS5-6 AMP was dissolve in the SDS micelle environment. The major component remained as a random coil 49.6%, while the beta-sheet (antiparallel) structure was 37.4%, and the turn conformatio was 13.0% (Figure 2b). The findings indicated that the NNS5-6 AMP primarily exhibite a beta-sheet (antiparallel) secondary structure with a random coil conformation. In th presence of SDS micelles, the interaction between NNS5-6 AMP and SDS micelles slightl altered the secondary structure components of the AMP but did not change the overa type of secondary structures compared with the native environment. The 3D molecula model further supported the analyzed CD spectra results. The NNS5-6 AMP displaye the beta-sheet conformation with an antiparallel orientation, with a turn at the prolin amino acid and a random coil at the N- and C-termini. The predicted molecular surfac revealed the hydrophobic and hydrophilic regions of NNS5-6 AMP (Figure 2c). The result suggest that the NNS5-6 AMP likely maintains a stable antiparallel beta-sheet structur at membrane surfaces, supporting the hypothesis that the cationic side of the AMP firs interacts electrostatically with the anionic membrane surface and destabilizes the lipi bilayer of the cell membrane with the hydrophobic part of the peptide [12].

2.5. Investigation of the Antibacterial Activities of the AMP

The antibacterial activities of the NNS5-6 AMP against *P. aeruginosa* TISTR 357 an *K. pneumoniae* TISTR 1383 were studied using a microdilution assay, with colistin as th standard. Both *P. aeruginosa* TISTR 357 and *K. pneumoniae* TISTR 1383 were inhibite by NNS5-6 AMP at the same MIC of 4 µg/mL, compared with an MIC of 1 µg/mL fo colistin. However, the concentrations required for the bactericidal effect of NNS5-6 AM differed between the two pathogens. The bactericidal activity of NNS5-6 AMP required concentration with a two-fold higher MIC concentration for killing *K. pneumoniae* TIST 1383, while for *P. aeruginosa* TISTR 357, the bactericidal concentration was equal to the MIC Colistin exhibited bactericidal activity at the MIC for both bacterial pathogens (Table 3).

Table 3. The MIC and MBC values of the purified AMP of NNS5-6 were determined using th microdilution method against two bacterial pathogens. Colistin was used as the positive control.

Active Compounds	Tested Strains	MIC (µg/mL)	MBC (µg/mL)
NNS5-6 AMP	*P. aeruginosa* TISTR 357	4	4
	K. pneumoniae TISTR 1383	4	8
Colistin	*P. aeruginosa* TISTR 357	1	1
	K. pneumoniae TISTR 1383	1	1

2.6. The Antibacterial Activity of NNS5-6 Derived AMP on Bacterial Pathogens Observed Usin Scanning Electron Microscopy (SEM) Analysis

Alterations in cell morphology and membrane integrity of *P. aeruginosa* TISTR 357 an *K. pneumoniae* TISTR 1383 treated with antibacterial compounds were observed using SEM The NNS5-6 AMP and colistin demonstrated antibacterial activity against these bacteria pathogens at 1× MIC. The cell surfaces of both pathogens responded differently to the antibacterial compounds. In untreated *P. aeruginosa*, the intact cell membrane retained its original rod shape with a smooth surface, indicating membrane integrity (Figure 3a). NNS5 6 AMP caused the *P. aeruginosa* cells to develop a rough surface and shrink. The cente region of the treated cells appeared intruded, indicating cell membrane breakage and the release of cytoplasm after the cell ruptured (Figure 3b). The action of colistin is well-know for its antibacterial activity through membrane disruption. Colistin-treated *P. aeruginosa* cells exhibited a killing effect, observed as holes at the center of the cells and deflated cells, confirming membrane disruption (Figure 3c). For *K. pneumoniae*, untreated cells maintained their shape and showed some exopolysaccharide deposits around them, indicating vigorous growth (Figure 3d). Upon treatment with NNS5-6 AMP, *K. pneumoniae* cells exhibited signs of rupture and cytoplasmic leakage, characterized by a porous morphology (Figure 3e). Similarly, colistin-treated cells displayed membrane damage, leading to pore formation (Figure 3f). The observed differences in morphological changes between treated *P. aeruginosa* cells and *K. pneumoniae* compared with untreated controls highlight the distinct membrane disruption characteristics induced by NNS5-6 AMP and colistin. This investigation suggests that NNS5-6 AMP effectively targets and damages bacterial membranes.

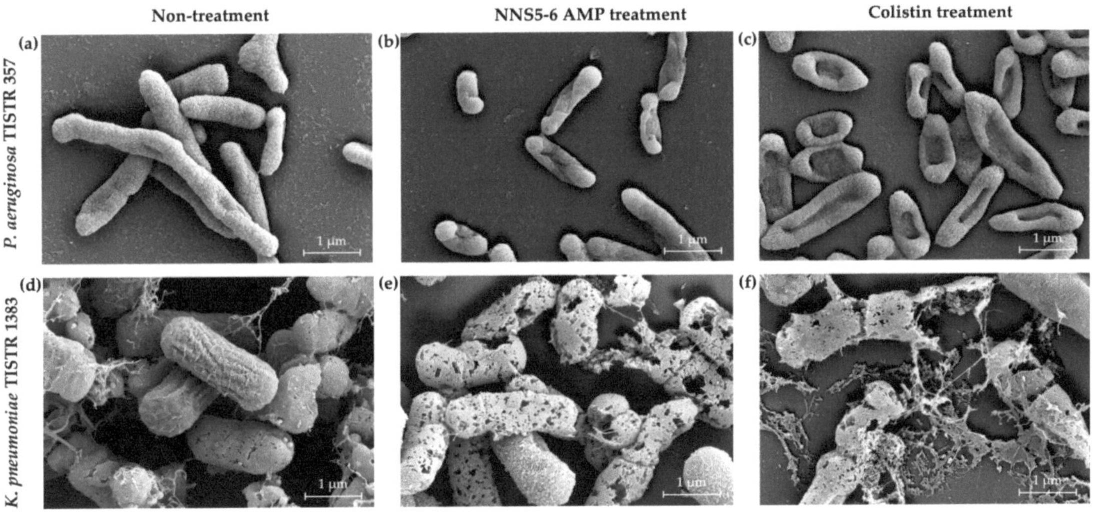

Figure 3. The effects of NNS5-6 AMP and colistin on *P. aeruginosa* TISTR 357 and *K. pneumoniae* TISTR 1383 were observed as morphological changes visualized under a scanning electron microscope at 20,000× magnification. The untreated condition of both bacteria was incubated with 0.85% NaCl, *P. aeruginosa* TISTR 357 (**a**) and *K. pneumoniae* TISTR 1383 (**d**). Effects of 1× MIC of NNS5-6 AMP on *P. aeruginosa* TISTR 357 (**b**) and *K. pneumoniae* TISTR 1383 (**e**). Effects of 1× MIC of colistin on *P. aeruginosa* TISTR 357 (**c**) and *K. pneumoniae* TISTR 1383 (**f**).

2.7. Time-Kill Assay of NNS5-6 AMP

The time-kill assay was conducted to determine the mode of antibacterial action of NNS5-6 AMP. The NNS5-6 AMP exhibited different inhibitory effects on *P. aeruginosa* TISTR 357 compared with *K. pneumoniae* TISTR 1383. For *P. aeruginosa* TISTR 357, a significant reduction in viable cells was observed within the first hour of treatment with 1× MIC and 2× MIC of NNS5-6 AMP (Figure 4a). The eradication effect of NNS5-6 AMP on viable cells was achieved when treatment time reached 8 h with concentrations of 1× and 2× MIC. From the initial treatment until the viable cells were eradicated, the reduction rate of viable cells was similar for 1× MIC (0.6297 log CFU/h) and 2× MIC (0.6421 log CFU/h). In contrast, *K. pneumoniae* TISTR 1383 showed a delayed response, with bacterial suppression beginning 2 h after treatment with 1× MIC. The reduction rate of viable cells was slower with 1× MIC (0.05790 log CFU/h) compared with 2× MIC (0.5715 log CFU/h). In addition, the eradication effect was observed at different concentrations of NNS5-6 AMP. The treatment with 2× MIC achieved eradication of viable cells within 8 h, whereas 1× MIC only suppressed cell growth and reduced viable cells by up to 2 log (CFU/mL) until 24 h of treatment (Figure 4b). The differences in antibacterial activity observed in the time-kill assay were consistent with the results from the microdilution assay and SEM micrograp both of which demonstrated the eradication effect. The killing curve for *K. pneumoni* TISTR 1383 was both concentration- and time-dependent, whereas, for *P. aeruginosa* TIST 357, it was time-dependent.

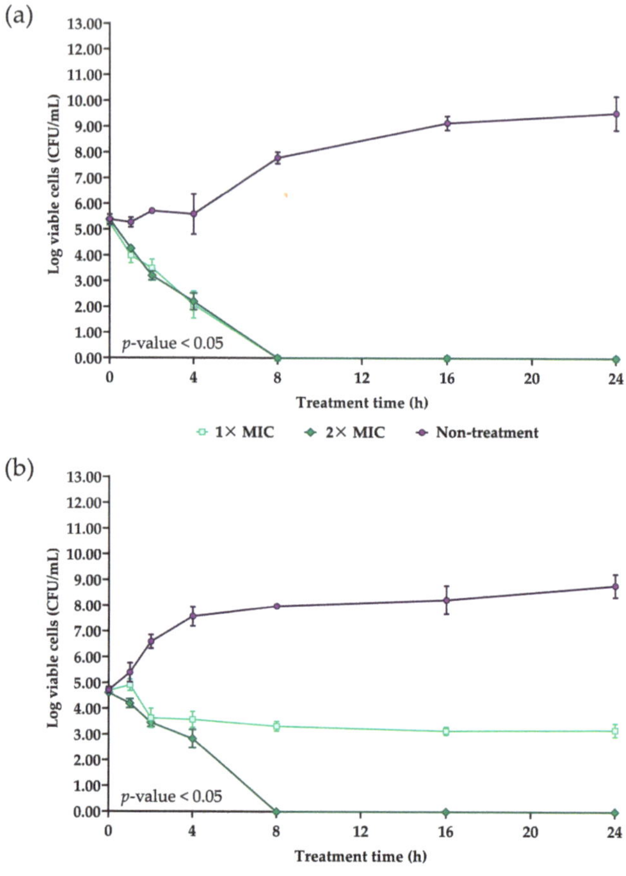

Figure 4. The time-kill assay was performed to determine the time- and concentration-depender effects of NNS5-6 AMP. The experiments used 1× and 2× MIC of NNS5-6 AMP for *P. aeruginosa* TIST 357 (**a**) and *K. pneumoniae* TISTR 1383 (**b**). The efficacy of the AMP was monitored by counting viab cells during the treatment period of up to 24 h. The experiment was performed in triplicate, and th log of viable cells is expressed as the mean with standard deviation (SD). The statistical significanc between different concentrations in treatment and non-treatment conditions was determined usin two-way ANOVA (p-value < 0.05).

2.8. Studies of Cell Permeability

The Sytox Green uptake assay was used to investigate cell permeabilization of bacteri pathogens after treatment with the AMP. The loss of cell membrane function due to th AMP disturbance was monitored by the fluorescence intensity, which reflects the bindin of Sytox Green to DNA as a result of membrane malfunction [13]. For *P. aeruginosa* TIST 357, the addition of NNS5-6 AMP caused an immediate increase in fluorescence intensit Fluorescence was found to have a plateau characteristic until 24 h of treatment. Differen levels of fluorescence intensity were observed at different concentrations of NNS5-6 AM

(0.125× MIC to 2× MIC) and Triton X-100 (0.125% to 2% *w*/*v*). Higher fluorescence intensity was observed with NNS5-6 AMP at concentrations of 0.5×, 1×, and 2× MIC compared with lower concentrations of 0.25× and 0.125× MIC. Additionally, the fluorescence intensity of NNS5-6 AMP treatment at concentrations of 0.5×, 1×, and 2× MIC showed no significant differences and was comparable with the fluorescence intensity observed with Triton X-100 treatment in the concentration range of 0.125% to 0.5% (Figure 5a). The fold increase in fluorescence intensity caused by NNS5-6 AMP treatment compared with non-treatment ranged from 1.95 ± 0.28 to 5.87 ± 0.31 when using 0.125× to 2× MIC of NNS5-6 AMP, while 5.93 ± 0.34 to 6.34 ± 0.39 was observed with 0.125% to 2% Triton X-100 treatment (Figure 5c). The increased cell permeability due to NNS5-6 AMP and Triton X-100 treatment against *P. aeruginosa* TISTR 357 can be attributed to a concentration-dependent action. However, the increased fluorescence intensity distinguished the degree of permeability into two groups, with one group including concentrations of 0.5×, 1×, and 2× MIC, and the other group including 0.25× and 0.125× MIC. The 0.25× MIC was identified as the cut-off concentration for the different degrees of cell permeability caused by NNS5-6 AMP treatment in *P. aeruginosa* TISTR 357. The different permeability was observed in *K. pneumoniae* TISTR 1383 treated with NNS5-6 AMP and Triton X-100. The concentration-dependent manner was found in both NNS5-6 AMP and Triton X-100 (Figure 5b). The similar levels of fluorescence intensity in the range of 0.125×, 0.25×, 0.5×, and 1× MIC of NNS5-6 AMP showed 1.66 ± 0.37 to 1.94 ± 0.29-fold increase. Moreover, 2× MIC of NNS5-6 AMP treatment resulted in a distinguishable and the highest fluorescence intensity, with a 3.21 ± 0.29-fold increase compared with other concentrations. Triton X-100 showed increases of 4.64 ± 0.62 to 8.12 ± 2.86-fold in fluorescence intensity, which was higher than those observed with NNS5-6 AMP for the treatment of *K. pneumoniae* TISTR 1383 (Figure 5d). These findings suggest that increased cell membrane permeability due to NNS5-6 AMP is concentration-dependent in both *P. aeruginosa* TISTR 357 and *K. pneumoniae* TISTR 1383. Although the cell permeability measurement during the treatment could not distinguish between live and dead cells, the results from the Sytox Green uptake assay support the concentration- and time-dependent effects observed in the time-kill assay. These results confirm the eradication of viable cells through membrane disruption, as captured by SEM micrography.

(a)

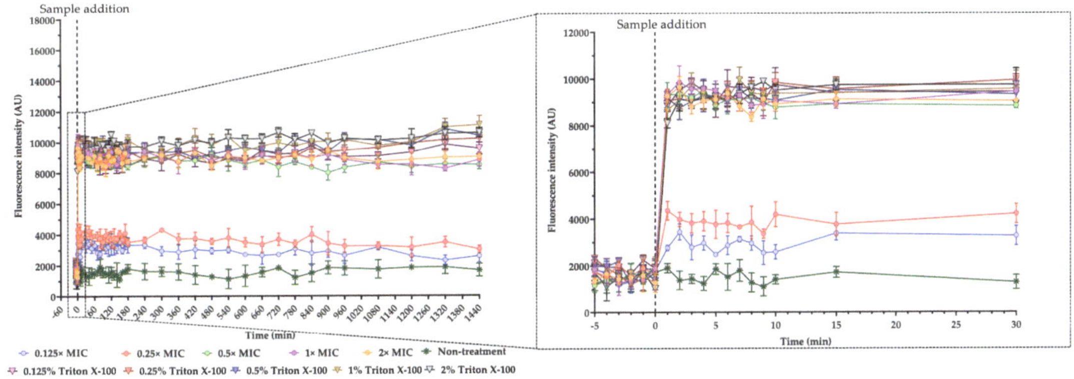

Figure 5. *Cont.*

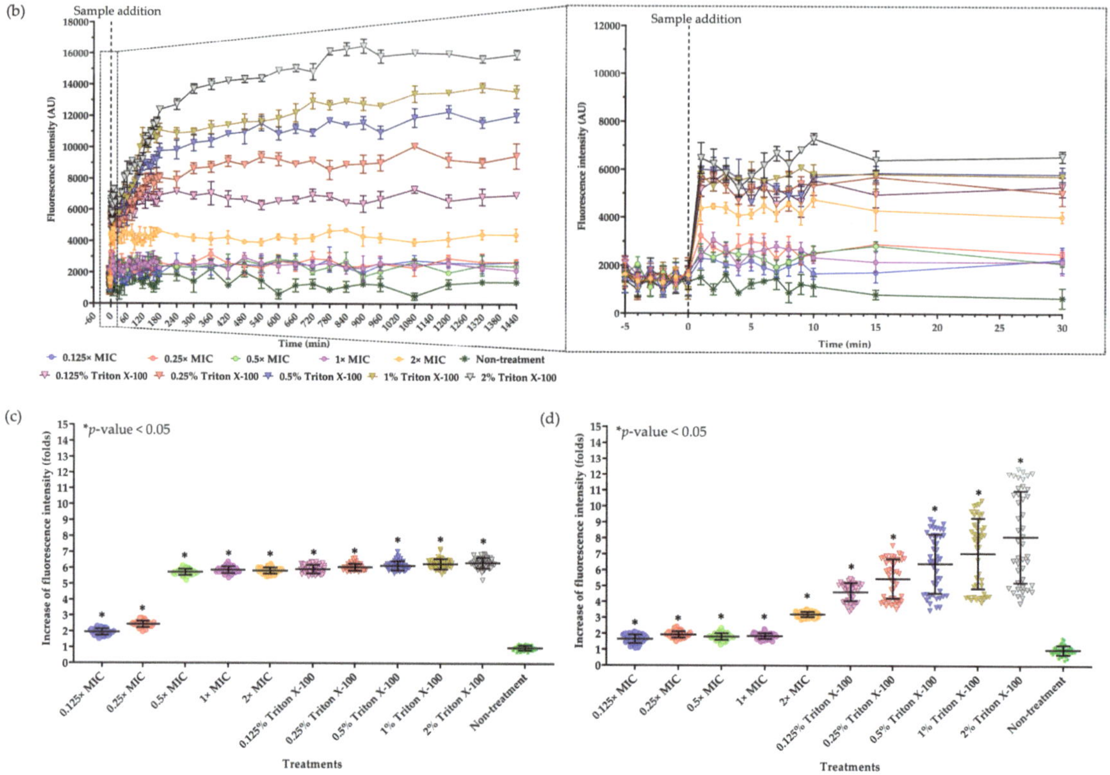

Figure 5. The Sytox Green uptake assay of NNS5-6 AMP-treated bacterial cells demonstrated the effec of NNS5-6 AMP on cell permeability. Cell cultures of *P. aeruginosa* TISTR 357 (**a**) and *K. pneumonia* TISTR 1383 (**b**) were incubated with different concentrations of AMP (0.125×, 0.25×, 0.5×, 1×, an 2× MIC). Various concentrations of Triton X-100 (0.125%, 0.25%, 0.5%, 1%, and 2% w/v) were used a positive controls to indicate the levels of cell membrane permeability. Membrane permeabilization o bacterial cells was monitored by the increase in fluorescence intensity caused by the Sytox Green-DN complex. The fluorescence intensity was observed over 24 h. The inset graph provides an expande view of the fluorescence baseline before sample addition and the fluorescence intensity in the initia 30 min after sample addition. The increase in fluorescence intensity during the entire treatmen period was expressed in folds, comparing different concentrations of samples to the non-treatmen condition of *P. aeruginosa* TISTR 357 (**c**) and *K. pneumoniae* TISTR 1383 (**d**). The experiments wer conducted in triplicate, with the mean and standard deviation presented. The statistical analysis wa performed to compare the overall time points between treatments and non-treatments using one-wa ANOVA (p-value < 0.05) shown in (**c**,**d**).

2.9. Stability Studies of NNS5-6 AMP under Various Conditions

The stability of NNS5-6 AMP was evaluated under various conditions (Table 4). Th antibacterial activity of NNS5-6 AMP against *P. aeruginosa* TISTR 357 was used as th primary indicator of its stability. The AMP demonstrated thermostability at temperature up to 40 °C for 12 h (99.07 ± 0.72% to 99.80 ± 0.94% residual activity). However, a 50 °C, the activity of AMP was reduced by almost half within 1 h and was completely los after 6 h. Furthermore, the antibacterial efficacy of NNS5-6 AMP was entirely lost whe exposed to 60 °C, 80 °C, and 100 °C for 1 h and under autoclave conditions. The residua activity of the AMP was significantly reduced when incubated with proteinase K fro 1 h to 12 h (83.62 ± 1.34% to 96.08 ± 3.41% residual activity). The AMP was stable whe incubated with trypsin for up to 6 h (99.38 ± 1.99% residual activity). Additionally, the AMP maintained its activity for up to 12 h when incubated with α-chymotrypsin, indicating its stability under α-chymotrypsin incubation (98.46 ± 0.88% to 99.42 ± 1.22% residual activity). The results indicated the proteinaceous nature of the AMP, making it susceptible to digestion by proteinase K and trypsin while not being digested by α-chymotrypsin. The overall residual activity after digestion with proteinase K, trypsin, and α-chymotrypsin was up to 80%.

Table 4. The stability of NNS5-6 AMP against temperatures, proteolytic enzymes, surfactants, and pH treatments was evaluated. The residual activity after each treatment is reported along with the incubation times (mean ± SD; n = 3).

Conditions	% Residual Activity of NNS5-6 AMP against *P. aeruginosa* TISTR 357		
	1 h	**6 h**	**12 h**
Effect of Temperatures			
Non-treated NNS5-6 AMP	100.00 ± 1.23	100.00 ± 0.70	100.00 ± 0.62
37 °C	99.39 ± 1.98	99.39 ± 0.61	99.69 ± 0.95
40 °C	99.80 ± 0.94	99.19 ± 0.93	99.07 ± 0.72
50 °C	65.16 ± 3.07 *	0.00 ± 0.00 *	0.00 ± 0.00 *
60 °C	0.00 ± 0.00 *	0.00 ± 0.00 *	0.00 ± 0.00 *
80 °C	0.00 ± 0.00 *	0.00 ± 0.00 *	0.00 ± 0.00 *
100 °C	0.00 ± 0.00 *	0.00 ± 0.00 *	0.00 ± 0.00 *
121 °C, 15 psi, 15 min		0.00 ± 0.00 *	
121 °C, 15 psi, 30 min		0.00 ± 0.00 *	
Effect of Proteolytic enzymes			
Non-treated NNS5-6 AMP	100.00 ± 0.95	100.00 ± 0.34	100.00 ± 0.58
NNS5-6 AMP with Proteinase K (1 mg/mL)	96.08 ± 3.41 *	86.58 ± 1.22 *	83.62 ± 1.34 *
NNS5-6 AMP with Trypsin (1 mg/mL)	99.38 ± 1.99	90.47 ± 2.02 *	88.90 ± 1.46 *
NNS5-6 AMP with α-chymotrypsin (1 mg/mL)	99.38 ± 0.95	99.42 ± 1.22	98.46 ± 0.88
Effect of Surfactants			
Non-treated NNS5-6 AMP	100.00 ± 0.60	100.00 ± 0.93	100.00 ± 0.61
NNS5-6 AMP with 1% SDS	116.77 ± 1.80 *	122.52 ± 1.86 *	119.19 ± 1.53 *
NNS5-6 AMP with 1% Triton X-100	116.17 ± 0.60 *	120.28 ± 1.53 *	120.81 ± 1.53 *
1% SDS alone		121.76 ± 2.42 *	
1% Triton X-100 alone		126.83 ± 2.72 *	
Effect of pH variation			
Non-treated NNS5-6 AMP	100.00 ± 1.19	100.00 ± 1.28	100.00 ± 0.72
pH 1.2	89.62 ± 1.85 *	83.83 ± 1.72 *	81.14 ± 1.71 *
pH 4.5	97.31 ± 0.33	95.69 ± 1.75	95.62 ± 1.19
pH 6.8	97.31 ± 3.33	96.46 ± 0.57	96.00 ± 0.57
pH 7.4	98.65 ± 0.58	98.18 ± 1.15	98.48 ± 0.66
pH 8.0	99.23 ± 1.15	97.22 ± 0.98	98.48 ± 0.87
pH 10.0	98.85 ± 1.20	97.42 ± 0.82	98.10 ± 1.44
pH 12.0	94.31 ± 0.68 *	93.40 ± 0.78 *	90.10 ± 2.38 *
pH 14.0	90.77 ± 0.88 *	84.40 ± 1.15 *	79.05 ± 1.75 *

* Significance according to Student's *t*-test at a p-value < 0.05 compared with non-treated NNS5-6 AMP.

Surfactants above the critical micelle concentration were used to assess their interference with NNS5-6 AMP activity. The combination of the AMP with SDS or Triton X-100 resulted in increased antibacterial activity compared with the AMP alone. However, SDS or Triton X-100 alone showed higher antibacterial activity (121.76 ± 2.42% to 126.83 ± 2.72% residual activity) compared with the AMP combined with these surfactants (116.17 ± 0.60% to 122.52 ± 1.86% residual activity) and the AMP alone (100.00 ± 0.60% to 100.00 ± 0.93% residual activity). The results from combining the AMP with surfactants at a concentration above the critical micelle concentration suggested that interactions between NNS5-6 AMP and surfactants occurred, but further studies are needed to understand these interactions fully. The activity of the AMP after incubation at pH 1.2 was significantly lower than th of non-treated AMP at all time points (81.14 ± 1.71% to 89.62 ± 1.85% residual activit In contrast, the activity of AMP remained above 95% when treated in a pH range of 4.5 10.0 (95.62 ± 1.19% to 99.23 ± 1.15% residual activity). The activity of AMP decreased in pH- and time-dependent manner in a pH range of 12–14 (94.31 ± 0.68% to 79.05 ± 1.75 residual activity). These findings indicated that the AMP was less tolerant to extreme acidic and alkaline environments but retained activity in the pH range of 4.5–10.0. Overa NNS5-6 AMP exhibited a stable profile against proteolytic enzymes, surfactants, and p with up to 80% residual activity over 12 h of treatment.

2.10. Phenotypic Characterization of NNS5-6

A single colony of NNS5-6 appeared as a circular, cream-colored colony with a undulating margin and a smooth surface (Figure S2a). The bacterial cells were observe under a light microscope at 1000× magnification. The one-day-old vegetative cells appeare Gram-positive-stained and rod-shaped bacilli (Figure S2b). Three-day-old endospores we oval-shaped, as confirmed by malachite green staining (Figure S2c). High-resolutic SEM images showed rod-shaped bacilli with dimensions of 0.4–0.5 µm in diameter an 2.0–2.3 µm in length (Figure S2d). The endospores had ridge-like characteristics wi dimensions of 0.8–1.0 µm in diameter and 1.8–2.0 µm in length (Figure S2e). The lig microscope and SEM images provided consistent morphological results for both vegetati cells and spores. The reduction in cell dimensions from vegetative cells to spores can b attributed to the dormant process, which helps the species tolerate stressful environmen and enhances survival adaptability.

2.11. Genome Insight for Coding Sequence Annotation and Whole-Genome Phylogenetic Analys

The read quality was high, and the assembled genome had a size of 6,522,808 bp wit a 139× sequencing depth, constructed by de novo assembly using the Velvet version 1.2.1 The assembled genome contained 283 contigs with an average contig length of 74,721 b The GC content was 53.20%. The genome presented a completeness of 99.68%. Th genome assembly had 0.82% contamination. A single chromosome sequence of *Paenibacill thiaminolyticus* NNS5-6 was deposited in the National Center for Biotechnology Informatio (NCBI) database under the accession number CP160395.

The taxonomy of NNS5-6 was predicted using genome-based sequencing and analyze by the Type (Strain) Genome Server (TYGS). NNS5-6 was found to show the closest relatio to *Paenibacillus thiaminolyticus* NRRL B-4156. The pairwise-comparison calculated by th Genome BLAST Distance Phylogeny (GBDP) method against reference genomes in th TYGS database showed similarity scores of 74.9% (95% CI, 70.9–78.5%), 61.7% (95% C 58.8–64.5%), and 74.8% (95% CI, 71.3–78.0%) as d_0, d_4, and d_6, respectively. The GC conter of the type strain *Paenibacillus thiaminolyticus* NRRL B-4156 is 53.64%, which is similar to tha of *Paenibacillus thiaminolyticus* NNS5-6 (53.20%), supporting the classification. TYGS result indicated that the NNS5-6 strain was most closely related to *Paenibacillus thiaminolyticu* NRRL B-4156 at the species level. The identity score (%) from FastANI analysis betwee *Paenibacillus thiaminolyticus* NRRL B-4156 and *Paenibacillus thiaminolyticus* NNS5-6 wa 95.02%, supporting the GBDP results and confirming that these organisms belong to th same species. The pairwise comparison between the two species was performed by genom sequence breakdown and mapped with the fragments of orthologous DNA sequence. Th similarity of each pairwise DNA fragment was visualized using FastANI v1.1.0 (Figure 6a

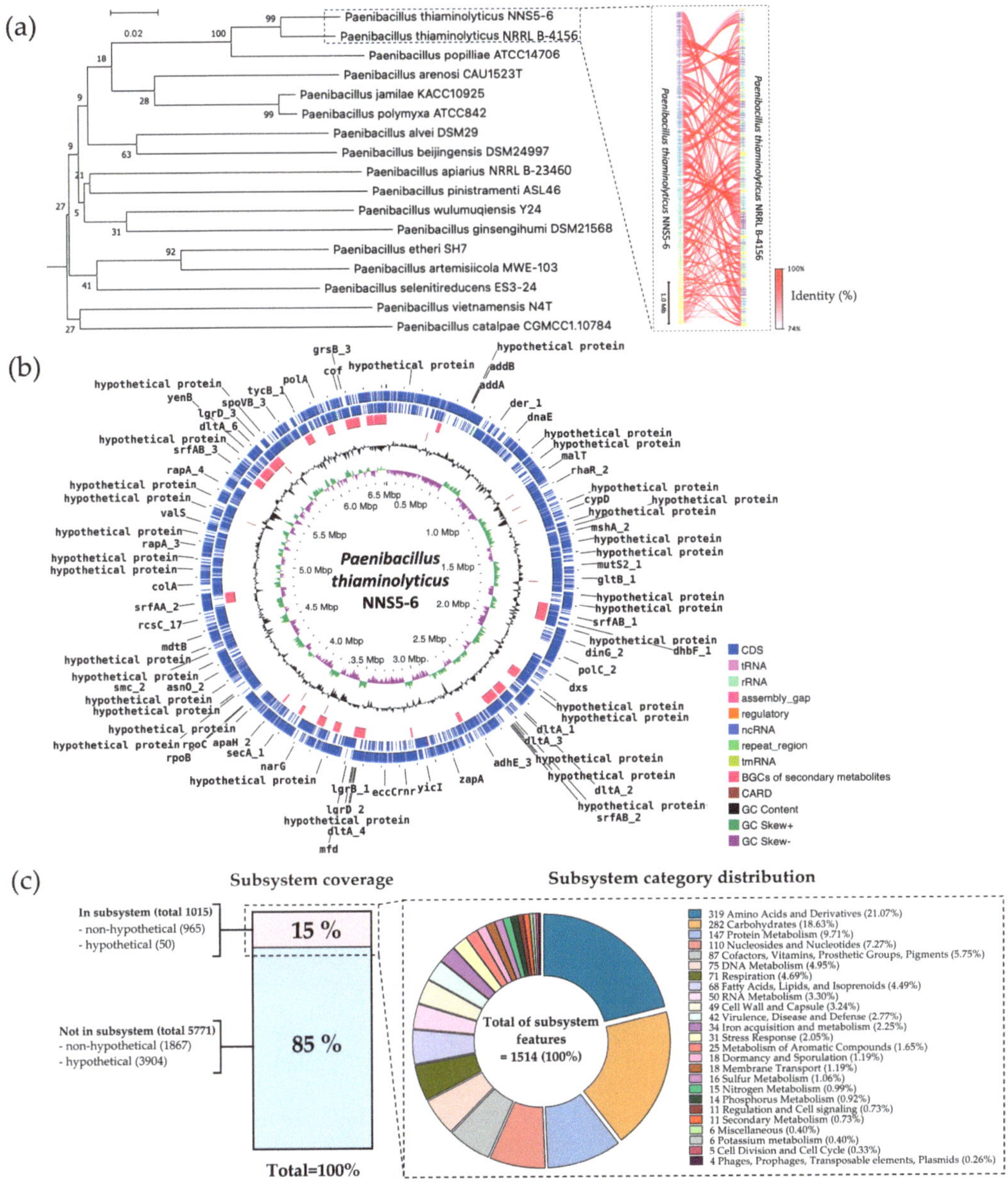

Figure 6. The genome-based phylogenetic tree of NNS5-6 was constructed using the GBDP method to identify the closest species and strains, via the TYGS web service. The inset visualized by FastANI demonstrates that the DNA fragments in the NNS5-6 genome are similar to the orthologous DNA fragment against the closest related genome, *Paenibacillus thiaminolyticus* NRRL B-4156 (**a**). The circular map of the NNS5-6 genome (6.5 Mb) displays the predicted coding sequences relevant to biological processes, including BGCs of secondary metabolites and antibiotic resistance genes (**b**). The subsystem technology via RAST covers 15% of the predicted subsystems, revealing the cellular machinery. The inset shows the distribution of subsystem categories from the 15% subsystem coverage, identifying 1514 features responsible for the biological processes of the bacterium (**c**).

The protein-coding sequences (CDS) were annotated using Rapid Prokaryotic Genom Annotation (Prokka) version 1.14.6, including biosynthetic gene clusters (BGCs) of se ondary metabolites, which were annotated by Antibiotics and Secondary Metaboli Analysis Shell (antiSMASH) version 7.0. The prediction of antibiotic resistance gene localized in the genome was performed using the Resistance Gene identifier (RGI) predi tion in the Comprehensive Antibiotic Resistance Database (CARD). The circular genom map and gene annotations were visualized using Proksee (CGViewBuilder version 1.1. https://proksee.ca, accessed on 14 July 2024), showing the number of different function of genes, which are classified and expressed by various colors (Figure 6b). There wer 6091 features annotated by Prokka, including 6024 CDS, 4 rRNA, 59 tRNA, and 4 ncRNA i the genome. The BGCs of secondary metabolites as antimicrobial compounds were counte by antiSMASH, which provided 19 BGCs with varying similarity based on orthologs i the Minimum Information about Biosynthetic Gene cluster (MIBiG) databases. The RG prediction identified 11 possible antibiotic-resistance genes.

Rapid Annotations using Subsystems Technology (RAST) was used to predict cellula machinery based on genomic information. The genes were categorized into subsyster and non-subsystem classifications. Of the total 6786 coding sequences (CDS), 15% wer categorized into subsystems and 85% into non-subsystems. The encoding proteins an notated by subsystem functionalization included 1015 genes, of which 965 genes wer encoded to non-hypothetical proteins and 50 genes to hypothetical proteins. In contras 5771 genes annotated by familial gene evidence in the database were non-subsystem. Ther were 1867 genes encoding non-hypothetical proteins and 3904 genes encoding hypothetica proteins. The gene functions in the subsystem were categorized into 1514 features tha ensemble for the cellular machinery (Figure 6c).

2.12. Comparative Analysis of Biosynthetic Gene Clusters in NNS5-6

The secondary metabolites of the NNS5-6 genome were predicted using antiSMASH which identified 19 BGCs that were thoroughly mapped in the genome. The known cluste BLAST was used to search for similar orthologs in the MIBiG database for each identifie BGC, which could indicate potential secondary metabolite productions. The results fron the known cluster BLAST search revealed that fifteen BGCs matched with similaritie ranging from 8% to 100% based on similar orthologs, whereas four BGCs did not matcl any BGCs in the database. The types of secondary metabolites produced by these BGC were categorized as non-ribosomal peptides (NRPs), ribosomally synthesized and pos translationally modified peptides (RiPPs), polyketides (PKs), co-processed productio of non-ribosomal peptide and polyketide (NRP + Polyketide), and others. The identica orthologs matched to BGCs of the antimicrobial peptides with NRP and NRP + Polyketid types such as paeninodin, paenibactin, and colistin A/B exhibited 100% similarity. Th high similarity of the matched orthologs ranged from 60% to 85%, including relevant BGC for ulbactin, paenibacterin, and colistin as NRPs, while ectoin was identified as an amin acid derivative. Orthologs with less than 50% similarity could represent less explore BGCs. Among those with 8–37% similarity matched to reference BGCs in the databas were polyketides (pellasoren, myxothiazole, and chejuenolide), NRPs (pelgipeptide an laterocidin), and NRP + Polyketide (paenilamicin). The matched secondary metabolite support the secondary metabolism in the Paenibacillus genus. However, BGCs 6, 8, 9, and 1 did not match any similar reference BGCs in the database, but antiSMASH predicted thes types of secondary metabolites as thiopeptide, ranthipeptide, linear azol(in)e-containin peptides, and NRP, respectively (Figure 7).

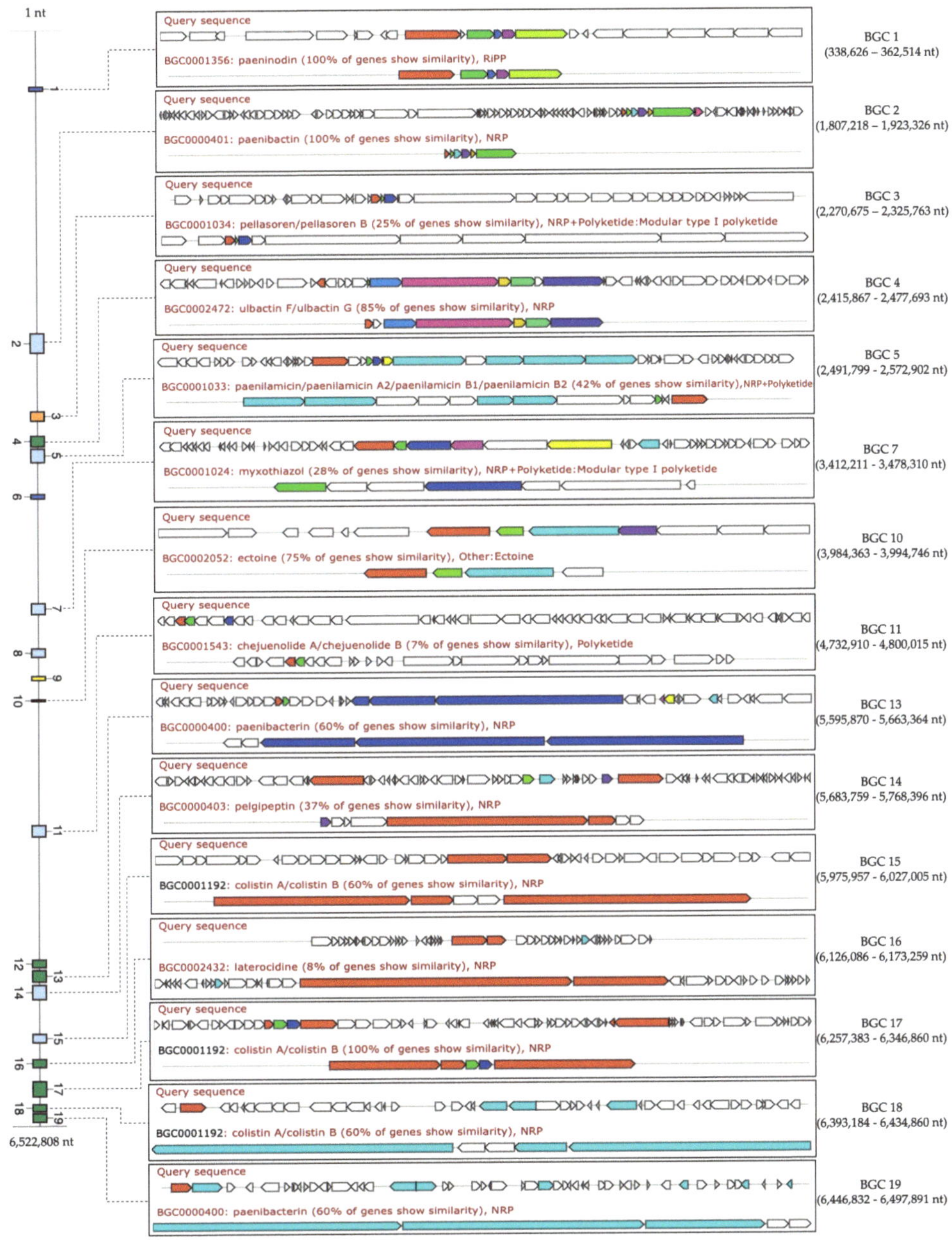

Figure 7. The predicted biosynthetic gene clusters (BGCs) of secondary metabolism in the NNS5-6 genome (6,522,808 nucleotides; nt) were investigated using antiSMASH. The known cluster BLAST was performed to match the reference BGCs in the MIBiG database. Nineteen BGCs were predicted and annotated for BGCs with the most similar orthologs, whereas 4 BGCs (6, 8, 9, and 12) did not match any BGCs found. The matched BGCs revealed a variety of secondary metabolites produced by the NNS5-6 genome. The color-coding of genes in the cluster followed the visualization provided by antiSMASH.

The NNS5-6 AMP shared an amino acid sequence similar to fusaricidins, whi were reported AMPs derived from Paenibacillus polymyxa. All predicted BGCs from t antiSMASH results were analyzed to find the biosynthetic gene relevant to fusaricid production in the NNS5-6 genome. The translated protein from the core biosynthetic gene BGC 13 (9142 amino acid residues) showed the highest similarity to fusaricidin syntheta (8524 amino acid residues; sequence ID: SUA94926.1) using the Domain Enhanced Looku Time Accelerated BLAST (DELTA-BLAST) algorithm in the NCBI database. The resu showed that the protein identity and coverage were 72.79% and 75.45%, respectively, ar the mismatches were 534 amino acid residues. The comparison of translated prote sequences between the NNS5-6 core biosynthetic gene and fusaricidin synthetase show the conserved region pattern (indicated by grey color). In contrast, the variable regio contained differences in amino acid sequences (indicated by red color) (Figure 8a). Th compared amino acid sequences between the two proteins support the reason that th AMP derived from NNS5-6 shares a similar amino acid sequence to fusaricidin conserve sequences (Thr-Val-Tyr-Thr) (Figure 8b). Therefore, NNS5-6 AMP could potentially be novel AMP related to fusaricidin derivatives. However, the production mechanisms NNS5-6 AMP and the fusaricidin-like core biosynthetic gene require further confirmatio through a molecular genetic approach.

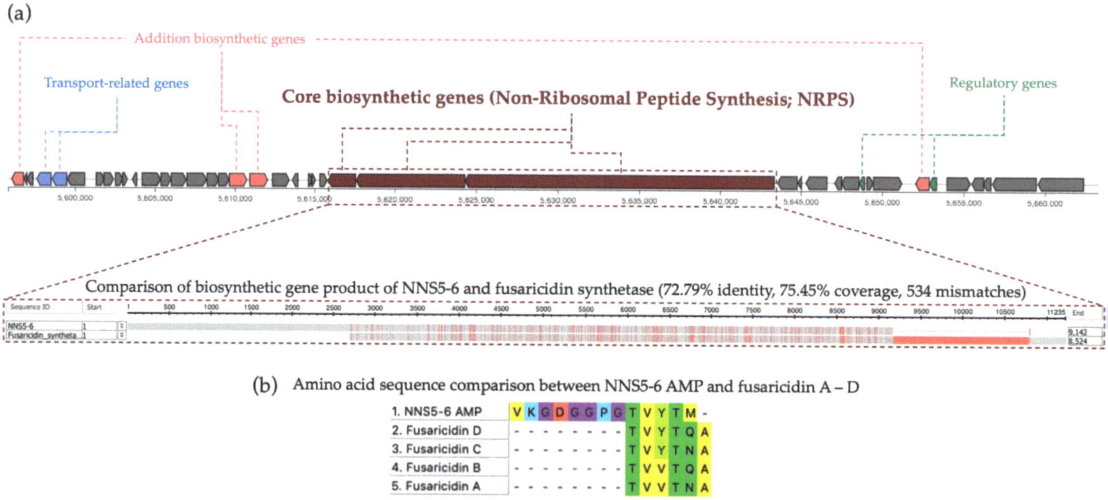

Figure 8. The comparative BGC of NNS5-6 AMP was proposed by the non-ribosomal peptid synthesis in antiSMASH. The BGC was composed of core biosynthetic genes, regulatory genes, an transport-related genes. The translated protein from the core biosynthetic genes showed a simila amino acid sequence to fusaricidin synthetase in the DELTA-BLAST search within the NCBI databas (red and grey colors indicate different and identical amino acids, respectively). The different amin acids in the gene product comparison are indicated by red color, as visualized by the NCBI MS viewer (**a**). The similar amino acid sequences between NNS5-6 AMP and fusaricidin A–D wer compared using multiple sequence alignment with MEGA X software (version 10.1.8) (**b**).

2.13. Prediction of Antibiotic Resistance Genes in the NNS5-6 Genome and Determination of Antibiotic Susceptibility

Assessing antimicrobial resistance is crucial when considering the use of newly dis covered bacteria in various industries and healthcare sectors. To ensure safety for furthe utilization, it is important to investigate the NNS5-6 strain for antibiotic resistance gene and susceptibility. The whole genome sequence of NNS5-6 was analyzed for antibioti resistance genes using the RGI prediction in the CARD database. The analysis identifie 11 predicted antibiotic-resistance genes. Among these, seven were related to glycopep tide antibiotic resistance genes with various percentages of identity and coverage when compared with reference genes in the database. Two vanY genes in the vanB cluster, two vanW genes in the vanI cluster, a vanT in the vanG cluster, a vanXY in the vanG cluster, and a vanG were identified. These predicted glycopeptide resistance genes had low percentages of identity, ranging from 33.04% to 53.47%, with 88.62% to 128.15% reference gene coverage. Additionally, other predicted antibiotic resistance genes were Otr(A) for tetracyclines, potxA for oxazolidinone antibiotics, norC for fluoroquinolone antibiotics, and qacG for antibiotic efflux pump. The identity ranged from 36.48% to 59.69%, with coverages ranging from 99.25% to 127.10% compared with the reference genes (Table 5). The antibiotic susceptibility test of NNS5-6 was conducted using the disk diffusion method to verify the predicted antibiotic resistance genes corresponding to phenotypic expression. The NNS5-6 bacterial strain exhibited high susceptibility to the combination of piperacillin (100 μg) and tazobactam (10 μg), ceftriaxone (30 μg), imipenem (10 μg), ciprofloxacin (5 μg), erythromycin (15 μg), doxycycline (30 μg), gentamicin (10 μg), cefoxitin (30 μg), and vancomycin (30 μg) (Table 6). The antibiotic susceptibility results contrasted with the predicted antibiotic resistance genes, which could be attributed to the low percentage of gene identity or reduced gene expression in the culture environment. Thus, while the resistance gene predictions provide important insights, they must be corroborated by susceptibility tests. The results indicate that NNS5-6 exhibits low resistance to commonly used antibiotics in both healthcare and industry settings.

Table 5. The prediction of antibiotic-resistance genes in the NNS5-6 genome was determined by their similarity to genetic sequences in the CARD database.

Antibiotic Resistance Gene	Antibiotic Resistance Gene Family	Resistance Mechanism	Position in Genome	Identity (%)	Coverage Length (%)
*van*Y gene in *van*B cluster	*van*Y, glycopeptide resistance gene cluster	antibiotic target alteration	291,432 to 292,277	33.04	104.85
Otr(A)	tetracycline-resistant ribosomal protection protein of tetracycline antibiotics	antibiotic target protection	1,037,370 to 1,039,346	45.40	99.25
*van*W gene in *van*I cluster	*van*W, glycopeptide resistance gene cluster	antibiotic target alteration	1,132,154 to 1,133,269	34.43	99.46
*qac*G	small multidrug resistance (SMR) antibiotic efflux pump	antibiotic efflux	1,236,341 to 1,236,751	42.86	127.10
*potx*A	Miscellaneous ABC-F subfamily ATP-binding cassette ribosomal protection proteins of oxazolidinones antibiotics	antibiotic target protection	1,668,816 to 1,670,789	36.48	121.22
*nor*C	major facilitator superfamily (MFS) antibiotic efflux pump of fluoroquinolone antibiotics	efflux pump complex or subunit conferring antibiotic resistance	3,712,623 to 3,714,038	59.69	101.95
*van*W gene in *van*I cluster	*van*W, glycopeptide resistance gene cluster	antibiotic target alteration	3,898,660 to 3,900,096	36.69	128.15
*van*Y gene in *van*B cluster	*van*Y, glycopeptide resistance gene cluster	antibiotic target alteration	5,289,399 to 5,290,199	33.56	99.25
*van*T gene in *van*G cluster	*van*T, glycopeptide resistance gene cluster	antibiotic target alteration	5,756,379 to 5,758,274	48.63	88.62
*van*XY gene in *van*G cluster	*van*XY, glycopeptide resistance gene cluster	antibiotic target alteration	5,758,264 to 5,759,118	43.70	111.81
*van*G	Van ligase, glycopeptide resistance gene cluster	antibiotic target alteration	5,759,115 to 5,760,176	53.47	101.15

Table 6. Antibiotic susceptibility test results for NNS5-6, following the CLSI M100 guidance protoc

Antibiotics	Zone of Inhibition (mm ± SD); n = 3
Ciprofloxacin (5 μg)	36.24 ± 0.63
Piperacillin (100 μg) and Tazobactam (10 μg)	50.29 ± 0.41
Imipenem (10 μg)	38.52 ± 0.52
Ceftriaxone (30 μg)	42.50 ± 0.48
Cefoxitin (30 μg)	25.40 ± 0.36
Doxycycline (30 μg)	31.07 ± 0.73
Vancomycin (30 μg)	21.51 ± 1.25
Erythromycin (15 μg)	34.88 ± 0.73
Gentamicin (10 μg)	25.57 ± 0.32

3. Discussion

WHO published a critical list of pathogens requiring urgent antimicrobial develo ment, which aimed to guide research priorities in 2017. The ESKAPE pathogens (*Enteroco cus faecium*, *Staphylococcus aureus*, *Klebsiella pneumoniae*, *Acinetobacter baumannii*, *Pseudomona aeruginosa*, and *Enterobacter* species) were designated as top priority. These bacterial specie are notorious for their antibiotic resistance and present the greatest challenge in combatin infectious diseases [14]. Among these, *Pseudomonas aeruginosa* and *Klebsiella pneumoniae* ar of particular concern, prompting the exploration of new antimicrobial strategies.

Our research identified a novel bacterial strain, NNS5-6, from mangrove sedimen samples, which exhibited strong antibacterial activity against *P. aeruginosa* TISTR 357 an *K. pneumoniae* TISTR 1383. This strain was characterized morphologically and geneticall confirming its identity as *Paenibacillus thiaminolyticus*.

The CFS of *Paenibacillus thiaminolyticus* NNS5-6 showed effective antibacterial activit against Gram-negative bacteria such as *P. aeruginosa* TISTR 357, *K. pneumoniae* TISTR 138 and *E. coli* TISTR 887, but not against Gram-positive bacteria such as *S. aureus* TISTR 51 and MRSA strain 2468. The antibacterial components of NNS5-6 were found among th early stationary phase of the growth curve. The active components were purified usin ammonium sulfate precipitation, cation-exchange chromatography, and size-exclusio chromatography. The components precipitated at 75% ammonium sulfate saturatio suggesting that the antibacterial compounds were strongly hydrophilic [15].

Our analysis identified the AMP from *Paenibacillus thiaminolyticus* NNS5-6 as having conserved amino acid sequence similar to fusaricidins produced by *Paenibacillus polymyx* Fusaricidins are known for their activity against Gram-positive bacteria and fungi. Fusar cidin C and D have been reported to have similar amino acid sequences, particularly i the Thr-Val-Tyr-Thr region [16–19]. Secondary structure determination of NNS5-6 AMP, a indicated by CD studies, revealed that when dissolved in purified water or SDS micelle the conformation of NNS5-6 AMP slightly changed, increasing the antiparallel beta-shee and random coil structures upon contact with the lipid membrane of SDS micelles. Th percentage of secondary structure proportions showed that the antiparallel beta-shee and random coil structures were in equilibrium in both solvents. The forces governin self-assembly processes, which allow peptides to achieve a stable low-energy state, includ weak interactions such as hydrogen bonding, electrostatic attractions, and Van der Waal forces. These interactions underpin the formation of secondary structures, such as alpha helices and beta-sheets, which are crucial in all biological processes. The self-assembl processes are influenced by several factors: (i) the amino acid sequence, (ii) the degree o hydrophobicity, (iii) the length of the peptides, and (iv) the self-assembly duration. Fo beta-sheet structure, each beta-strand is connected laterally by hydrogen bonds, creatin a pleated sheet structure that is rigidified through interpeptide and interchain hydroger bonds. These hydrogen bonding patterns can form two different structures, parallel or an tiparallel beta-sheets. Antiparallel beta-sheets are energetically more favored than paralle ones because their hydrogen bonds are better aligned [20].

NNS5-6 AMP demonstrated strong antibacterial activity with an MIC of 4 µg/mL against both *P. aeruginosa* TISTR 357 and *K. pneumoniae* TISTR 1383. Time-kill kinetics showed rapid eradication of *P. aeruginosa* TISTR 517 and *K. pneumoniae* TISTR 1383 within 8 h at 1× and 2× MIC, respectively. NNS5-6 AMP could be effective in limiting infections by these pathogens during the initial hours of bacterial colonization. Sytox Green penetration through the bacterial membrane indicates compromised cellular membrane integrity. This dye interacts with nucleic acids, leading to intense fluorescence, which reflects increased cellular permeability following the AMP action. In this study, the cell permeability test revealed that in *P. aeruginosa* TISTR 357, the AMP exhibited similar fluorescence intensity at concentrations of 2× and 1× MIC. This finding aligned with the results of time-kill kinetics, which showed comparable killing rates for 2× and 1× MIC. The AMP at a concentration of 2× MIC exhibited greater killing activity than at 1× MIC in *K. pneumoniae* TISTR 1383, which was consistent with the significantly higher fluorescence intensity observed at 2× MIC compared with 1× MIC. This study on cell permeability indicated that *P. aeruginosa* TISTR 357 exhibited greater sensitivity to NNS5-6 AMP compared with *K. pneumoniae* TISTR 1383. This observation suggests that the bactericidal effect of NNS5-6 AMP is mediated through membrane permeabilization, which correlates with the findings from time-kill kinetics and the antibacterial activity assessed using the agar well diffusion assay. However, the findings from the cell permeability experiment and the time-kill assay showed differences in the characteristics of action. This suggests that when the AMP contacts the cell membrane, it disrupts the cell membrane immediately, resulting in substantial fluorescence detection right after the addition of Sytox Green. However, the pathogen cells require sufficient time to be killed. Additionally, the differences observed in the time-kill assay and cell permeability between *P. aeruginosa* TISTR 357 and *K. pneumoniae* TISTR 1383 could be attributed to the distinct chemical compositions of their cell membranes. The outer membrane (OM) of Gram-negative bacteria is a unique structure serving as a permeability barrier against antibiotics. It consists of phospholipids, lipopolysaccharides (LPS), outer membrane beta-barrel proteins (OMP), and lipoproteins [21]. The lipid compositions of the cell membranes in *P. aeruginosa* contain 21% 1-palmitoyl-2-oleoyl-*sn*-glycero-3-phospho-(1′-*rac*-glycerol) (POPG), 60% 1-palmitoyl-2-oleoyl-*sn*-glycero-3-phosphoethanolamine (POPE), and 11% cardiolipin. The lipid compositions of the cell membrane in *K. pneumoniae* consist of 5% POPG, 82% POPE, and 6% cardiolipin. Furthermore, the lipid composition ratios of 1-palmitoyl-2-oleoyl-*sn*-glycero-3-phosphocholine (POPC):POPG: Cardiolipin in *P. aeruginosa* and *K. pneumoniae* was 1:9:1 and 7:3:1, respectively [22]. The reasons and mechanisms underlying why *P. aeruginosa* TISTR 357 is more sensitive to NNS5-6 AMP than *K. pneumoniae* TISTR 1383 requires further investigation. It is proposed that this sensitivity could be attributable to differences in the interaction between the AMP and the cell membranes of *P. aeruginosa* and *K. pneumoniae*. Therefore, the approach to determining the mechanism of action of the AMP is to investigate the AMP–membrane interaction, as bacterial membranes consist of various types of lipids and proteins [23]. Negative GRAVY values indicate a hydrophilic peptide [24]. This finding correlated with the results of NNS5-6 AMP, which has a GRAVY value of −0.231. The NNS5-6 AMP was composed of hydrophilic amino acids such as Lys, Asp, Thr, and Tyr. Hydrophobicity and hydrophilicity are crucial for understanding peptide–membrane interactions, which play a key role in permeation through target cells. Typically, highly hydrophobic peptides are more likely to form pores, whereas more polar peptides tend to interact with the negative charges on membranes. The ratio of polar to non-polar amino acids determines the interaction mechanism, thereby affecting the ability of the peptides to cross the membrane [24,25]. The positive charge of NNS5-6 AMP, containing lysine, could bind to the negatively charged bacterial membranes via electrostatic interactions. Subsequently, the hydrophobic side chains insert into the lipid bilayer, causing membrane disruption [10]. AMPs generally exert their activity by interacting with and disrupting cell membranes. This disruption can occur through several mechanisms, including pore formation (either barrel–stave or toroidal) or a carpet-like action. The secondary structure of NNS5-6 AMP was an antiparallel beta-sheet conformation, which plays an important role in forming pores in bacterial c membranes [26]. Nevertheless, the structure–activity relationship of beta-sheet structur has been less studied than alpha-helix structures. SEM studies revealed that NNS5-6 AM affected the surface morphology and membrane integrity of *P. aeruginosa* TISTR 357 a *K. pneumoniae* TISTR 1383. The bactericidal mechanism of NNS5-6 AMP differed based the bacterial species. Consequently, the effect of NNS5-6 AMP on bacterial cell morpholo differed distinctly between the two tested species. This variation could be attributed differences in bacterial cell membrane compositions [27]. After treatment with NNS5 AMP, *P. aeruginosa* TISTR 357 cells tended to shrink and displayed noticeable holes, wh *K. pneumoniae* TISTR 1383 showed signs of cell rupture with pore formation and leakage cytoplasmic contents. *P. aeruginosa* TISTR 357 and *K. pneumoniae* TISTR 1383 cells treat with NNS5-6 AMP showed distinct characteristics to those observed in cells treated wi colistin. However, these SEM results suggest the mechanisms by which the antibacter compound interacts with the membrane.

The AMP of *Paenibacillus thiaminolyticus* NNS5-6 was sensitive to proteinase K an trypsin, confirming its proteinaceous nature. The AMP activity was found to be stable temperatures below 40 °C. The AMP was also stable across a wide pH range (pH 4.5–1 compared with extremely acidic and alkaline environments. This information could he inform temperature conditions and buffer choices during purification and guide the form lation development of drug delivery systems for suitable administration routes. Previo research has demonstrated that fusaricidin A, produced by *Paenibacillus bovis* sp. nc BD3526, exhibits good heat stability even at 121 °C for 15 min. These compounds reta their antimicrobial activity across a broad pH range (2.0 to 9.0) and are insensitive to pr tease treatment [28]. The findings of this report differ from our observations regarding th fusaricidin-like AMP derived from *Paenibacillus thiaminolyticus* NNS5-6.

Phylogenetic analysis indicated that NNS5-6 had the closest taxonomic relatio to *Paenibacillus thiaminolyticus* NRRL B-4156 based on genome comparison. *Paenibac lus thiaminolyticus* NRRL B-4156 was isolated from a high salinity area (5% *w*/*v* NaC in Japan and had a single linear chromosome with a size of 6,537,496 bp (accessio number: NDGK00000000.1), 53.64% GC content, 5756 CDS, and 15 BGCs of secondar metabolisms [29]. The investigating strain, NNS5-6, showed a genome composition simil to *Paenibacillus thiaminolyticus* NRRL B-4156. *Paenibacillus* belongs to the genus withi the family Paenibacillaceae. The genus is known for its aerobic or facultatively anaerob rod-shaped, endospore-forming, Gram-positive bacilli [30]. *Paenibacillus thiaminolyticus* wa initially included in the genus *Bacillus* and was later reclassified into the genus *Paenibacill* based on the results of 16S rRNA gene and cellular fatty acid composition analyses [31]. Th vegetative cells and endospores of *Paenibacillus thiaminolyticus* NNS5-6 resembled those c *Paenibacillus thiaminolyticus* NRRL B-4156. The size of *Paenibacillus thiaminolyticus* NNS5- was similar to that of *Paenibacillus thiaminolyticus* NRRL B-4156 [29]. It has been reporte that *Paenibacillus* species had the ability to synthesize a variety of antimicrobial compound including antimicrobial peptides. *Paenibacillus polymyxa* OSY-DF exhibits a promisin broad spectrum of antimicrobial activity [32]. The strain produces polymyxin E1 an paenibacillin, which belong to the group of lantibiotics [33]. The antimicrobial lipopeptid paenibacterin, derived from a soil isolate, *Paenibacillus* OSY-SE, exhibits antibacterial acti ity against Gram-positive and Gram-negative bacteria. Paenibacterin is a cyclic lipopeptid consisting of 13 amino acids and a C15 fatty acyl moiety [34]. There have been report that *Paenibacillus* sp. produces polymyxins. *Paenibacillus thiaminolyticus* SY20 has bee reported to produce polymyxin A1 [15]. Antimicrobial assays showed that they coul inhibit numerous Gram-negative species, including *Escherichia coli* ATCC 25922, *Salmonell enteritidis* CCTCC AB 94018, *Klebsiella pneumoniae* ATCC 10031, *Enterobacter sakazakii* ATC 29544, *Vibrio parahaemolyticus* ATCC 10031, *Psychrobacter pulmonis*, *Pseudomonas aerugino* PAO1, and *Salmonella typhimurium* ATCC 14028. However, the compound was ineffectiv against Gram-positive bacteria. On the other hand, one of the polymyxins of *Paenibacill* sp. strain B2 was reported to be active against Gram-positive bacteria, which might b related to the presence of the unusual amino acid [35]. The predicted BGCs of the NNS5-6 genome showed paeninodin, paenibacterin, and polymyxin biosynthesis, which were consistent with the secondary metabolites reported in the *Paenibacillus* genus. *Paenibacillus thiaminolyticus* NRRL B-4156 has some BGCs similar to the NNS5-6 genome. There is a report of a fusaricidin A synthase enzyme produced by *Paenibacillus thiaminolyticus* (sequence ID: SUA94926.1) in the NCBI database. The NNS5-6 AMP is active against Gram-negative bacteria; this finding was not in accordance with previous reports of fusaricidin from *Bacillus polymyxa* KT-8, which exhibited antibacterial activity against *S. aureus* but not against Gram-negative bacteria [16]. However, a novel cyclic lipopeptide analog of fusaricidin has demonstrated antibacterial activity against Gram-negative bacteria, such as *P. aeruginosa*. The addition of positively charged exocyclic residue, such as diaminobutyric acid, has been shown to enhance activity against Gram-negative bacterial strains [36]. The predicted BGC that appeared to be relevant to NNS5-6 AMP showed high similarity to the fusaricidin synthetase found in the *Paenibacillus thiaminolyticus* NCTC11027 (accession number: UGRZ01000003.1). The fusaricidin-like core biosynthesis gene found in the *Paenibacillus thiaminolyticus* NNS5-6 genome contained a different amino acid sequence, which could be responsible for the different synthesis of the NNS5-6 AMP structure. The antibacterial effect against Gram-negative bacteria of the fusaricidin-like peptide of *Paenibacillus thiaminolyticus* NNS5-6 is hypothesized to be related to the additional amino acid sequence of the NNS5-6 AMP and could potentially alter the typical activity spectrum of the peptide [37]. The exact mechanism of action of fusaricidins and their analogs remains unclear. Studies on cyclic lipopeptides derived from the fusaricidin family have shown that they can depolarize the cytoplasmic membranes of Gram-positive bacteria in a concentration-dependent manner. However, membrane depolarization does not necessarily correlate with bacterial cell death, suggesting that membrane-targeting activity may not be the primary mode of action for fusaricidins [38]. Other research has shown that fusaricidin inhibits purine and pyrimidine synthesis. Fusaricidin treatment leads to increased degradation of nucleotide precursors, suggesting that this can reduce the availability of nucleic acid-related substances in *Bacillus subtilis*. Moreover, fusaricidin causes membrane destruction. An increase in OH production interferes with protein and nucleic acid biosynthesis in the cells [39]. Genome annotation of *Paenibacillus thiaminolyticus* NNS5-6 against the CARD database using the RGI feature revealed 11 antibiotic resistance genes, conferring resistance to four classes of antibiotics, with a low percentage of identity to the reference genes. These included glycopeptide antibiotic resistance genes, fluoroquinolone antibiotic resistance genes, tetracycline antibiotic resistance genes, oxazolidinone antibiotic resistance genes, and antibiotic efflux pump genes. The results were relevant to antibiotic-resistance genes found in *Paenibacillus thiaminolyticus* PATH554 [40]. However, antibiotic susceptibility test of *Paenibacillus thiaminolyticus* NNS5-6 verified that this strain was susceptible to commonly used antibiotics, including carbapenems, macrolides, glycopeptides, penicillins, tetracyclines, fluoroquinolones, cephalosporins, and aminoglycosides. This verification indicated that the antibiotic resistance genes identified in the RGI results showed low identity or were present at low expression levels, affecting the translation of antibiotic resistance gene products.

Although NNS5-6 AMP shows potential for killing Gram-negative bacterial pathogens through membrane disruption, the AMP would require further improvement to enhance its stability through structure optimization and delivery systems. These enhancements would assist in the successful delivery to the target site and prevent degradation due to physiological barriers. For clinical application, the study of structure-activity relationships is mandatory. However, oral and parenteral administration requires further investigation to verify therapeutic efficacy, toxicity, pharmacokinetic and pharmacodynamic properties [41]. Future research should identify the molecular mechanisms of NNS5-6 AMP and verify the predicted genes involved in its production. Safety assessments, including cytotoxicity tests In Vitro, and toxicity evaluations in vivo are essential for its applications in the health sectors. The NNS5-6 AMP represents a valuable candidate for addressing infections caused by *P. aeruginosa* and *K. pneumoniae*, potentially offering an alternative to current treatments.

4. Materials and Methods

4.1. Sample Collection and Bacterial Isolation

Mangrove sediment was collected from Thasala district, Nakhon Si Thammarat, Tha land. The five samples were randomly collected from mangrove areas at a depth 10–15 cm. The surface part of the sediment (1–2 cm) was removed before collecting th samples. The collected samples were then placed in clean polyethylene bags and packe into an ice box. Subsequently, ten grams of sediment were transferred into a sterile flas and diluted with 90 mL of 0.85% NaCl solution (RCI Labscan Ltd., Bangkok, Thailand). A ditionally, a 10% *w*/*v* sediment suspension was prepared with deionized water. Then, th pH and salinity were measured. The sediment underwent agitation in a shaking incubat at 150 rpm for 30 min under ambient conditions, followed by an additional incubation 60 °C for 30 min. The samples were then subjected to 10-fold serial dilutions up to 10^{-} Each dilution (100 µL) was then spread onto Mueller Hinton (MH) agar, Zobell Marin (ZM) agar, and Starch Casein (SC) agar supplemented with 1.5% NaCl (Titan Biotech Ltd Rajasthan, India). The plates were incubated at 30 °C for a week before being re-streaked obtain pure isolates [42].

4.2. Antibacterial Screening Using the Soft Agar Overlay Method against P. aeruginosa TISTR 357

A single colony of the isolate was transferred to the same solid medium used for th initial isolation and incubated at 30 °C for 3 days. *P. aeruginosa* TISTR 357 was used as tes strain. The test strain was subcultured in MH agar and incubated at 37 °C for 18 h. Th suspension of *P. aeruginosa* TISTR 357 was prepared by dispersing a single colony in a 0.85 sterile NaCl solution and adjusting the turbidity to be equivalent to 0.1 optical densit (OD) at 625 nm (1.5 × 10^{8} CFU/mL) using a UV-visible spectrophotometer (Genesys 2 Thermo Scientific, Waltham, MA, USA). One mL of *P. aeruginosa* TISTR 357 was uniform mixed into 9 mL of molten 0.7% agar containing MH medium and then overlaid onto th bacterial isolate-seeded plates. The soft agar overlaid plates were incubated at 37 °C fo 24 h. The antibacterial compounds-producing isolates were observed by the appearance o an inhibition zone [43].

4.3. Verification of Antibacterial Activity Using the Agar Well Diffusion Technique

The active isolates from the agar overlay assay were transferred into broth medi identical to that used during the initial isolation. The cultures were incubated at 30 °C with shaking at 150 rpm for 18 h. The optical density of the starter culture was adjuste to a turbidity of 0.1 OD at 625 nm using a 0.85% sterile NaCl solution before transferrin 1 mL of the culture into 49 mL of fresh broth. The inoculum was then incubated at 30 °C with shaking at 150 rpm for 24 h. The CFS was obtained by centrifugation at 10,000× at 4 °C for 15 min and filtered through a 0.2 µm sterile cellulose acetate syringe filte The antibacterial activity of the CFS derived from each isolate was investigated using *P. aeruginosa* TISTR 357, *K. pneumoniae* TISTR 1383, *E. coli* TISTR 887, *S. typhimurium* TISTR 1469, *V. parahaemolyticus* TISTR 1596, and *S. aureus* TISTR 517 from the Thailand Institut of Scientific and Technological Research (TISTR), Thailand. The antibiotic-resistant strain MRSA strain 2468, was kindly provided by the medical technology laboratory at the Schoo of the Allied Health Sciences, Walailak University, Thailand. The microbial indicators wer incubated at 37 °C for 18 h on MH agar. Then, the bacterial indicators were prepared to a turbidity equivalent to 0.1 OD at 625 nm before being spread on MH agar. The CFS (100 µL was aseptically transferred to 9 mm diameter wells, followed by an incubation period a 37 °C for 18 h. Colistin (1 µg) and vancomycin (30 µg) (Sigma-Aldrich Co., St. Louis, MO USA) were used as a positive control, while broth media was used as a negative contro The experiment was conducted in triplicate, and the mean ± SD of the diameters of the inhibition zones was measured [44].

4.4. Investigation of the Production Kinetics of Antimicrobial Compounds of NNS5-6

The preculture of NNS5-6 was adjusted to an OD of 0.1 at 625 nm. It was then inoculated into 50 mL of MH broth at a 2% concentration. The mixture was placed in an incubator at 30 °C with agitation at 150 rpm for 7 days. Samples were aseptically collected at intervals of 2, 4, 6, 8, 12, 16, 20, 24, 48, 72, 96, 120, 144, and 168 h. Bacterial growth was monitored by the OD at 625 nm. The CFS was collected using the same procedure as the previous experiment. The agar well diffusion assay was conducted with bacterial pathogens as described in the previous experiment. This experiment was performed in triplicate. Statistical analysis was used to compare the antibacterial activity of CFS at each incubation time. The presence of significant differences was determined by Student's t-test at a p-value < 0.05. The kinetic of antibacterial compounds production curve with mean $\pm$ SD was presented [45].

4.5. Purification of the Antimicrobial Peptide

A single colony of NNS5-6 was suspended in a 0.85% sterile NaCl solution. The bacterial suspension was adjusted to a turbidity of 0.1 OD at 625 nm. The OD-adjusted suspension was used to prepare a 2% inoculum in 200 mL of MH broth in a 1 L sterile Erlenmeyer flask. The 1 L total culture was incubated at 30 °C with shaking at 150 rpm for 20 h. The CFS was collected as described in the previous experiment. Ammonium sulfate was then added stepwise to the CFS to achieve saturation levels of 25%, 50%, and 75%. The precipitates formed at each saturation level were collected by centrifugation at 18,000× g at 4 °C for 15 min. The collected precipitates were then dissolved in 50 mM ammonium acetate buffer solution, pH 5.0, and desalted using a dialysis bag with a 3.5 kDa molecular weight cut-off membrane (SnakeSkin membrane, Pierce, Rockford, IL, USA). The dissolved protein was dialyzed in the same buffer solution at 4 °C for 16 h. Each dialyzed fraction was tested for antibacterial activity against *P. aeruginosa* TISTR 357 using the agar well diffusion assay. Fractions exhibiting antibacterial activity were subjected to cation-exchange chromatography using HiTrap SP column (GE Healthcare Bio-Sciences AB, Uppsala, Sweden). The samples were equilibrated with a 50 mM ammonium acetate buffer solution at pH 5.0 to allow binding to the column. A gradient elution was performed using the same buffer solution with 1 M NaCl, gradually increasing from 0% to 100% over 25 mL. The eluates, detected at 214 nm absorbance, were collected and tested for antibacterial activity. Fractions containing the antibacterial compound were then injected into a size-exclusion chromatography column (Superdex™ 75 10/300 GL, GE Healthcare Bio-Sciences AB, Uppsala, Sweden). The samples were separated by molecular sieving using a mobile phase of 50 mM ammonium acetate buffer at pH 5.0. The active compound was eluted over two column volumes (50 mL) at a flow rate of 0.5 mL/min. The eluted samples were monitored by UV absorption at 214 nm, and each 1 mL fraction was collected and evaporated using a speed vacuum concentrator (RVC 2-25 CDplus, Martin Christ, Osterode am Harz, Germany). The dried residues were reconstituted in purified water to their original fraction volumes and subsequently assayed for antibacterial activity against *P. aeruginosa* TISTR 357 using the agar well diffusion method. Active fractions were then lyophilized (Gamma 2-16 LCSplus, Martin Christ, Osterode am Harz, Germany), and their masses were determined. These dried fractions were reconstituted in sterile purified water to their pre-lyophilization volumes. The reconstituted samples were subjected to serial two-fold dilutions and evaluated for antibacterial activity against *P. aeruginosa* TISTR 357. The arbitrary activity of each active fraction was calculated by taking the final dilution showing an inhibition zone to the power of 2 and multiplying it by 10.

4.6. Sodium Dodecyl Sulfate-Polyacrylamide Gel Electrophoresis (SDS-PAGE) and Agar Overlay Assay

The fractions purified by size-exclusion chromatography were subjected to analysis using 15% SDS-PAGE with two sets of samples to confirm the purity of the peptide by estimating the molecular weight and determining the active protein band [46]. After electrophoresis was completed, the gel was split into two parts. One part was staine with Coomassie brilliant blue G-250 for visualization of protein bands, while the other w treated with a mixture of 25% ethanol and 5% glacial acetic acid for 1 h and then washe with purified water for 3 h. The protein-fixed gel was subsequently overlaid with soft M agar containing *P. aeruginosa* TISTR 357 (10^6 CFU/mL) and incubated at 37 °C for 18 h determine the position of the inhibition zone.

4.7. Peptide Sequencing

Peptide sequencing was performed following previously reported methods [47]. T purified sample was analyzed using an UltiMate 3000 liquid chromatography (LC) sy tem coupled with high-resolution mass spectrometry (MS) (Thermo Fisher Scientific In Waltham, MA, USA). The peptide was dissolved in 0.1% formic acid and 1% aceto trile before being injected into a reversed-phase UHPLC column (4.6 mm × 30 mm; C Hypersil Gold, Thermo Fisher Scientific Inc., Waltham, MA, USA). The sample was sep rated by gradient elution from buffer A (0.1% formic acid) to buffer B (0.1% formic ac in acetonitrile) over 40 min at a flow rate of 300 µL/min. The separated peptides we ionized using an electrospray ionization (ESI) source with a capillary voltage of 3.2 kV a temperature of 300 °C. De novo sequencing was conducted to determine the peptic sequences using LC-MS data in full mass scanning mode. The MS parameters for detectir the peptide fragmentation mass were as follows: resolution of 120,000, automatic ga control (AGC) target of 1×10^6, maximum injection time of 100 m s, and scanning rang of m/z 400–2200. The results of the full MS scan were processed using Freestyle softwa (version 1.4) (Thermo Fisher Scientific Inc., Waltham, MA, USA) for peak identification. Th identified peaks from the parallel reaction monitoring system were analyzed for fragment tion by the second mass spectrometry (MS2). MS2 parameters were as follows: resolution 30,000, AGC target of 1×10^6, maximum injection time of 100 ms, and isolation window m/z 1.4. The collected mass data were used to predict the amino acid sequences using Pea Studio X (Bioinformatics Solutions Inc., Waterloo, ON, Canada). The predicted peptid sequences were included in the analysis if the ALC score was above 70%. The prediction physicochemical parameters was performed using ProtParam on the Expasy server [48].

4.8. Determination of the Peptide Secondary Structure

The secondary structure of the purified AMP was determined using CD spectroscop (JASCO Corporation, Tokyo, Japan) over a wavelength range of 190–250 nm. The purifie AMP (1 mg/mL) was dissolved in purified water or 50 mM SDS to determine the nativ conformation and to simulate the bacterial membrane environment, respectively. C spectra were analyzed to determine the secondary structure, and the components wer calculated using the BeStSel method via a web-based service [49]. Molecular modeling the AMP structure in a three-dimensional conformation was predicted using PEP-FOLD for visualizing and determining molecular surface area [50].

4.9. Determination of Minimum Inhibitory Concentration (MIC) and Minimum Bactericidal Concentration (MBC) of NNS5-6 AMP

The MIC and MBC of the purified AMP were determined using the broth microdilutio method following the guidance provided by the Clinical and Laboratory Standards Institut (CLSI) [51]. *P. aeruginosa* TISTR 357 and *K. pneumoniae* TISTR 1383 were cultured on MI agar and incubated at 37 °C for 18 h. A single colony of the tested bacteria was suspende in 0.85% NaCl until the suspension turbidity reached an OD of 0.1 at 625 nm. The cells we then diluted to 5×10^6 CFU/mL using cation-adjusted Mueller Hinton broth (CAMHB The diluted cell suspension of 10 µL was transferred into each well of a 96-well plate wit a final volume of 100 µL per well. The AMP was added to achieve final concentration ranging from 0.125 to 64 µg/mL. Colistin was used as the positive control, while antibioti free samples were used as negative controls. The 96-well plate was incubated at 37 °C fo 24 h. Each strain was tested in triplicate. The MIC was defined as the lowest concentratio of the AMP that showed no observable growth of bacteria. Subsequently, 100 μL of each dilution was spread on MH agar and then incubated at 37 °C for 24 h. The MBC was identified as the lowest concentration at which no bacterial colony growth was observed on the agar plate.

4.10. Scanning Electron Microscopy (SEM) of Cells Treated with NNS5-6 AMP

The effect of NNS5-6 AMP on *P. aeruginosa* TISTR 357 and *K. pneumoniae* TISTR 1383 was assessed by examining morphological alterations under SEM. *P. aeruginosa* TISTR 357 and *K. pneumoniae* TISTR 1383 were cultured in MH broth for 18 h and subsequently centrifuged to harvest the cell pellets. The cell pellets were washed with 0.85% sterile NaCl solution. The washed cells were resuspended in CAMHB to achieve a turbidity equivalent to 0.1 OD at 625 nm. The prepared cells were diluted to a concentration of 5×10^5 CFU/mL in CAMHB before being treated with NNS5-6 AMP at 1× MIC for 12 h. The sample for SEM micrography was prepared by fixing the bacterial sample in 2.5% glutaraldehyde in 0.1 M phosphate buffer pH 7.2 for 24 h before ethanol dehydration of the sample. Complete ethanol removal was achieved using a critical point drying machine (Quorum Technologies Ltd., Lewes, East Sussex, UK). The dehydrated sample was coated with gold using a sputter coater machine. The micrograph image was displayed by SEM at 20,000× magnification [52]. The morphological changes in *P. aeruginosa* TISTR 357 and *K. pneumoniae* TISTR 1383 cells treated with NNS5-6 AMP were compared with those treated with 1× MIC of colistin.

4.11. Time-Kill Kinetics of NNS5-6 AMP

P. aeruginosa TISTR 357 and *K. pneumoniae* TISTR 1383 were separately adjusted to an initial treatment cell density of 5×10^5 CFU/mL with CAMHB. The experiment preparation followed the microdilution assay procedures. Briefly, the NNS5-6 AMP concentrations were prepared by diluting to 1× and 2× MIC using CAMHB, while AMP-free CAMHB served as the non-treatment condition. The reactions were carried out at 37 °C, followed by spreading the entire volume from each well in each treatment condition on MH agar at specific time intervals (0–24 h). The plates were placed in a 37 °C incubator for 24 h before counting the colonies. The trends in bacterial reduction were graphically depicted using a logarithmic scale of viable cells. The bacterial cell reduction was determined within 24 h. Each experiment was performed in triplicate on the 96-well plates [53]. The presence of significant differences (p-value < 0.05) was assessed by two-way ANOVA, followed by post hoc Tukey's test for conducting multiple comparisons between treated and non-treated samples at each time interval.

4.12. Stability Studies of NNS-5-6 AMP

The stability of the NNS5-6 AMP under various conditions at different exposure times of 1, 6, and 12 h was examined [54]. The AMP was dissolved in sterile purified water and adjusted a final concentration to 16 μg/mL. The sensitivity of the AMP to temperatures of 37, 40, 50, 60, 80, and 100 °C, as well as autoclaving at 121 °C for 15 and 30 min was evaluated. The sensitivity of the AMP to proteolytic enzymes with different digesting characteristics was studied by incubating with a concentration of 1 mg/mL of proteinase K, trypsin, and α-chymotrypsin (Sigma-Aldrich, Warren, MI, USA). The compatibility of the AMP with surfactants was evaluated by exposing the AMP to 1% SDS or 1% Triton X-100 (AppliChem GmbH, Darmstadt, Germany). The degradation of the AMP under various pH conditions, including physiological and extended alkaline conditions, was evaluated by adjusting the pH of the solution to 1.2, 4.5, 6.8, 7.4, 8.0, 10.0, 12.0, and 14.0. After incubation, the pH of the AMP solution was neutralized to the original pH before being assessed for antibacterial activity. The antibacterial assay against *P. aeruginosa* TISTR 357 was conducted using the agar well diffusion assay in triplicate. The stability profiles are presented as the percentage of the residual activity compared with the non-treatment conditions (mean ± SD), and statistical significance was determined by the Student's *t*-te at a *p*-value < 0.05.

4.13. Effect of the AMP on Cell Membrane Permeability

The overnight precultured inoculum of *P. aeruginosa* TISTR 357 and *K. pneumoni* TISTR 1383 was collected and washed three times with sterile phosphate-buffered salir (PBS) pH 7.4 supplemented with 0.2% CAMHB as a diluent. The cell suspension wa diluted with the diluent until the OD at 625 nm was equal to 0.1 (1.5×10^8 CFU/mL). A aliquot of the diluted cells (100 μL) was introduced into each well and incubated wit 10 μM of Sytox Green (Thermo Fisher Scientific Inc., Waltham, MA, USA) in the dark fo 15 min. The AMP was diluted with sterile PBS pH 7.4 supplemented with 0.2% CAMH and 100 μL was added to the wells to achieve final concentrations of 0.125×, 0.25×, 0.5 1×, and 2× MIC. After the introduction of the AMP, the cell membrane permeability c treated bacteria was evaluated by monitoring the fluorescence intensity resulting fro the penetration of Sytox Green across the compromised cell membrane and subsequent binding to nucleic acid. The fluorescence intensity was measured using a microplate reade (Thermo Scientific Inc., Waltham, MA, USA) with excitation and emission wavelengths c 504 and 523 nm, respectively. Significant differences in fluorescence intensity of overa time points were analyzed by one-way ANOVA (*p*-value < 0.05) and post hoc Tukey's te for multiple comparisons between treated and non-treated conditions [52].

4.14. Characterization of Bacterial Morphology

A single colony of NNS5-6 was cultured on MH agar for 1 and 3 days to obtai vegetative cells and spores, respectively. The general appearance of colony morpholog was determined under a stereo microscope (Carl Zeiss, Oberkochen, Germany). Gra staining and malachite green staining were performed to determine the vegetative ce morphology and spore-forming capability, respectively, under a light microscope at 1000 magnification (Carl Zeiss, Oberkochen, Germany). The SEM micrography (Carl Zeis Oberkochen, Germany) was employed to examine the bacterial and spore morpholog with high-resolution images at 20,000× magnification [55,56].

4.15. Whole Genome Sequencing and Bioinformatic Analysis

The chromosome of NNS5-6 was extracted before sequencing by Illumina Hise (PE150 mode, Illumina, San Diego, CA, USA) using the service of U2Bio Co., Ltd. (Seou Republic of Korea). Data cleaning of raw reads and genome assembly were performe using the Galaxies Australia platform version 24.0 [57]. The quality of raw reads before an after trimming the adapters was checked by FastQC version 0.12.1 [58]. The raw reads wer trimmed by Fastp version 0.23.4 to remove the adapter sequences and filter out low-qualit reads and reads that had a base length below 30 bases [59]. The bacterial genome wa assembled by Shovill version 1.1.0 using Velvet version 1.2.10 as the assembler [60]. Th quality and genome parameters of the assembled genome were evaluated by QUAS version 5.2.0 [61]. The genome completeness and contamination in the genome sequenc were assessed using CheckM version 1.0.18 [62]. The known genes were annotated an predicted using Prokka version 1.14.6, and the cellular machinery was predicted usin RAST [63,64]. The prediction of BGCs of secondary metabolism of antibacterial agents wa carried out using antiSMASH version 7.0 [65]. The predicted secondary metabolites fro the NNS5-6 BGCs were compared with the reported reference BGCs in the database tha was connected to MIBiG version 3.1 and GenBank in the NCBI database [66,67]. NNS 6 was identified at the genus and species levels using genomic data to determine th closest taxonomic relationship. Digital DNA-DNA hybridization was performed usin the GBDP method to provide the similarity of the genome-based comparison betwee the NNS5-6 genome and the reference genome. The genome-based phylogenetic tree wa constructed using the TYGS genome server [68]. FastANI version 1.1.0 in the Proksee web based service was used to support the phylogenetic result. The sequence identity of DN fragments in the NNS5-6 genome was calculated using the average nucleotide identity method, comparing it to the closest species identified from the GBDP result. FastANI also provided the visualization of the position of the matched DNA fragments between the two compared genomes [69]. The insights into the BGC responsible for NNS5-6 AMP production were proposed by comparing the encoded protein of the core biosynthetic gene in the NNS5-6 genome with the conserved domains of the biosynthetic enzymes in the NCBI database. The search method was performed using the Domain Enhanced Lookup Time Accelerated BLAST (DELTA-BLAST) algorithm. The matched protein from the database and the encoded protein of the core biosynthetic gene of NNS5-6 were inspected for amino acid sequences, which were aligned using the Multiple Sequence Comparison by Log-Expectation (MUSCLE) algorithm in MEGA X software version 10.1.8 [70]. The NCBI Multiple Sequence Alignment (MSA) viewer version 1.25.0 was used to determine the similarity percentage and visualize the different amino acid sequences between the encoded protein of the core biosynthetic gene in the NNS5-6 genome and database-matched protein. The antibiotic resistance genes (ARGs) in the NNS5-6 genome were assessed to support one of the safety requirements for the future utilization of the NNS5-6 isolate. The ARGs were predicted by the RGI feature provided by the CARD database [71]. The ARGs were presented with percentages of identity and coverage compared with the matched reference ARGs in the database [72]. The NNS5-6 genomic information was visualized by the Proksee web-based service [73]. The circular genome map displayed general information, including the in-depth characterization of annotated genes and predicted biosynthetic gene clusters of secondary metabolisms. The genome map information was classified using different colored tracks to better understanding and readability.

4.16. Antibiotic Susceptibility Studies of NNS5-6

The susceptibility of NNS5-6 to standard antibiotics was assessed using a disc diffusion assay [74]. A single colony of 18 h-precultured NNS5-6 was adjusted to an OD of 0.1 at 625 nm before being spread onto MH agar plates. Antibiotic discs (Oxoid Ltd., Hampshire, UK) of ciprofloxacin (5 µg), piperacillin (100 µg) combined with tazobactam (10 µg), imipenem (10 µg), ceftriaxone (30 µg), cefoxitin (30 µg), doxycycline (30 µg), vancomycin (30 µg), erythromycin (15 µg), and gentamicin (10 µg) were placed on the culture-seeded agar and then incubated at 30 °C for 18 h. The susceptibility tests were conducted in triplicate. The inhibition zones were reported as the mean ± SD.

5. Conclusions

The AMP from mangrove-derived *Paenibacillus thiaminolyticus* NNS5-6 was effective against *P. aeruginosa* TISTR 357 and *K. pneumoniae* TISTR 1383. This novel AMP had a similar amino acid sequence to fasaricidins and exhibited potent antibacterial activity by disrupting the cell membrane. The stability profile of NNS5-6 AMP showed tolerance across a wide range of pH levels, proteolytic enzymes, and surfactants, but the temperature above 40 °C should be concerned about losing activity. Genetic analysis of the NNS5-6 genome identified BGCs responsible for various secondary metabolite productions. The high susceptibility of NNS5-6 to commonly used antibiotics could serve as preliminary safety data. Future research will explore the molecular mechanisms of NNS5-6 AMP that contribute to antibacterial activity, including verification of proposed genes responsible for secondary metabolite production. The pharmacodynamics and pharmacokinetics of NNS5-6 AMP should be further studied to evaluate the efficacy of AMP, along with the comprehensive safety assessments, which are vital requirements for its clinical applications and utilizations.

Supplementary Materials: The following supporting information can be downloaded at: https://www.mdpi.com/article/10.3390/antibiotics13090846/s1, Figure S1: The purification of NNS5-6 AMP was performed using size-exclusion chromatography. Figure S2. The physical characteristics of NNS5-6.

Author Contributions: Conceptualization, N.S. (Nuttapon Songnaka), N.S. (Namfa Sermkaew) a A.A.; methodology, N.S. (Nuttapon Songnaka), N.S. (Namfa Sermkaew) and A.A.; validation, N (Nuttapon Songnaka); formal analysis, N.S. (Nuttapon Songnaka), N.S. (Namfa Sermkaew), A. S.K., C.A., Y.Y. and J.U.; investigation, N.S. (Nuttapon Songnaka), N.S. (Namfa Sermkaew), A.A., S. C.A., Y.Y. and J.U.; resources, N.S. (Nuttapon Songnaka), N.S. (Namfa Sermkaew), A.A., S.K., C. Y.Y. and J.U.; data curation, N.S. (Nuttapon Songnaka) and N.S. (Namfa Sermkaew); writing—origir draft preparation, N.S. (Nuttapon Songnaka), N.S. (Namfa Sermkaew), A.A., S.K., C.A., Y.Y. and J. writing—review and editing, N.S. (Nuttapon Songnaka), N.S. (Namfa Sermkaew), A.A., S.K., C. Y.Y. and J.U.; visualization, N.S. (Nuttapon Songnaka) and N.S. (Namfa Sermkaew); supervision, N (Nuttapon Songnaka); project administration, N.S. (Nuttapon Songnaka) and N.S. (Namfa Sermkae funding acquisition, N.S. (Nuttapon Songnaka) and A.A. All authors have read and agreed to t published version of the manuscript.

Funding: This research was funded by Walailak University's annual government statement expenditure under the Plant Genetic Conservation Project Under The Royal Initiative of Her Ro Highness Princess Maha Chakri Sirindhorn, grant number RSPG-WU-09/2567.

Institutional Review Board Statement: The study was approved by the Institutional Biosafe Committee, Walailak University (WU-IBC-67-005).

Informed Consent Statement: Not applicable.

Data Availability Statement: Data are contained within the article and Supplementary Materials.

Acknowledgments: We acknowledge the Center of Scientific and Technological Equipment Walailak University for its research facilities. Chanat Aonbangkhen would also like to thank t Office of the Ministry of Higher Education, Science, Research, and Innovation.

Conflicts of Interest: The authors declare no conflict of interest.

References

1. O'Neill, J. *Review on Antimicrobial Resistance, Tackling Drug-resistant Infections Globally: Final Report and Recommendations*; Wellcon Trust and HM Government: London, UK, 2016; pp. 10–16.
2. Botelho, J.; Grosso, F.; Peixe, L. Antibiotic resistance in *Pseudomonas aeruginosa*–Mechanisms, epidemiology and evolution. *Dru Resist. Updat.* **2019**, *44*, 100640. [CrossRef] [PubMed]
3. Pang, Z.; Raudonis, R.; Glick, B.R.; Lin, T.J.; Cheng, Z. Antibiotic resistance in *Pseudomonas aeruginosa*: Mechanisms and alternati therapeutic strategies. *Biotechnol. Adv.* **2019**, *37*, 177–192. [CrossRef]
4. Yasir, M.; Dutta, D.; Willcox, M.D.P. Activity of antimicrobial peptides and ciprofloxacin against *Pseudomonas aeruginosa* biofilr *Molecules* **2020**, *25*, 3843. [CrossRef]
5. Wang, G.; Zhao, G.; Chao, X.; Xie, L.; Wang, H. The characteristic of virulence, biofilm and antibiotic resistance of *Klebsiel pneumoniae*. *Int. J. Environ. Res. Public Health* **2020**, *17*, 6278. [CrossRef] [PubMed]
6. Bhatnagar, I.; Kim, S.K. Immense essence of excellence: Marine microbial bioactive compounds. *Mar. Drugs* **2010**, *8*, 2673–270 [CrossRef] [PubMed]
7. Karthik, Y.; Ishwara Kalyani, M.; Krishnappa, S.; Devappa, R.; Anjali Goud, C.; Ramakrishna, K.; Wani, M.A.; Alkafafy, M Hussen Abduljabbar, M.; Alswat, A.S.; et al. Antiproliferative activity of antimicrobial peptides and bioactive compounds fro the mangrove *Glutamicibacter mysorens*. *Front. Microbiol.* **2023**, *14*, 1096826. [CrossRef]
8. Karthik, Y.; Kalyani, M.I.; Sheetal, K.; Rakshitha, D.; Bineesha, B.K. Cytotoxic and antimicrobial activities of microbial protein from mangrove soil actinomycetes of Mangalore, Dakshina Kannada. *Biomedicine* **2020**, *40*, 59–67.
9. Núñez-Montero, K.; Barrientos, L. Advances in Antarctic research for antimicrobial discovery: A comprehensive narrative revie of bacteria from Antarctic environments as potential sources of novel antibiotic compounds against human pathogens an microorganisms of industrial importance. *Antibiotics* **2018**, *7*, 90. [CrossRef]
10. Bin Hafeez, A.; Jiang, X.; Bergen, P.J.; Zhu, Y. Antimicrobial peptides: An update on classifications and databases. *Int. J. Mol. Sc* **2021**, *22*, 11691. [CrossRef]
11. Ben Hur, D.; Kapach, G.; Wani, N.A.; Kiper, E.; Ashkenazi, M.; Smollan, G.; Keller, N.; Efrati, O.; Shai, Y. Antimicrobial peptide against multidrug-resistant *Pseudomonas aeruginosa* biofilm from cystic fibrosis patients. *J. Med. Chem.* **2022**, *65*, 9050–906 [CrossRef]
12. Li, X.; Zuo, S.; Wang, B.; Zhang, K.; Wang, Y. Antimicrobial mechanisms and clinical application prospects of antimicrobia peptides. *Molecules* **2022**, *27*, 2675. [CrossRef] [PubMed]
13. Roth, B.L.; Poot, M.; Yue, S.T.; Millard, P.J. Bacterial viability and antibiotic susceptibility testing with SYTOX green nucleic aci stain. *Appl. Environ. Microbiol.* **1997**, *63*, 2421–2431. [CrossRef]
14. De Oliveira, D.M.P.; Forde, B.M.; Kidd, T.J.; Harris, P.N.A.; Schembri, M.A.; Beatson, S.A.; Paterson, D.L.; Walker, M.J. Antimicr bial resistance in ESKAPE pathogens. *Clin. Microbiol. Rev.* **2020**, *33*, e00181-19. [CrossRef] [PubMed]

. Wu, Y.P.; Liu, D.M.; Liang, M.H.; Huang, Y.Y.; Lin, J.; Xiao, L.F. Genome-guided purification and characterization of polymyxin A1 from *Paenibacillus thiaminolyticus* SY20: A rarely explored member of polymyxins. *Front. Microbiol.* **2022**, *13*, 962507. [CrossRef] [PubMed]

. Kajimura, Y.; Kaneda, M. Fusaricidin A, a new depsipeptide antibiotic produced by *Bacillus polymyxa* KT-8. Taxonomy, fermentation, isolation, structure elucidation and biological activity. *J. Antibiot.* **1996**, *49*, 129–135. [CrossRef] [PubMed]

'. Lee, S.H.; Cho, Y.E.; Park, S.H.; Balaraju, K.; Park, J.W.; Lee, S.W.; Park, K. An antibiotic fusaricidin: A cyclic depsipeptide from *Paenibacillus polymyxa* E681 induces systemic resistance against Phytophthora blight of red-pepper. *Phytoparasitica* **2013**, *41*, 49–58. [CrossRef]

. Jeong, H.; Choi, S.K.; Ryu, C.M.; Park, S.H. Chronicle of a soil bacterium: *Paenibacillus polymyxa* E681 as a tiny guardian of plant and human health. *Front. Microbiol.* **2019**, *10*, 467. [CrossRef]

. Vater, J.; Niu, B.; Dietel, K.; Borriss, R. Characterization of novel fusaricidins produced by *Paenibacillus polymyxa*-M1 using MALDI-TOF mass spectrometry. *J. Am. Soc. Mass Spectrom.* **2015**, *26*, 1548–1558. [CrossRef]

. Caporale, A.; Adorinni, S.; Lamba, D.; Saviano, M. Peptide-protein interactions: From drug design to supramolecular biomaterials. *Molecules* **2021**, *26*, 1219. [CrossRef]

. Choi, U.; Lee, C.R. Antimicrobial agents that inhibit the outer membrane assembly machines of Gram-negative bacteria. *J. Microbiol. Biotechnol.* **2019**, *29*, 1–10. [CrossRef]

2. Chai, H.; Allen, W.E.; Hicks, R.P. Spectroscopic investigations of the binding mechanisms between antimicrobial peptides and membrane models of *Pseudomonas aeruginosa* and *Klebsiella pneumoniae*. *Bioorg. Med. Chem.* **2014**, *22*, 4210–4222. [CrossRef] [PubMed]

3. Cheng, J.T.; Hale, J.D.; Elliott, M.; Hancock, R.E.; Straus, S.K. The importance of bacterial membrane composition in the structure and function of aurein 2.2 and selected variants. *Biochim. Biophys. Acta* **2011**, *1808*, 622–633. [CrossRef] [PubMed]

4. Panayi, T.; Diavoli, S.; Nicolaidou, V.; Papaneophytou, C.; Petrou, C.; Sarigiannis, Y. Short-chained linear scorpion peptides: A pool for novel antimicrobials. *Antibiotics* **2024**, *13*, 422. [CrossRef]

5. Uggerhøj, L.E.; Poulsen, T.J.; Munk, J.K.; Fredborg, M.; Sondergaard, T.E.; Frimodt-Moller, N.; Hansen, P.R.; Wimmer, R. Rational design of alpha-helical antimicrobial peptides: do's and don'ts. *Chembiochem* **2015**, *16*, 242–253. [CrossRef] [PubMed]

5. Feijoo-Coronel, M.L.; Mendes, B.; Ramírez, D.; Peña-Varas, C.; de Los Monteros-Silva, N.Q.E.; Proaño-Bolaños, C.; de Oliveira, L.C.; Lívio, D.F.; da Silva, J.A.; da Silva, J.M.S.F.; et al. Antibacterial and antiviral properties of chenopodin-derived synthetic peptides. *Antibiotics* **2024**, *13*, 78. [CrossRef] [PubMed]

7. Hincapié, O.; Giraldo, P.; Orduz, S. In silico design of polycationic antimicrobial peptides active against *Pseudomonas aeruginosa* and *Staphylococcus aureus*. *Antonie. Van Leeuwenhoek* **2018**, *111*, 1871–1882. [CrossRef]

8. Hua, B.; Feng, H.; Han, J.; Qiao, Z.; Wang, X.; Zhang, Q.; Liu, Z.; Wu, Z. Isolation and characterization of a new fusaricidin-type antibiotic produced by *Paenibacillus bovis* sp. nov BD3526. *Curr. Microbiol.* **2020**, *77*, 3990–3999. [CrossRef]

9. Sannino, D.; Angert, E.R. Genomic insights into the thiamin metabolism of *Paenibacillus thiaminolyticus* NRRL B-4156 and *P. apiarius* NRRL B-23460. *Stand. Genomic Sci.* **2017**, *12*, 59. [CrossRef]

0. Sáez-Nieto, J.A.; Medina-Pascual, M.J.; Carrasco, G.; Garrido, N.; Fernandez-Torres, M.A.; Villalón, P.; Valdezate, S. *Paenibacillus* spp. isolated from human and environmental samples in Spain: Detection of 11 new species. *New Microbes. New Infect.* **2017**, *19*, 19–27. [CrossRef]

1. Shida, O. Transfer of *Bacillus alginolyticus*, *Bacillus chondroitinus*, *Bacillus curdlanolyticus*, *Bacillus glucanolyticus*, *Bacillus kobensis*, and *Bacillus thiaminolyticus* to the genus *Paenibacillus* and emended description of the genus *Paenibacillus*. *Int. J. Syst. Bacteriol.* **1997**, *47*, 289–298. [CrossRef]

2. Huang, E.; Yousef, A.E. Draft genome sequence of *Paenibacillus polymyxa* OSY-DF, which coproduces a lantibiotic, paenibacillin, and polymyxin E1. *J. Bacteriol.* **2012**, *194*, 4739–4740. [CrossRef] [PubMed]

3. He, Z.; Kisla, D.; Zhang, L.; Yuan, C.; Green-Church, K.B.; Yousef, A.E. Isolation and identification of a *Paenibacillus polymyxa* strain that coproduces a novel lantibiotic and polymyxin. *Appl. Environ. Microbiol.* **2007**, *73*, 168–178. [CrossRef]

4. Guo, Y.; Huang, E.; Yuan, C.; Zhang, L.; Yousef, A.E. Isolation of a *Paenibacillus* sp. strain and structural elucidation of its broad-spectrum lipopeptide antibiotic. *Appl. Environ. Microbiol.* **2012**, *78*, 3156–3165. [CrossRef] [PubMed]

5. Selim, S.; Negrel, J.; Govaerts, C.; Gianinazzi, S.; van Tuinen, D. Isolation and partial characterization of antagonistic peptides produced by *Paenibacillus* sp. strain B2 isolated from the sorghum mycorrhizosphere. *Appl. Environ. Microbiol.* **2005**, *71*, 6501–6507. [CrossRef] [PubMed]

6. Gil, J.; Pastar, I.; Houghten, R.A.; Padhee, S.; Higa, A.; Solis, M.; Valdez, J.; Head, C.R.; Michaels, H.; Lenhart, B.; et al. Novel cyclic lipopeptides fusaricidin analogs for treating wound infections. *Front. Microbiol.* **2021**, *12*, 708904. [CrossRef]

7. Lebedeva, J.; Jukneviciute, G.; Čepaitė, R.; Vickackaite, V.; Pranckutė, R.; Kuisiene, N. Genome mining and characterization of biosynthetic gene clusters in two cave strains of *Paenibacillus* sp. *Front. Microbiol.* **2021**, *11*, 612483. [CrossRef]

8. Bionda, N.; Fleeman, R.M.; Shaw, L.N.; Cudic, P. Effect of ester to amide or N-methylamide substitution on bacterial membrane depolarization and antibacterial activity of novel cyclic lipopeptides. *Chem. Med. Chem.* **2013**, *8*, 1394–1402. [CrossRef]

9. Yu, W.B.; Yin, C.Y.; Zhou, Y.; Ye, B.C. Prediction of the mechanism of action of fusaricidin on *Bacillus subtilis*. *PLoS ONE* **2012**, *7*, e50003. [CrossRef]

0. Lu, J.; Sha, Y.; Gao, M.; Shi, W.; Lin, X.; Li, K.; Bao, Q.; Feng, C. Identification and characterization of a novel aminoglycoside *O*-nucleotidyltransferase ANT(6)-If from *Paenibacillus thiaminolyticus* PATH554. *Front. Microbiol.* **2023**, *14*, 1184349. [CrossRef]

41. Mazurkiewicz-Pisarek, A.; Baran, J.; Ciach, T. Antimicrobial peptides: Challenging journey to the pharmaceutical, biomedic and cosmeceutical use. *Int. J. Mol. Sci.* **2023**, *24*, 9031. [CrossRef]
42. Kashfi, R.; Kelsey, C.; Gang, D.J.; Call, D.R.; Gang, D.R. Metabolomic diversity and identification of antibacterial activities bacteria isolated from marine sediments in Hawai'i and Puerto Rico. *Front. Mol. Biosci.* **2020**, *7*, 23. [CrossRef]
43. Hockett, K.L.; Baltrus, D.A. Use of the soft-agar overlay technique to screen for bacterially produced inhibitory compounds. *J. V Exp.* **2017**, *14*, 55064.
44. Balouiri, M.; Sadiki, M.; Ibnsouda, S.K. Methods for in vitro evaluating antimicrobial activity: A review. *J. Pharm. Anal.* **2016**, 71–79. [CrossRef]
45. Regmi, S.; Choi, Y.S.; Choi, Y.H.; Kim, Y.K.; Cho, S.S.; Yoo, J.C.; Suh, J.W. Antimicrobial peptide from *Bacillus subtilis* CSB13 Characterization, killing kinetics, and synergistic potency. *Int. Microbiol.* **2017**, *20*, 43–53. [PubMed]
46. Xu, C.; Fu, Y.; Liu, F.; Liu, Z.; Ma, J.; Jiang, R.; Song, C.; Jiang, Z.; Hou, J. Purification and antimicrobial mechanism of a nov bacteriocin produced by *Lactobacillus rhamnosus* 1.0320. *LWT* **2021**, *137*, 110338. [CrossRef]
47. Sermkaew, N.; Atipairin, A.; Wanganuttara, T.; Krobthong, S.; Aonbangkhen, C.; Yingchutrakul, Y.; Uchiyama, J.; Songnaka, N. Novel bacitracin-like peptide from mangrove-isolated *Bacillus paralicheniformis* NNS4-3 against MRSA and its genomic insigh *Antibiotics* **2024**, *13*, 716. [CrossRef]
48. Gasteiger, E.; Hoogland, C.; Gattiker, A.; Duvaud, S.; Wilkins, M.R.; Appel, R.D.; Bairoch, A. Protein Identification and Analys Tools on the Expasy Server. In *The Proteomics Protocols Handbook*, 1st ed.; Walker, J.M., Ed.; Humana Press: Totowa, NJ, USA, 200 pp. 571–607.
49. Micsonai, A.; Wien, F.; Kernya, L.; Lee, Y.H.; Goto, Y.; Réfrégiers, M.; Kardos, J. Accurate secondary structure prediction and fo recognition for circular dichroism spectroscopy. *Proc. Natl. Acad. Sci. USA* **2015**, *112*, E3095–E3103. [CrossRef] [PubMed]
50. Rey, J.; Murail, S.; de Vries, S.; Derreumaux, P.; Tuffery, P. PEP-FOLD4: A pH-dependent force field for peptide structure predictic in aqueous solution. *Nucleic Acids Res.* **2023**, *51*, W432–W437. [CrossRef]
51. Clinical and Laboratory Standards Institute. *Methods for Dilution Antimicrobial Susceptibility Tests for Bacteria That Grow Aerobicall* 11th ed.; Clinical and Laboratory Standards Institute: Wayne, PA, USA, 2018; pp. 15–52.
52. Songnaka, N.; Lertcanawanichakul, M.; Hutapea, A.M.; Krobthong, S.; Yingchutrakul, Y.; Atipairin, A. Purification and characte zation of novel anti-MRSA peptides produced by *Brevibacillus* sp. SPR-20. *Molecules* **2022**, *27*, 8452. [CrossRef]
53. Uzair, B.; Menaa, F.; Khan, B.A.; Mohammad, F.V.; Ahmad, V.U.; Djeribi, R.; Menaa, B. Isolation, purification, structural elucidatic and antimicrobial activities of kocumarin, a novel antibiotic isolated from actinobacterium *Kocuria marina* CMG S2 associate with the brown seaweed *Pelvetia canaliculata*. *Microbiol. Res.* **2018**, *206*, 186–197. [CrossRef]
54. Choyam, S.; Jain, P.M.; Kammara, R. Characterization of a potent new-generation antimicrobial peptide of *Bacillus*. *Front. Microbi* **2021**, *12*, 710741. [CrossRef] [PubMed]
55. Tripathi, N.; Sapra, A. Gram Staining. In *StatPearls [Internet]*; StatPearls Publishing: Treasure Island, FL, USA, 2023. Availab online: https://www.ncbi.nlm.nih.gov/books/NBK562156/ (accessed on 18 May 2024).
56. Hussey, M.A.; Zayaitz, A. Endospore stain protocol. *Am. Soc. Microbiol.* **2007**, *8*, 1–11.
57. Galaxy Community. The Galaxy platform for accessible, reproducible, and collaborative data analyses: 2024 update. *Nucleic Acid Res.* **2024**, *52*, gkae410.
58. Babraham Bioinformatics: FastQC A Quality Control Tool for High Throughput Sequence Data. Available online: http://ww bioinformatics.babraham.ac.uk/projects/fastqc/ (accessed on 18 June 2024).
59. Chen, S.; Zhou, Y.; Chen, Y.; Gu, J. fastp: An ultra-fast all-in-one FASTQ preprocessor. *Bioinformatics* **2018**, *34*, i884–i890. [CrossRe [PubMed]
60. Zerbino, D.R.; Birney, E. Velvet: Algorithms for de novo short read assembly using de Bruijn graphs. *Genome Res.* **2008**, *1* 821–829. [CrossRef]
61. Gurevich, A.; Saveliev, V.; Vyahhi, N.; Tesler, G. QUAST: Quality assessment tool for genome assemblies. *Bioinformatics* **2013**, *2* 1072–1075. [CrossRef] [PubMed]
62. Parks, D.H.; Imelfort, M.; Skennerton, C.T.; Hugenholtz, P.; Tyson, G.W. CheckM: Assessing the quality of microbial genome recovered from isolates, single cells, and metagenomes. *Genome Res.* **2015**, *25*, 1043–1055. [CrossRef] [PubMed]
63. Seemann, T. Prokka: Rapid prokaryotic genome annotation. *Bioinformatics* **2014**, *30*, 2068–2209. [CrossRef]
64. Overbeek, R.; Olson, R.; Pusch, G.D.; Olsen, G.J.; Davis, J.J.; Disz, T.; Edwards, R.A.; Gerdes, S.; Parrello, B.; Shukla, M.; et a The SEED and the Rapid Annotation of microbial genomes using Subsystems Technology (RAST). *Nucleic Acids Res.* **2014**, *4* D206–D214. [CrossRef]
65. Blin, K.; Shaw, S.; Augustijn, H.E.; Reitz, Z.L.; Biermann, F.; Alanjary, M.; Fetter, A.; Terlouw, B.R.; Metcalf, W.W.; Helfrich, E.J.N et al. antiSMASH 7.0: New and improved predictions for detection, regulation, chemical structures and visualisation. *Nuclei Acids Res.* **2023**, *51*, W46–W50. [CrossRef]
66. Terlouw, B.R.; Blin, K.; Navarro-Muñoz, J.C.; Avalon, N.E.; Chevrette, M.G.; Egbert, S.; Lee, S.; Meijer, D.; Recchia, M.J.J.; Reit Z.L.; et al. MIBiG 3.0: A community-driven effort to annotate experimentally validated biosynthetic gene clusters. *Nucleic Acid Res.* **2023**, *51*, D603–D610. [CrossRef] [PubMed]
67. Clark, K.; Karsch-Mizrachi, I.; Lipman, D.J.; Ostell, J.; Sayers, E.W. GenBank. *Nucleic Acids Res.* **2016**, *44*, D67–D72. [CrossRef [PubMed]

Meier-Kolthoff, J.P.; Göker, M. TYGS is an automated high-throughput platform for state-of-the-art genome-based taxonomy. *Nat. Commun.* **2019**, *10*, 2182. [CrossRef] [PubMed]

Jain, C.; Rodriguez-R, L.M.; Phillippy, A.M.; Konstantinidis, K.T.; Aluru, S. High throughput ANI analysis of 90K prokaryotic genomes reveals clear species boundaries. *Nat. Commun.* **2018**, *9*, 5114. [CrossRef]

Kumar, S.; Stecher, G.; Li, M.; Knyaz, C.; Tamura, K. MEGA X: Molecular Evolutionary Genetics Analysis across Computing Platforms. *Mol. Biol. Evol.* **2018**, *35*, 1547–1549. [CrossRef] [PubMed]

Alcock, B.P.; Huynh, W.; Chalil, R.; Smith, K.W.; Raphenya, A.R.; Wlodarski, M.A.; Edalatmand, A.; Petkau, A.; Syed, S.A.; Tsang, K.K.; et al. CARD 2023: Expanded curation, support for machine learning, and resistome prediction at the Comprehensive Antibiotic Resistance Database. *Nucleic Acids Res.* **2023**, *51*, D690–D699. [CrossRef]

Jia, B.; Raphenya, A.R.; Alcock, B.; Waglechner, N.; Guo, P.; Tsang, K.K.; Lago, B.A.; Dave, B.M.; Pereira, S.; Sharma, A.N.; et al. CARD 2017: Expansion and model-centric curation of the comprehensive antibiotic resistance database. *Nucleic Acids Res.* **2017**, *45*, D566–D573. [CrossRef]

Grant, J.R.; Enns, E.; Marinier, E.; Mandal, A.; Herman, E.K.; Chen, C.Y.; Graham, M.; Van Domselaar, G.; Stothard, P. Proksee: In-depth characterization and visualization of bacterial genomes. *Nucleic Acids Res.* **2023**, *51*, W484–W492. [CrossRef]

Clinical and Laboratory Standards Institute. *M100 Performance Standards for Antimicrobial Susceptibility Testing*, 13th ed.; Clinical and Laboratory Standards Institute: Wayne, PA, USA, 2020; pp. 58–66.

antibiotics

MDP

Article

Effect of Antimicrobial Peptide BiF2_5K7K on Contaminated Bacteria Isolated from Boar Semen and Semen Qualities during Preservation and Subsequent Fertility Test on Pig Farm

Krittika Keeratikunakorn [1], Panida Chanapiwat [1], Ratchaneewan Aunpad [2], Natharin Ngamwongsatit [3,4] and Kampon Kaeoket [1,*]

1 Semen Laboratory, Department of Clinical Sciences and Public Health, Faculty of Veterinary Science, Mahidol University, 999 Phuttamonthon 4 Rd., Salaya, Phuttamonthon, Nakhon Pathom 73170, Thailand; krittika.ker@student.mahidol.edu (K.K.); panida.chn@mahidol.edu (P.C.)

2 Graduate Program in Biomedical Sciences, Faculty of Allied Health Sciences, Thammasat University, Rangsit Campus, Klong Luang, Pathumthani 12120, Thailand; aratchan@tu.ac.th

3 Department of Clinical Sciences and Public Health, Faculty of Veterinary Science, Mahidol University, 999 Phuttamonthon 4 Rd., Salaya, Phuttamonthon, Nakhon Pathom 73170, Thailand; natharin.nga@mahidol.edu

4 Laboratory of Bacteria, Veterinary Diagnostic Center, Faculty of Veterinary Science, Mahidol University, 999 Phuttamonthon 4 Rd., Salaya, Phuttamonthon, Nakhon Pathom 73170, Thailand

* Correspondence: kampon.kae@mahidol.edu

Abstract: The purpose of this study was to determine the impact of an antimicrobial peptic BiF2_5K7K, on semen quality and bacterial contamination in boar semen doses used for artifici insemination. A key factor affecting semen quality and farm production is bacterial contaminatio in semen doses. Using antibiotics in a semen extender seems to be the best solution for minimizi bacterial growth during semen preservation. However, concern regarding antibiotic-resistant m croorganisms has grown globally. As a result, antimicrobial peptides have emerged as interestir alternative antimicrobial agents to replace the current antibiotics used in semen extenders. BiF2_5K7 is an antimicrobial peptide that can inhibit Gram-negative and Gram-positive bacteria isolated fro boar semen and sow vaginal discharge. In this study, ten fresh boar semen samples were collected an diluted with one of two types of semen extender: with (positive control) or without (negative contro an antibiotic (i.e., gentamicin). The semen extender without an antibiotic contained antimicrobi peptide BiF2_5K7K at different concentrations (15.625, 31.25, 62.5, and 125 μg/mL). The sample were stored at 18 °C until use. Semen quality parameters were assessed on days 0, 1, 3, and 5, an the total bacterial count was also evaluated at 0, 24, 36, 48, and 72 h after storage. A fertility test c a pig farm was also performed via sow insemination with a commercial extender plus BiF2_5K7 at a concentration of 31.25 μg/mL. No significant difference was found in terms of semen qualit on days 0 or 1. On days 3 and 5, the total motility, progressive motility, and viability remaine normal in the 15.625 and 31.25 μg/mL groups. However, the sperm parameters decreased startin on day 3 for the 125 μg/mL group and on day 5 for the 62.5 μg/mL group. For total bacteri count at 0, 24, 36, 48, and 72 h, the lowest bacterial count was found in the positive control grou and the highest bacterial count was found in the negative control group compared with the othe groups. Comparing antimicrobial peptide groups from 0 to 48 h, the lowest bacterial count wa found in the 125 μg/mL group, and the highest bacterial count was found in the 15.625 μg/m group. For the fertility test, artificial insemination was conducted by using a commercial extende plus BiF2_5K7K at a concentration of 31.25 μg/mL. The results showed a superior pregnancy rat farrowing rate, and total number of piglets born compared with artificial insemination conducte using a commercial extender plus antibiotic. In conclusion, BiF2_5K7K can inhibit bacterial growth i extended boar semen for 24 h, and thereafter, the bacterial count slightly increases. However, th increase in the number of bacterial counts from days 0 to 3 had no negative effect on sperm qualit in the positive control, 15.625, or 31.25 μg/mL groups. This indicates that BiF2_5K7K might be a antimicrobial peptide candidate with potential for use as an alternative antimicrobial agent to replac the conventional antibiotic used in boar semen extenders.

Citation: Keeratikunakorn, K.; Chanapiwat, P.; Aunpad, R.; Ngamwongsatit, N.; Kaeoket, K. Effect of Antimicrobial Peptide BiF2_5K7K on Contaminated Bacteria Isolated from Boar Semen and Semen Qualities during Preservation and Subsequent Fertility Test on Pig Farm. *Antibiotics* **2024**, *13*, 579. https://doi.org/10.3390/antibiotics13070579

Academic Editors: Jean-Marc Sabatier, Marisa Di Pietro and Piyush Baindara

Received: 9 May 2024
Revised: 19 June 2024
Accepted: 21 June 2024
Published: 22 June 2024

Keywords: antimicrobial peptide; boar semen; semen extender; semen quality

1. Introduction

Liquid boar semen preservation is routinely used in artificial insemination (AI) in the swine industry [1,2], as AI can reduce the risk of disease transmission and improve genetics, as well as increase the production or quality of piglets [1,3]. In the modern pig business, more than 93% of sows are inseminated via artificial insemination, and boar semen diluted with semen extenders is mostly used in the breeding herd [2]. The use of a semen extender is necessary to support the longevity and quality of sperm; that is, the extender protects sperm from cold shock, controls pH and osmotic pressure, and inhibits bacterial growth [1].

Although AI can reduce the transmission of disease from boar to gilt/sow, bacterial contamination in semen may affect their reproductive performance [3]. Bacteriospermia in humans and animals results in reduced quality, quantity, and longevity of spermatozoa [3,4]. In addition to reduced semen quality, it can result in embryonic or fetal death, endometritis, and vaginal discharge in sows [3]. Fresh boar sperm contain both Gram-negative and Gram-positive bacteria, including *Staphylococcus* spp., *Streptococcus* spp., *E. coli*, *Klebsiella* spp., *Aeromonas* spp., *Pseudomonas* spp., *Proteus* spp., and *Providencia* spp. [5–8]. It has been reported that contamination of boar semen with *E. coli*, *Proteus mirabilis*, *Pseudomonas aeruginosa*, *Clostridium perfringens*, and *Staphylococcus aureus* not only causes poor motility, but also reduces the integrity of the sperm membrane and acrosome [2]. Antibiotics play an important role in boar semen extenders, controlling bacterial contamination, reducing transmission of pathogens to the gilt/sow, and increasing the longevity of spermatozoa during storage [4,9].

It has been reported in tropical countries, including Thailand, that bacteria found in boar semen have developed resistance to multiple antibiotics commonly used in pig farms and added to boar semen extender. These antibiotics include amoxicillin, gentamicin, and colistin [8,10]. Consequently, bacteria isolated from boar semen have shown critical antibiotic resistance genes such as *mcr-3* and *int1* [8,10]. The medical world is concerned about antibiotic resistance, as many antibiotics are liberally used not only in humans, but also in livestock, and the pace of new antibiotic discoveries is slow [11]. In practice, many antibiotics are mixed into semen extenders to inhibit bacterial growth and limit the deleterious effects of contamination [4,9,12]. Gentamicin, neomycin, streptomycin, and other antibiotics are commonly used in boar semen extenders [13–15]. In many cases, more than one antibiotic is mixed into the boar semen extender; for example, combinations of gentamicin and florfenicol as well as gentamicin and polymyxin B have been used [16]. Since the emergence of antibiotic-resistant bacteria, many alternative antibacterial agents have been studied to reduce the use of antibiotics, including antimicrobial peptides (AMPs) [17]. To date, more than 2500 AMPs have been deposited in the Antimicrobial Peptide Database (APD) [18]. Antimicrobial peptides (AMPs), including proline-rich antimicrobial peptides (PrAMPs), tryptophan- and arginine-rich antimicrobial peptides, histidine-rich antimicrobial peptides, and glycine-rich antimicrobial peptides, have been identified as potential antimicrobial agents. These peptides show potential in combating antibiotic-resistant bacteria [19–21]. AMPs provide an alternative option to reduce or replace antibiotics used in swine and poultry production [22]. As observed since 2004, the number of publications on the topic of AMPs has increased every year [23]. In human medicine, AMPs, including Nisin A S26A, S29D, and S29E, have been applied for the prevention of food-borne pathogens such as *E. coli* and *Salmonella* Typhimurium [24]. In addition to using AMPs to prevent food-borne disease, AMPs are utilized as additional medical treatments, for example, in the treatment of burn wound infections using PXL150 and D2A21 [25]. With regard to the broad-spectrum antimicrobial activities of antimicrobial peptides, they are used to replace antibiotics in pig farms as growth promoters [26]. In a meta-analysis study, it was found that AMPs can improve average daily gain (ADG) and decrease the diarrhea rate in piglets [26]. The primary characteristic of AMPs is their ability to eliminate bacteria whi minimizing harm to the host cell. This makes them a compelling option for reducing substituting antibiotics in semen extenders [27].

The objective of this study was to investigate the antimicrobial peptide properti of BiF2_5K7K against bacteria isolated from sow vaginal discharge and boar semen. Fu thermore, we tested the bacterial inhibition efficiency of the BiF2_5K7K antimicrobi peptide when mixed with boar semen extender to investigate its potential as an antibiot replacement and its effect on extended boar semen quality.

2. Results

2.1. Minimum Inhibitory Concentration (MIC) and Minimum Bactericidal Concentration (MBC) Assay

The results of MIC and MBC assays of BiF2_5K7K against pathogens isolated from bo semen and sow vaginal discharge are presented in Table 1. Except for *Klebsiella pneumoni Morganella morganii, Proteus mirabilis,* and *Providencia rettgeri* (MIC > 250 µg/mL), MI values of BiF2_5K7K between 15.625 and 250 µg/mL inhibited the growth of bacter isolated from boar semen and sow vaginal discharge (Table 1). Meanwhile, MBC value of BiF2_5K7K between 15.625–250 µg/mL showed bactericidal effects. However, sever bacteria, including *Klebsiella pneumoniae, Morganella morganii, Proteus mirabilis, Providenc rettgeri,* and *Staphylococcus hyicus,* showed MBC values of more than 250 µg/mL (Table 1

Table 1. MIC and MBC values of BiF2_5K7K tested against 12 bacteria isolated from boar semen an sow vaginal discharge.

Gram	ID	Sample	Bacteria	BiF2_5K7K	
				MIC (µg/mL)	MBC (µg/mL)
Negative	S1LLF	Boar semen	*Citrobacter koseri*	15.625	31.25
	S8-6LF	Boar semen	*Enterobacter hormaechei*	250	250
	S5LF3	Boar semen	*Escherichia coli*	15.625	15.625
	MI912-2LF/62	Sow vaginal discharge	*Klebsiella pneumoniae*	>250	>250
	V5-6	Sow vaginal discharge	*Morganella morganii*	>250	>250
	S6-4	Boar semen	*Providencia alcalifaciens*	15.625	62.5
	S3	Boar semen	*Proteus mirabilis*	>250	>250
	V2-5	Sow vaginal discharge	*Providencia rettgeri*	>250	>250
	S2NLF	Boar semen	*Pseudomonas aeruginosa*	31.25	125
	V4-3	Sow vaginal discharge	*Pasteurella aerogenes*	125	125
Positive	S7-5W	Boar semen	*Staphylococcus sciuri*	15.625	15.625
	V2-3	Sow vaginal discharge	*Staphylococcus hyicus*	125	>250

2.2. Total Bacterial Count

The mean total bacterial count of fresh boar semen was log2.27 ± 0.80 CFU/m (ranged from log1.81 to log2.98 CFU/mL) (Table 2). After incubation at 18 °C, seme samples with different concentrations of BiF2_5K7K were measured at 0, 24, 36, 48, an 72 h, and the results of the total bacterial count are presented in Table 3. With increasin incubation time, the total bacterial count increased. At 0 h after incubation, the lowes total bacterial count was found in the positive control group (log1.22 CFU/mL, BTS plu antibiotic) when compared with other groups. Comparing treatment groups (BTS withou antibiotic plus BiF2_5K7K), the total bacterial count varied from log1.33 to log1.47 CFU/m which was lower than in the negative control group (log1.79 CFU/mL, BTS without antib otic). At 24 h after incubation, the pattern of total bacterial count in all groups was more o less the same compared with the slightly increased bacterial count at 0 h. The lowest an highest total bacterial counts were found in the positive control group (log0.67 CFU/mL and the negative control group (log2.35 CFU/mL), respectively. Meanwhile, the tota bacterial count in the treatment groups depended on the concentrations of BiF2_5K7 and varied from log1.51 to log1.84 CFU/mL. The total bacterial count of BiF2_5K7K at concentrations of 62.5 and 31.25 µg/mL at 24 h after incubation were the lowest in the treatment group, and were not significantly different from the positive control group (BTS with antibiotic) (Table 3). The total bacterial count in treatment groups decreased with an increased concentration of BiF2_5K7K (Table 3).

Table 2. Descriptive statistics for sperm parameter measurements of fresh boar semen (n = 10).

Semen Parameters	Mean ± S.D.	Range
Concentration ($\times 10^6$ spz/mL)	211.50 ± 71.10	146–345
Osmolality (mOsm/kg)	304.80 ± 8.50	288–315
Total motility (%)	94.40 ± 3.90	86.60–99.40
Progressive motility (%)	90.90 ± 5.80	80.00–98.50
Sperm viability (%)	88.3 0± 2.80	85–93
Intact acrosome (%)	85.40 ± 2.90	80–91
MMP (%)	82.50 ± 2.70	80–89
Total bacterial count (CFU/mL)	log1.81 ± 0.80	log1.81–log2.98

MMP: Sperm with high mitochondrial membrane potential.

Table 3. Total bacteria count (mean ± SEM) from boar semen samples (n = 10) at 0, 24, 36, 48, and 72 h after incubation at 18 °C.

Group	Concentration (µg/mL)	Total Bacterial Count (log) (CFU/mL)				
		Incubation Time				
		0 h	24 h	36 h	48 h	72 h
BTS	-	1.79 ± 0.23	2.35 ± 0.26 ^{b}	3.47 ± 0.58 ^{b}	3.98 ± 0.76 ^{b}	6.25 ± 1.75 ^{b}
BTS + ABO	-	1.22 ± 0.52	0.67 ± 0.33 ^{a}	0.85 ± 0.15 ^{a}	1.12 ± 0.42 ^{a}	0.00 ± 0.00 ^{a}
BiF2_5K7K *	125	1.33 ± 0.30	1.78 ± 0.31 ^{b}	2.58 ± 1.06 ^{b}	3.55 ± 0.97 ^{b}	6.14 ± 2.79 ^{b}
BiF2_5K7K *	62.5	1.42 ± 0.34	1.53 ± 0.31 ^{a}	3.36 ± 0.75 ^{b}	3.72 ± 0.83 ^{b}	6.23 ± 2.68 ^{b}
BiF2_5K7K *	31.25	1.39 ± 0.27	1.51 ± 0.29 ^{a}	3.21 ± 0.89 ^{b}	4.32 ± 1.15 ^{b}	6.59 ± 2.82 ^{b}
BiF2_5K7K *	15.625	1.47 ± 0.19	1.84 ± 0.33 ^{b}	3.82 ± 0.94 ^{b}	3.02 ± 1.03 ^{b}	7.09 ± 2.16 ^{b}

* Combination of BiF2_5K7K and BTS without antibiotics. ^{a,b} Significant difference among groups at the same incubation time ($p < 0.05$). ABO: antibiotic; BTS: Beltsville Thawing Solution.

2.3. Sperm Quality Parameter Analysis

The sperm quality of fresh boar semen samples is presented in Table 2. On day 1, the sperm quality parameters remained normal and there were no significant differences in all sperm parameters among the six groups, except in the STR (straightness) and LIN (linearity) parameters. The straightness and linearity values of the 15.625 µg/mL group were significantly different from the other groups (75.2% and 33.3%) (Table 4). On day 3, inferior progressive motility occurred in the negative control, 125 µg/mL, and 62.5 µg/mL groups compared with the other groups. The 125 µg/mL dose of BiF2_5K7K had significantly lower effects on sperm motility patterns, including the VCL, VSL, VAP, and ALH parameters, compared with the other groups, especially the negative control group (Table 5). In addition, lower percentages of total motility, viability, intact acrosome, and sperm with high MMP were found in the 125 µg/mL group compared with the other control and treatment groups (Table 5), and these decreased as incubation times increased in some parameters on day 5 (Table 6). On day 5, significantly superior percentages of total motility and progressive motility were found in the 15.625 and 31.25 µg/mL groups compared with the 125 µg/mL group ($p < 0.05$, Table 6). There were significantly lower percentages of sperm motility patterns in the 125 µg/mL group compared with the other groups ($p < 0.05$), except in the STR and LIN parameters. However, there were no significant differences in sperm viability or intact acrosomes among the six groups (Table 6).

Table 4. Mean ± SEM of semen quality parameters on day 1 after incubation at 18 °C (n = 10).

Sperm Parameters	Group					
	BTS	BTS + ABO	BiF2_5K7K 125 µg/mL *	BiF2_5K7K 62.50 µg/mL *	BiF2_5K7K 31.25 µg/mL *	BiF2_5K7K 15.625 µg/mL *
MOT (%)	90.5 ± 1.6	90.7 ± 1.5	90.3 ± 1.3	92.3 ± 1.0	92.6 ± 1.0	90.5 ± 1.1
PMOT (%)	81.8 ± 2.6	81.23 ± 2.8	81.3 ± 2.0	84.1 ± 2.0	83.8 ± 1.9	81.6 ± 2.2
VCL (µm/s)	120.4 ± 9.4	125.4 ± 8.2	188.1 ± 6.9	116.1 ± 6.3	113.8 ± 6.9	110.6 ± 5.6
VSL (µm/s)	37.4 ± 3.1	38.8 ± 3.7	37.2 ± 3.4	37.7 ± 3.1	37.3 ± 3.1	36.8 ± 2.4
VAP (µm/s)	51.4 ± 3.6	53.9 ± 3.9	50.9 ± 3.6	50.8 ± 3.3	49.9 ± 3.3	48.6 ± 2.5
ALH (µm)	1.18 ± 0.09	1.21 ± 0.67	1.14 ± 0.05	1.11 ± 0.04	1.08 ± 0.04	1.06 ± 0.04
STR (%)	72.1 ± 1.7 ^{a}	71.2 ± 1.8 ^{a}	72.2 ± 1.7 ^{a}	73.89 ± 1.5 ^{a}	74.0 ± 1.3 ^{a}	75.2 ± 1.3 ^{b}
LIN (%)	31.1 ± 1.2 ^{a}	30.7 ± 1.4 ^{a}	31.1 ± 1.3 ^{a}	32.2 ± 1.1 ^{a}	32.6 ± 1.0 ^{a}	33.3 ± 0.8 ^{b}
Viability (%)	85.4 ± 0.9	86.3 ± 0.8	84.0 ± 0.7	85.5 ± 0.5	86.3 ± 0.8	84.8 ± 0.5
Intact acrosome (%)	83.5 ± 0.8	83.8 ± 0.7	82.4 ± 0.9	82.4 ± 0.8	82.5 ± 1.0	82.7 ± 0.9
MMP (%)	77.2 ± 1.6	78.3 ± 1.2	77.7 ± 1.0	77.9 ± 1.2	79.9 ± 1.4	78.7 ± 0.9

* Combination of BiF2_5K7K and BTS without antibiotics. ^{a,b} Significant difference among groups at the sar incubation time (p < 0.05). ABO: antibiotic; BTS: Beltsville Thawing Solution; MOT: total motility; PMC progressive motility; VCL: curvilinear velocity; VSL: straight-line velocity; VAP: average pathway velocity; AL amplitude of lateral head displacement; STR: straightness; LIN: linearity; MMP: sperm with high mitochondr membrane potential.

Table 5. Mean ± SEM of semen quality parameters on day 3 after incubation at 18 °C (n = 10).

Sperm Parameters	Group					
	BTS	BTS + ABO	BiF2_5K7K 125 µg/mL *	BiF2_5K7K 62.50 µg/mL *	BiF2_5K7K 31.25 µg/mL *	BiF2_5K7K 15.625 µg/mL *
MOT (%)	85.9 ± 2.7	86.1 ± 2.6	73.3 ± 6.9	85.5 ± 3.4	87.5 ± 2.9	86.3 ± 2.9
PMOT (%)	73.3 ± 3.9 ^{a}	73.4 ± 4.1 ^{a}	60.8 ± 7.3 ^{b}	73.6 ± 4.6 ^{a}	76.4 ± 3.9 ^{a}	75.1 ± 4.2 ^{a}
VCL (µm/s)	106.3 ± 10.9 ^{a}	109.9 ± 12.1 ^{a}	89.8 ± 13.9 ^{b}	107.1 ± 11.9 ^{a}	102.2 ± 11.7 ^{a,b}	100.4 ± 11.5 ^{a,b}
VSL (µm/s)	33.0 ± 3.5 ^{a}	34.1 ± 4.1 ^{a}	27.8 ± 4.7 ^{b}	35.0 ± 4.5 ^{a}	33.6 ± 4.2 ^{a}	33.4 ± 4.3 ^{a}
VAP (µm/s)	45.3 ± 4.7 ^{a}	47.4 ± 5.4 ^{a}	38.2 ± 5.9 ^{b}	46.0 ± 5.5 ^{a}	44.3 ± 5.3 ^{a,b}	44.0 ± 5.1 ^{a,b}
ALH (µm)	1.09 ± 0.09 ^{a}	1.09 ± 0.10 ^{a}	0.94 ± 0.12 ^{b}	1.05 ± 0.09 ^{a,b}	1.01 ± 0.09 ^{a,b}	0.99 ± 0.09 ^{a,b}
STR (%)	72.9 ± 1.0 ^{a}	71.8 ± 1.3 ^{a}	72.1 ± 1.5 ^{a}	75.0 ± 1.4 ^{a,b}	75.8 ± 1.7 ^{b}	75.4 ± 1.2 ^{b}
LIN (%)	31.2 ± 0.0 ^{a,b}	31.0 ± 0.9 ^{a,b}	30.8 ± 0.8 ^{a}	32.7 ± 1.1 ^{a,b}	32.9 ± 1.1 ^{a,b}	33.1 ± 1.0 ^{b}
Viability (%)	81.6 ± 1.5	82.1 ± 1.4	77.6 ± 2.3	80.3 ± 1.7	82.8 ± 1.1	81.3 ± 1.1
Intact acrosome (%)	80.0 ± 1.5	80.3 ± 0.9	75.3 ± 0.3	78.6 ± 1.5	79.4 ± 1.2	78.6 ± 1.2
MMP (%)	71.3 ± 1.8	72.0 ± 2.3	66.4 ± 3.8	72.4 ± 2.3	73.9 ± 2.5	74.1 ± 2.0

* Combination of BiF2_5K7K and BTS without antibiotics. ^{a,b} Significant difference among groups at the san incubation time (p < 0.05). ABO: antibiotic; BTS: Beltsville Thawing Solution; MOT: total motility; PMO progressive motility; VCL: curvilinear velocity; VSL: straight-line velocity; VAP: average pathway velocity; AL amplitude of lateral head displacement; STR: straightness; LIN: linearity; MMP: sperm with high mitochondri membrane potential.

Table 6. Mean ± SEM of semen quality parameters on day 5 after incubation at 18 °C (n = 10).

Sperm Parameters	Group					
	BTS	BTS + ABO	BiF2_5K7K 125 µg/mL *	BiF2_5K7K 62.50 µg/mL *	BiF2_5K7K 31.25 µg/mL *	BiF2_5K7K 15.625 µg/mL *
MOT (%)	72.6 ± 7.2 ^{a,b}	79.3 ± 4.7 ^{a}	57.3 ± 10.1 ^{b}	75.5 ± 6.5 ^{a,b}	80.0 ± 5.6 ^{a}	80.3 ± 4.8 ^{a}
PMOT (%)	60.2 ± 7.6 ^{a,b}	66.0 ± 5.3 ^{a,b}	45.9 ± 10.1 ^{b}	63.9 ± 7.7 ^{a,b}	69.8 ± 6.5 ^{a}	71.1 ± 6.0 ^{a}
VCL (µm/s)	79.7 ± 11.8 ^{a}	94.9 ± 10.7 ^{a}	71.2 ± 16.3 ^{b}	90.6 ± 14.1 ^{b}	95.2 ± 14.4 ^{b}	94.4 ± 12.5 ^{b}
VSL (µm/s)	24.4 ± 3.9 ^{a}	29.9 ± 3.7 ^{b}	23.5 ± 6.1 ^{a}	28.1 ± 5.2 ^{a,b}	30.3 ± 5.5 ^{a,b}	29.8 ± 4.9 ^{a,b}
VAP (µm/s)	33.6 ± 5.1 ^{a}	41.0 ± 4.7 ^{b}	30.9 ± 7.4 ^{c}	38.0 ± 6.4 ^{a,b}	40.7 ± 6.8 ^{a,b}	39.6 ± 5.9 ^{a,b}
ALH (µm)	0.86 ± 0.11 ^{a}	0.98 ± 0.09 ^{a}	0.76 ± 0.13 ^{b}	0.93 ± 0.12 ^{a}	0.96 ± 0.11 ^{a}	0.95 ± 0.09 ^{a}
STR (%)	71.4 ± 1.3	72.4 ± 1.7	72.9 ± 2.0	72.8 ± 1.3	73.0 ± 1.5	73.6 ± 1.6
LIN (%)	30.0 ± 0.6	31.4 ± 1.1	31.1 ± 1.3	30.0 ± 1.1	30.8 ± 1.5	30.4 ± 1.3
Viability (%)	79.2 ± 2.0	79.6 ± 1.0	73.1 ± 3.3	78.9 ± 1.2	79.6 ± 1.8	79.6 ± 1.4
Intact acrosome (%)	74.8 ± 1.4	77.1 ± 0.9	73.0 ± 3.0	77.1 ± 0.8	77.8 ± 1.5	78.1 ± 1.3
MMP (%)	64.4 ± 2.5 ^{a}	66.0 ± 2.7 ^{a}	51.0 ± 6.7 ^{b}	64.2 ± 3.5 ^{a}	64.6 ± 3.0 ^{a}	67.9 ± 1.7 ^{a}

* Combination of BiF2_5K7K and BTS without antibiotics. ^{a,b,c} Significant difference among groups at the sam incubation time (p < 0.05). ABO: antibiotic; BTS: Beltsville Thawing Solution; MOT: total motility; PMO progressive motility; VCL: curvilinear velocity; VSL: straight-line velocity; VAP: average pathway velocity; ALI amplitude of lateral head displacement; STR: straightness; LIN: linearity; MMP: sperm with high mitochondria membrane potential.

2.4. Scanning Electron Microscopy

The sperm morphology evaluation, conducted using scanning electron microscopy (SEM), is presented in Figure 1. The sperm morphology in the positive control group (BTS with antibiotic) revealed normal morphology (Figure 1C,D), whereas abnormal sperm morphology, including membrane detachment, acrosomal damage, and sperm agglutination, as well as the attachment of bacteria on spermatozoa, was found in the negative control group (BTS without antibiotic) (Figure 1A,B). For the 62.5 μg/mL treatment groups (BTS plus BiF2_5K7K at different concentrations), the sperm morphology in these groups showed both normal and abnormal morphology (Figure 1E,F), with a lesser degree of abnormal morphology than in the negative control group (Figure 1E,F).

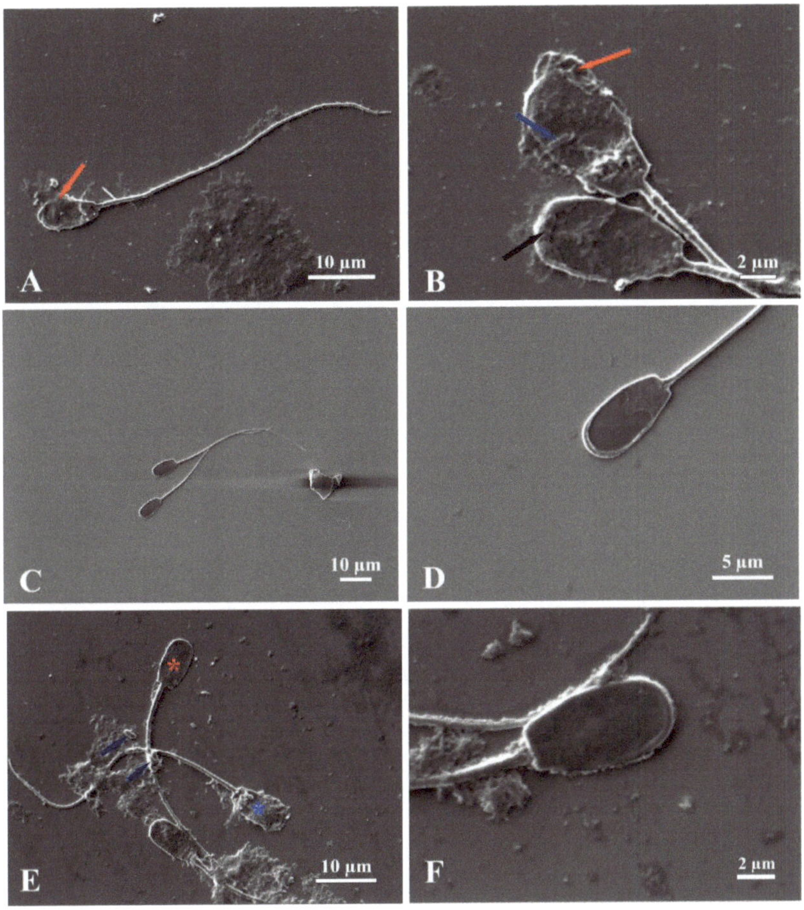

Figure 1. Scanning electron micrographs of boar sperm on day 3 after incubation with non-antibiotic BTS (**A**,**B**), BTS with antibiotic (**C**,**D**), and BiF2_5K7K at 62.5 μg/mL (**E**,**F**). The non-antibiotic BTS group (**A**,**B**) presented sperm with abnormal heads (red arrows), swelling acrosomes (black arrow), and rod bacteria (blue arrows). The BTS with antibiotic group (**C**,**D**) presented normal boar sperm. The BiF2_5K7K at 62.5 μg/mL group (**E**,**F**) showed both normal (red star) and abnormal boar sperm (blue star), and attracted bacteria (**E**) (blue arrow) as well as normal boar sperm at high magnification (**F**).

2.5. Fertility Test on the Pig Farm

For the fertility tests on the pig farm, the pregnancy rate, the percentage of pregnanc the percentage of farrowing rate, the total number of piglets born, the number of piglets bo alive, stillbirths, and mummies, as well as the litter birthweight, are presented in Table In the treatment group, BTS supplemented with BiF2_5K7K peptide at a concentration 31.25 µg/mL showed higher fertility results, such as pregnancy rate, farrowing rate, tot number of piglets born, and number of piglets born alive, than those of commercial bo semen extenders.

Table 7. Reproductive performance (mean ± SD) of sows inseminated using liquid stored semen wi commercial BTS with antibiotic (control) and BTS supplemented with 31.25 µg/mL of BiF2_5K7 peptide (treatment).

Parameters	Groups	
	Control (n = 20)	**Treatment (n = 20)**
Average parity	3.7 ± 0.4	3.7 ± 0.5
Pregnancy rate (%)	90.0 ± 0.3	100.0 ± 0.0
Farrowing rate (%)	80.0 ± 0.4	85.0 ± 0.4
Total number of piglets born	12.6 ± 3.0	14.1 ± 2.6
Number of piglets born alive	10.8 ± 3.1 [a]	13.1 ± 2.4 [b]
Stillborn piglets (%)	0.06 ± 0.25	0.17 ± 0.52
Mummified fetuses (%)	0.0 ± 0.0	0.0 ± 0.0
Litter birthweight (kg)	15.5 ± 4.9 [a]	18.8 ± 3.9 [b]

[a,b] Significant difference using Student's t-test between control and treatment groups ($p < 0.05$).

3. Discussion

The results of this study clearly show that BiF2_5K7K inhibits the growth of bot Gram-negative and Gram-positive bacteria isolated from boar semen and in extended boa semen. Most of the bacteria contaminating fresh boar semen were *E. coli*, *Pseudomona aeruginosa*, *Proteus mirabilis*, and *Staphylococcus* spp., which is in agreement with the mos common bacteria contaminated in fresh boar semen reported by previous studies [8,28–30 In contrast to Ngo et al. [31], Gram-positive bacteria such as *Staphylococcus* spp. were th most frequently contaminated in fresh boar semen. It has been documented that contan ination with *Pseudomonas aeruginosa* and *E. coli* negatively impacts boar sperm throug either causing sperm agglutination or decreasing sperm motility [5,29]. The presence o *E. coli* in boar semen prior to artificial insemination is primarily responsible for sow er dometritis and accounts for 72.3% of cases [5,32,33]. Clinical endometritis often present with vaginal discharge, which can be attributed to various factors, including hormona imbalance [34] or post-ovulatory insemination [32]. While a range of antibiotics can hel reduce the severity of acute endometritis, it is important to note that this condition ca worsen and develop into chronic endometritis, which can have a significant negative effec on pigs' reproductive performance [35]. BiF2_5K7K at concentrations of 15.625 and 31.2 showed an ability to inhibit bacteria isolated from boar semen for only 24 h; however, i did not show a deleterious effect on semen quality during storage for 3 days. As a result this peptide might be an alternative to antibiotics for supplementation into boar seme extenders in order to diminish the negative effect of bacterial contamination in fresh boa semen. In commercial semen extenders, antibiotics including amoxicillin, gentamicir neomycin, penicillin, and streptomycin are added and widely used to inhibit the growth o bacteria [12,13,29,36]. These antibiotics have also been routinely used for the treatment o bacterial infections via both injection and feed medication on many pig farms worldwide. I has been reported that bacteria isolated from boar semen [8] and diarrheic piglets [10] show high resistance to many antibiotics. This is in accordance with the present result, wher we found that some bacteria (i.e., *Enterobacter* spp., *Klebsiella* spp., *Proteus* spp., *Providenci* spp., and *Staphylococcus* spp.) had high levels of MIC and MBC. It is worth noting tha at concentrations of 250 µg/mL, BiF2_5K7K cannot inhibit *Proteus mirabilis*, whilst it ca inhibit *E. coli* and *Pseudomonas aeruginosa* at concentrations of 15.625 and 31.25 μg/mL, respectively. Similar outcomes were also reported when using synthetic cyclic hexapeptides c-WWW and c-WFW, which are unable to inhibit *Proteus* spp. [9,37]. Considering the results of the total bacterial count of fresh boar semen on day 0 and extended boar semen from days 1 to 5 in all groups, the BiF2_5K7K antimicrobial peptide at concentrations of 15.625, 31.25, 62.5, and 125 μg/mL is able to inhibit bacterial growth in extended boar semen stored at 18 °C for at least 24 h after incubation. The number of bacterial counts in those treatment groups was as low as log1.51 to log1.78 CFU/mL, respectively, when compared with the bacterial count of log2.35 CFU/mL in the negative control group. It has been reported that the total bacteria count in fresh boar semen should range from 22.40 to 188.20 ($\times 10^3$ CFU/mL) in order to optimize reproduction in pig farms [38]; moreover, a reduction in sperm viability of 6.4% has been documented, which corresponds to every log10 increase in the total bacterial count [31]. Previous research has indicated that the quality of boar semen is affected by the bacteria count. It has been reported that sperm viability decreases with an *E. coli* concentration of approximately 10^3 CFU/mL [39]. In addition, boar semen contaminated with an *E. coli* concentration greater than 3.5×10^3 CFU/mL resulted in inferior reproductive performance by reducing litter size in pig farms [35]. Fertilizing ability was also found to have decreased by 10^4–10^6 CFU/mL in boar semen contaminated with *Pseudomonas aeruginosa* [40]. Moreover, in a study of the effects of anaerobic bacteria, including *Clostridium perfringens*, the total motility of boar semen was reduced at a concentration of 10^7–10^8 CFU/mL [41]. Although, after storage for 36 h, the total bacterial count in treatment groups increased from log1.51 to log1.78 at 24 h to log0.85 to log3.82 CFU/mL, this total bacterial count was considerably lower than in the negative control group (log3.47 CFU/mL, BTS without antibiotic). This indicates that pig farmers can use the alternative BiF2_5K7K peptide as a replacement for antibiotics in semen extenders; however, this extended semen should be used for artificial insemination within 24 h and no later than 36 h in order to avoid deleterious effects from a high number of bacteria.

Considering semen quality from days 0 to 5 in different groups, despite the fact that the total bacterial count increased over time after storage of the extended boar semen at 18 °C, this negative effect on semen quality was not observed until day 3; in particular, a negative effect was observed in progressive motility values in a BiF2_5K7K concentration of 125 μg/mL. Only BiF2_5K7K at concentrations of 31.25 and 15.625 μg/mL was able to maintain all sperm parameters comparable to the positive control group (BTS with antibiotic). On day 5, a BiF2_5K7K concentration of 125 μg/mL showed lesser total motility, progressive motility, and mitochondrial membrane potential than other groups. This might be due to the fact that too high a concentration of antimicrobial peptide may cause deleterious effects on the spermatozoa [19]. During storage on days 3 and 5, besides the negative effect found for sperm parameters, as mentioned above, the sperm morphology determined via scanning electron microscopy revealed that most of the plasma membrane damage around the head and acrosome region was found in the BTS without an antibiotic group, which may have been caused by a high number of bacteria, as described earlier by Bonet et al. [42]. It is worth noting that the ability to inhibit bacterial growth without damaging spermatozoa is considered an imperative property of the antimicrobial peptide for supplementation in boar semen extenders [9,19,43]. It has been suggested that direct and rapid binding to the external bacterial cell wall, such as lipopolysaccharide (LPS) in Gram-negative bacteria or teichoic acid in Gram-positive bacteria, might be the mechanism through which antimicrobial peptides interact [22,44,45], due to the fact that there is a difference in charge between the bacterial and animal membranes. Although the total net charge of the sperm is negative on the sperm head, the sperm head position has a positive charge [46]. Consequently, the possibility of interacting with AMPs is lower; as a result, antimicrobial peptides act on the bacterial membrane rather than the sperm membrane [41,47]. The positive charge AMPs have strong interactions with the negative charge on the outermost bacterial cell surface due to the presence of lipopolysaccharides or teichoic acid [45,47–49]; however, the AMPs have weak interactions with the sperm membrane, which has a negative charge in the innermost region near the cytoplasm [11,44]. It has previously been reported th certain antimicrobial peptides such as Nisin and Protegrine 1 (PG 1) have a detriment effect on sperm [50,51]. Nisin has shown a prompt effect on spermicidal activities ar immobilization of spermatozoa within 20 s; moreover, the toxic dose of Nisin varies fro 50–400 μg depending on the animal species [50]. PG 1 has also shown a negative effe on sperm motility and viability, although it has demonstrated compromised antibacteri activity when compared with the antibiotic group [51]. Even though antimicrobial peptid have shown greater antimicrobial activity at high concentrations than at low concentratio they can also damage boar spermatozoa during storage at 17 °C, as reported by Shaoyo et al. [52]. A combination of antimicrobial peptides and antibiotics is occasionally use for reducing antibiotics and reducing the negative effect of too high a concentration antimicrobial peptide on boar sperm. It has been shown that a combination of 0.16 g/L epsilon-polylysine (ε-PL) and 0.125 g/L of gentamycin provides sperm quality equivale to adding 0.25 g/L of gentamycin alone in liquid-stored pig semen [52]. Neverthele utilizing BiF2_5K7K as a semen extender at a concentration of 31.25 μg/mL may lead increased production expenses. Hence, further investigation may be necessary to dimini the concentration of BiF2_5K7K, such as by combining minimal amounts of antibiotics wi BiF2_5K7K or using two or more AMPs as a cocktail peptide. Moreover, the use of a sing antimicrobial peptide, an antimicrobial peptide combined with a commercial antibiotic, a combination of antimicrobial peptides in order to cope with multidrug-resistant bacter has also been reported [53]. The short-term semen extender (BTS) utilized in the prese study can preserve semen quality for a maximum of three days after dilution [54]. Howev it may be stored for as long as five days [55]. For the reasons mentioned above, this stu evaluated the sperm quality at days 0, 1, 3, and 5 after storage to ensure sure the BT maintained the sperm quality ensured by the manufacturer. The total bacterial count wa determined after 0, 24, 36, 48, and 72 h of storage, as the bacteria increased rapidly durin storage and significantly after 72 h [2,5,56]. It is also important to emphasize that extende boar semen was generally utilized by the pig farms within 24 h of storage. Consequentl the present experimental design corresponded to standard clinical practice.

In the present study, the antimicrobial peptide BiF2_5K7K showed its effectivene against both Gram-positive and Gram-negative bacteria isolated from fresh boar semen an sow vaginal discharge. This antimicrobial peptide not only has an effect on antibacteri activity, but also causes less damage to boar sperm during storage at 17 °C compared wi antibiotics. In this study, the toxic impact of BiF2_5K7K was observed at high concentratio (62.5 and 125 μg/mL) and varied based on the incubation time during storage in bo sperm. Considering the sperm quality and the total bacterial count in each treatment grou it seems likely that BiF2_5K7K at concentrations of 15.625 and 31.25 μg/mL are the optim doses to replace antibiotics in boar semen extenders. For the fertility test, it has bee demonstrated in bovine sperm that beta-defensin 126 improves sperm motility but do not promote the fertilizing ability during in vitro tests [57]. BiF2_5K7K at a concentration 31.25 μg/mL was selected to test the impact on reproductive performance in relation to th standard farm condition (BTS with antibiotic) because of the bacterial inhibitory effect aft incubation at 24 h (log1.51 ± 0.29 CFU/mL) and the less negative effect on semen qualit The reproductive performance parameter for the above experiment was based on the wor of Koketsu et al. [58]. In agreement with the present results, the fertility test conducted on pig farm using BiF2_5K7K supplemented in boar semen extenders for artificial inseminatio showed a superior pregnancy rate, farrowing rate, total number of piglets born, and numb of piglets born alive compared with the commercial semen extender used at the pig farr BiF2_5K7K has the ability to inhibit bacterial growth without affecting the efficiency reproductive ability, which is no different from the use of BTS with antibiotics. The extende boar semen in the present study, which contained BiF2_5K7K, may be responsible for th positive effects observed on various reproductive parameters, including the increase i the number of piglets born alive, the improvement in the farrowing rate and pregnanc rate, and other related factors. Antimicrobial peptides, such as β-defensin, cathelicidi

PMAP23, and PMAT37, have been observed to be present in the endometrium of female pigs during the reproductive cycle and in the placental tissue of pregnant sows [59,60]. The mechanism underlying of the particular antimicrobial peptides could influence the number of piglets born need further studies. Therefore, BiF2_5K7K may be of interest to the swine industry in order to minimize the use of antibiotics in pig farms. Nevertheless, it is worth noted that field fertility test was performed by using 20 sows in each group, and as a result, a further experiment with a greater number of sows in each group may be needed.

4. Materials and Methods

4.1. Peptide Synthesis

The BiF2_5K7K antimicrobial peptide was inspired by natural AMPs, as previously described [61,62]. According to the 2020 study by Klubthawee et al. [62], peptide synthesis methods were used. As trifluoroacetate salts, the BiF2_5K7K components were purified using HPLC after being produced using solid-phase methods and 9-fluorenylmethoxycarbonyl (Fmoc) chemistry (ChinaPeptides, Shanghai, China). The content of residual TFA, quantified using 19F nuclear magnetic resonance (NMR), was less than 1.7% (wt/wt). Dehydration condensation was used for producing the TAMRA-labeled BiF2_5K7K, and an amide bond at the N-terminus was utilized for attaching TAMRA to BiF2_5K7K. Analytical reversed-phase HPLC determined that all of the peptides were more than 98% purified. Electrospray ionization mass spectrometry (ESI-MS) was used to identify the peptides.

The characteristics of BiF2_5K7K are presented in Table 8. BiF2_5K7K consisted of 12 amino acid sequence peptides with a molecular weight of 1541.07 g/mol. The net electric charge and hydrophobicity were +6 and 0.336, respectively [61].

Table 8. Physicochemical properties of the BiF2_5K7K peptide.

Peptide	Amino acid Sequence	Number of Amino Acids	Molecular Weight (g/mol)	Net Charge	Hydrophobicity	Percentage of Hydrophobic Residues
BiF2_5K7K	FLVKKIKKILRR	12	1541.07	+6	0.336	50%

4.2. Minimum Inhibitory Concentration (MIC) and Minimum Bactericidal Concentration (MBC) Test

The MIC assay of BiF2_5K7K was conducted using 12 bacteria isolated from boar semen and sow vaginal samples from the stock collection of the Bacterial Laboratory, Veterinary Diagnostic Center, Faculty of Veterinary Science, Mahidol University, Thailand. The bacterial stock was kept in glycerol at −80 °C. Bacteria from the culture stock were cultured on MacConkey agar (Difco, Reno, NV, USA) or sheep blood agar (Biomedia, Nonthaburi, Thailand), and then incubated for 18 to 24 h at 37 °C. After that, one to three bacterial colonies were transferred into a regular saline solution (0.85% NaCl) and thoroughly mixed. The turbidity of the bacterial sample was measured using a 0.5 McFarland standard, approximately 10^8 CFU/mL. Following guidelines from the Clinical and Laboratory Standards Institute (CLSI), the broth microdilution method was used to conduct the MIC assay. The assays were performed in triplicate using 96-well plates. In each well, 100 µL of the bacterial suspension, which had been diluted in Mueller–Hinton broth (Difco, USA) to 10^6 CFU/mL, was added to 100 µL of the appropriate dilutions of BiF2_5K7K at the selected concentrations using a two-fold dilution method (1.953–250 µg). As a control, a medium without BiF2_5K7K was used. After incubation, the MIC values were determined and defined as the lowest concentration of each BiF2_5K7K at which evidence of bacterial growth was absent. After the MIC assay, 100 µL aliquots from each well of bacterial growth representing the MIC values were streaked on the culture media agar plate and incubated for 18–24 h at 37 °C. The MBC value was defined as the lowest concentration of BiF2_5K7K at which nonbacterial colonies did not proliferate on the culture media agar.

4.3. Boar Semen Collection and Preparation

Ten individual adult boars of ages ranging from 1.5 to 3 years were chosen for seme collection. The gloved-hand method was used to collect sperm from each boar. Semen w filtered through gauze, and only sperm-rich fractions were collected during the collectic process. The sperm motility, concentration, percentage of viability, intact acrosomes, m tochondrial membrane potential, osmolality, and total bacterial concentration of the fres semen were measured after collection. Only semen ejaculates with progressive motili values of more than 70% and concentrations of more than 100 × 10^6 spermatozoa/mL we included in the experiment.

As shown in Table 9, the fresh boar semen was divided into 6 groups via dilution wi Beltsville Thawing Solution (BTS; Minitube, Tiefenbach, Germany), BTS with antibiot (Minitube, Tiefenbach, Germany), and BTS without antibiotic plus various concentratior of BiF2_5K7K. Each group's sperm concentration was 4.5 × 10^9 spermatozoa/100 mL. Th diluted semen samples were incubated at 18 °C until evaluation. After incubation, the tot bacterial concentration was assessed at 0, 24, 36, 48, and 72 h. The quality of sperm w evaluated on days 1, 3, and 5 after storage.

Table 9. Group of experiments with varying BiF2_5K7K concentrations.

Group	Antimicrobial Peptide	Concentration (μg/mL)
Group 1	Negative control (BTS)	-
Group 2	Positive control (BTS with gentamicin)	-
Group 3	BiF2_5K7K *	125
Group 4	BiF2_5K7K *	62.50
Group 5	BiF2_5K7K *	31.25
Group 6	BiF2_5K7K *	15.625

* Combination of BiF2_5K7K and BTS without antibiotics.

4.4. Total Bacterial Count

The spread plate technique was employed to ascertain the total bacterial *count* subs quent to the incubation of a boar semen sample at 18 °C. The semen samples were subjecte to ten-fold dilution with normal saline solution (0.85% NaCl). One hundred microliter (μL) of each semen sample dilution were evenly distributed on Plate Count Agar (PCA (Difco, Nevada, USA) and incubated at 37 °C. After 48 h of incubation, the colonies wer enumerated and converted into colony-forming units per milliliter (CFU/mL).

4.5. Sperm Parameter Analysis

4.5.1. Sperm Motility

Computer-assisted sperm motility analysis (CASA) was used to examine sperm moti ity (AndroVision®, Minitube, Tiefenbach, Germany). In brief, 3 μL of extended seme was pipetted into a pre-warmed counting chamber (Leja®, IMV Technologies, L'Aigl Basse-Normandie, France) and then immediately measured using CASA software (REF 12500/0000). Five fields of each sample were evaluated, and at least 600 cells were counte per analysis. The analysis results expressed the percentages of motile sperm and progre sive motile sperm, as well as motility patterns including curvilinear velocity (VCL, μm/s average pathway velocity (VAP, mm/s), straight-line velocity (VSL, mm/s), amplitude o lateral head displacement (ALH, mm), straightness (STR; VSL/VAP, %), and linearity (LIN VSL/VCL, %) [63,64].

4.5.2. Sperm Viability

The viability of the sperm was examined using Ethidium homodimer-1 (EthD-1, E116 Invitrogen, Waltham, MA, USA) and SYBR-14 (Sperm viability kit, Molecular probes, L701 Thermo Fisher Scientific, Waltham, MA, USA). SYBR-14 (0.54 μM in DMSO) and EthD- (1.17 μM in PBS) were combined with an aliquot of 10 μL of the semen samples and the incubated at 37 °C for 15 min. After incubation, 5 µL of the processed sample was placed onto a pre-warmed glass slide and covered with a coverslip. A total of 200 sperm were assessed under a fluorescence microscope at 1000× magnification and classified as live or dead sperm [63,64].

4.5.3. Sperm Acrosomal Integrity

The acrosomal integrity of the sperm was evaluated by using fluorescein isothiocyanate-labeled peanut (*Arachis hypogaea*) agglutinin (FITC-PNA) with EthD-1 staining. Next, 10 µL samples of the diluted semen were mixed with 10 µL of EthD-1 and incubated at 37 °C for 15 min. Five µL of the mixture was smeared on a glass slide and fixed with 95% ethanol for 30 s. Each glass slide was covered with 50 µL of FITC-PNA (diluted with PBS 1:10 v/v) and incubated in a moist chamber at 4 °C for 30 min. After incubation, each sample was rinsed with cold PBS and air-dried. A total of 200 sperm were assessed using a fluorescent microscope and classified as intact acrosomes or damaged acrosomes [54,63,65].

4.5.4. Sperm with High Mitochondrial Membrane Potential (MMP)

Fluorochrome 5,5′,6,6′-tetrachloro-1,1′,3,3′-tetraethylbenzimidazolyl-carbocyanine iodide (1.53 mM) (JC-1; T3168, Invitrogen, Waltham, MA USA) was used in the staining process to determine the mitochondrial membrane potential of the sperm. A sample of 50 µL of diluted semen was mixed with 3 µL of a 1.53 mM JC-1 solution and 3 µL of a 2.4 mM propidium iodide (PI) solution in DMSO. The mixture was then incubated for 10 min at 37 °C in a dark container. Two hundred sperm were analyzed and divided into groups according to their degree of mitochondrial membrane potential using a 400× magnification fluorescent microscope [63,66].

4.6. Scanning Electron Microscopy (SEM)

Sperm samples were subjected to evaluation for morphology under a scanning electron microscope by using the classical conventional procedure as follows: the semen samples were fixed with 2.5% glutaraldehyde (Electron Microscopy Sciences, Hatfield, UK) in PBS for 24 h. After fixation, a washing process with PBS was conducted for 15 min and repeated three times. The samples were then stained with 0.1% osmium tetroxide (Sigma-Aldrich, Darmstadt, Germany) for 1 h and washed three times for 15 min with PBS. In the dehydration step, the samples were dehydrated with a graded series of ethanol at concentrations of 70%, 80%, 90%, and 95% absolute ethanol. The semen samples were processed and then placed onto an SEM stub and coated with 50 nm platinum particles [42]. Finally, the sperm morphology was observed under the scanning electron microscope (JEOL, JSM-IT500LA, Tokyo, Japan).

4.7. Fertility Test on the Pig Farm

After weaning, estrus was detected twice a day by monitoring the vulva for swelling and redness, as well as by performing a back-pressure test when a boar was around [54]. All the sows were inseminated thrice with a conventional AI catheter at 12 h, 24 h, and 36 h after standing estrus with a dose of semen (boar of proven fertility). The semen dose contained 3×10^9 spermatozoa in 80 mL of BTS with gentamicin (control, $n = 20$) and BTS supplemented with 31.25 µg/mL of BiF2_5K7K peptide (treatment, $n = 20$), and was stored at 18 °C for no more than 24 h. The pregnancy tests were performed on days 23–24 of pregnancy via transabdominal ultrasonography, real-time B-mode (50STringa, sector probe with 5 MHz, ESAOTE Pie Medical, Maastricht, The Netherlands) [67]. The pregnancy rate; the percentage of farrowing rate; the total number of piglets born; and the number of piglets born alive, dead, and mummified, as well as litter birthweight, were recorded.

4.8. Statistical Analysis

For the MIC and MBC data, the descriptive statistic was applied. Using PASW Statistics for Windows, version 18.0 (SPSS Inc., Chicago, IL, USA), the statistical analysis was performed. The Shapiro–Wilk test was used to evaluate the data distribution, and t results showed a normal distribution ($p > 0.05$). The total bacteria count and the fertili data were presented as mean ± SD. The semen parameter data included total motili progressive motility, curvilinear velocity, straight-line velocity, average pathway veloci amplitude of lateral head displacement, straightness, and linearity, as well as sperm wi high mitochondrial membrane potential, and were presented as mean ± SEM. The bacter count and sperm parameter data analysis were performed using the one-way analysis variance (ANOVA) test, and mean values were compared using Duncan's test. Data c fertility were compared using the Student's t-test, and the Chi-square test was used f the pregnancy and farrowing rates. When dealing with non-normally distributed da the Mann–Whitney U (Wilcoxon's rank sum) test was applied. Statistical significance w determined at a p-value < 0.05.

5. Conclusions

In this study, the BiF2_5K7K antimicrobial peptide demonstrated the ability to inhil the growth of bacteria isolated from boar semen, and, thus, appears to be a worthy ternative to antibiotics in boar semen extenders. Nevertheless, the successful applicatic of this particular AMP depends on the concentration and incubation time during stc age. According to the present study's findings, adding BiF2_5K7K at a concentration 31.52 µg/mL in a BTS semen extender without antibiotics, with storage at 18 °C for 24 demonstrated the most effective bacterial-inhibitory effect. Furthermore, the 31.52 µg/m BiF2_5K7K concentration demonstrated the least harmful impact on boar semen paramete and revealed superior fertility when tested on a pig farm.

Author Contributions: K.K. (Krittika Keeratikunakorn) performed the experiments, collected ar analyzed the data, and wrote the first manuscript. R.A. conducted the antimicrobial peptide synthes designed and supervised the study, and edited the manuscript. P.C. performed the clinical tri and analysis and collected the data. N.N. designed and supervised the study, conceived the ide investigated the study, analyzed the data, and edited the manuscript. K.K. (Kampon Kaeok conceptualized and supervised the study and clinical trial, edited the manuscript, secured fundir and conducted project administration. All authors have read and agreed to the published version the manuscript.

Funding: This project is financially supported by the National Research Council of Thailand (NRC and Mahidol University (NRCT5-RSA63015-05).

Institutional Review Board Statement: This study was conducted in compliance with the ARRIV guidelines. The research ethics were approved by the Faculty of Veterinary Science, Mahidol Unive sity, Animal Care and Use Committee (FVS-MU-IACUC-Protocol No. MUVS-2021-10-41), animal u license No. U1-01281-2558.

Informed Consent Statement: Not applicable.

Data Availability Statement: The original contributions presented in the study are included in th article; further inquiries can be directed to the corresponding author.

Acknowledgments: The authors wish to express their appreciation to the Semen Laboratory, Ve erinary Diagnostic Center, Faculty of Veterinary Science, Mahidol University, and other supportiv staff for providing materials and kind assistance. The authors would like to thank Nawapol Udpua Chawalit Takoon, and Suwilai Chaveanghong, scientists of Mahidol University Frontier Researc Facility (MU-FRF), for their kind assistance, instrumental operation, and technical support durin the scanning electron microscopy work.

Conflicts of Interest: The authors declare no conflicts of interest.

References

1. Pezo, F.; Romero, F.; Zambrano, F.; Sanchez, R.S. Preservation of boar semen: An update. *Reprod. Domest. Anim.* **2019**, *54*, 423–434. [CrossRef] [PubMed]
2. Contreras, M.J.; Núñez-Montero, K.; Bruna, P.; García, M.; Leal, K.; Barrientos, L.; Weber, H. Bacteria and boar semen storage: Progress and challenges. *Antibiotics* **2022**, *11*, 1796. [CrossRef] [PubMed]
3. Maes, D.; Nauwynck, H.; Rijsselaere, T.; Mateusen, B.; Vyt, P.; de Kruif, A.; VanSoom, V. Diseases in swine transmitted by artificial insemination: An overview. *Theriogenology* **2008**, *70*, 1337–13345. [CrossRef] [PubMed]
4. Vickram, A.; Dhama, K.; Archana, K.; Parameswari, R.; Rameshpathy, M.; Iqbal, H.; Sridharan, T.B. Antimicrobial peptides in semen extenders: A valuable replacement option for antibiotics in cryopreservation- A prospective review. *J. Exp. Biol. Agric. Sci.* **2017**, *5*, 578–588.
5. Gączarzewicz, D.; Udała, J.; Piasecka, M.; Błaszczyk, B.; Stankiewicz, T. Bacterial contamination of boar semen and its relationship to sperm quality preserved in commercial extender containing gentamicin sulfate. *Pol. J. Vet. Sci.* **2016**, *19*, 451–459. [CrossRef] [PubMed]
6. Rodriguez, A.L.; Van Soom, A.; Arsenakis, I.; Maes, D. Boar management and semen handling factors affect the quality of boar extended semen. *Porc. Health Manag.* **2017**, *3*, 15. [CrossRef] [PubMed]
7. Bennemann, P.E.; Machado, S.A.; Girardini, L.K.; Sonalio, K.; Tonin, A.A. Bacterial contaminants and antimicrobial susceptibility profile of boar semen in Southern Brazil studs. *Rev. Mvz. Cordoba.* **2018**, *23*, 6637–6648.
8. Keeratikunakorn, K.; Kaewchomphunuch, T.; Kaeoket, K.; Ngamwongsatit, N. Antimicrobial activity of cell free supernatants from probiotics inhibits against pathogenic bacteria isolated from fresh boar semen. *Sci. Rep.* **2023**, *13*, 5995. [CrossRef]
9. Schulze, M.; Dathe, M.; Waberski, D.; Muller, K. Liquid storage of boar semen: Current and future perspectives on the use of cationic antimicrobial peptides to replace antibiotics in semen extenders. *Theriogenology* **2016**, *85*, 39–46. [CrossRef]
10. Nguyet, L.T.Y.; Keeratikunakorn, K.; Kaeoket, K.; Ngamwongsatit, N. Antibiotic resistant *Escherichia coli* from diarrheic piglets from pig farms in Thailand that harbor colistin-resistant *mcr* genes. *Sci. Rep.* **2022**, *12*, 9083. [CrossRef]
11. Kumar, R.; Ali, S.A.; Singh, S.K.; Bhushan, V.; Mathur, M.; Jamwal, S.; Mohanty, A.K.; Kaushik, K.J.; Kumar, S. Antimicrobial peptides in farm animals: An updated review on its diversity, function, modes of action and therapeutic prospects. *Vet. Sci.* **2020**, *7*, 206. [CrossRef] [PubMed]
12. Schulze, M.; Grobbel, M.; Riesenbeck, A.; Brüning, S.; Schaefer, J.; Jung, M.; Grossfeld, R. Dose rates of antimicrobial substances in boar semen preservation-time to establish new protocols. *Reprod. Domest. Anim.* **2017**, *52*, 397–402. [CrossRef] [PubMed]
13. Gadea, J. Review: Semen extenders used in the artificial insemination of swine. *Span. J. Agric. Res.* **2003**, *1*, 17–27. [CrossRef]
14. Morrell, J.M.; Wallgren, M. Alternatives to antibiotics in semen extenders: A review. *Pathogens* **2014**, *3*, 934–946. [CrossRef] [PubMed]
15. Santos, C.S.; Silva, A.R. Current and alternative trends in antibacterial agents used in mammalian semen technology. *Anim. Reprod.* **2020**, *17*, e20190111. [CrossRef] [PubMed]
16. Bryła, M.; Trzcińska, M. Quality and fertilizing capacity of boar spermatozoa during liquid storage in extender supplemented with different antibiotics. *Anim. Reprod. Sci.* **2015**, *163*, 157–163. [CrossRef] [PubMed]
17. Alaoui Mdarhri, H.; Benmessaoud, R.; Yacoubi, H.; Seffar, L.; Guennouni Assimi, H.; Hamam, M.; Boussettine, R.; Filali-Ansari, N.; Lahlou, F.A.; Diawara, I.; et al. Alternatives therapeutic approaches to conventional antibiotics: Advantages, limitations and potential application in medicine. *Antibiotics* **2022**, *11*, 1826. [CrossRef] [PubMed]
18. Zhang, L.J.; Gallo, R.L. Antimicrobial peptides. *Curr. Biol.* **2016**, *26*, R14–R19. [CrossRef]
19. Schulze, M.; Grobbel, M.; Muller, K.; Junkes, C.; Dathe, M.; Rudiger, K.; Jung, M. Challenges and limits using antimicrobial peptides in boar semen preservation. *Reprod. Domest. Anim.* **2015**, *50* (Suppl. S2), 5–10. [CrossRef]
20. Zhang, Q.Y.; Yan, Z.B.; Meng, Y.M.; Hong, Y.X.; Shao, G.; Ma, J.J.; Cheng, X.R.; Lui, J.; Kang, J.; FU, C.Y. Antimicrobial peptides: Mechanism of action, activity and clinical potential. *Mil. Med. Res.* **2021**, *8*, 48. [CrossRef]
21. Huan, Y.; Kong, Q.; Mou, H.; Yi, H. Antimicrobial peptides: Classification, design, application and research progress in multiple fields. *Front. Microbiol.* **2020**, *11*, 582779. [CrossRef]
22. Wang, S.; Zeng, X.F.; Yang, Q.; Qiao, S.Y. Antimicrobial peptides as potential alternatives to antibiotics in food animal industry. *Int. J. Mol. Sci.* **2016**, *17*, 630. [CrossRef]
23. Mahlapuu, M.; Hakansson, J.; Ringstad, L.; Bjorn, C. Antimicrobial peptides: An emerging category of therapeutic agents. *Front. Cell. Infect. Microbiol.* **2016**, *6*, 194. [CrossRef]
24. Sultana, A.; Luo, H.; Ramakrishna, S. Antimicrobial peptides and their applications in biomedical sector. *Antibiotics* **2021**, *10*, 1094. [CrossRef]
25. Björn, C.; Noppa, L.; Näslund Salomonsson, E.; Johansson, A.L.; Nilsson, E.; Mahlapuu, M.; Håkansson, J. Efficacy and safety profile of the novel antimicrobial peptide PXL150 in a mouse model of infected burn wounds. *Int. J. Antimicrob. Agents.* **2015**, *45*, 519–524. [CrossRef]
26. Xu, B.C.; Fu, J.; Zhu, L.Y.; Li, Z.; Wang, Y.Z.; Jin, M.L. Overall assessment of antimicrobial peptides in piglets: A set of meta-analyses. *Animal* **2020**, *14*, 2463–2471. [CrossRef]
27. Andersson, D.I.; Hughes, D.; Kubicek-Sutherland, J.Z. Mechanisms and consequences of bacterial resistance to antimicrobial peptides. *Drug Resist. Updat.* **2016**, *26*, 43–57. [CrossRef]

28. Okazaki, T.; Mihara, T.; Fujita, Y.; Yoshida, S.; Teshima, H.; Shimada, M. Polymyxin B neutralizes bacteria-released endotoxin ar improves the quality of boar sperm during liquid storage and cryopreservation. *Theriogenology* **2010**, *74*, 1691–1700. [CrossRe
29. Dalmutt, A.C.; Moreno, L.Z.; Gomes, V.T.M.; Cunha, M.P.V.; Barbsa, M.R.F.; Sato, M.I.Z.; Knöbl, T.; Pedroso, C.A.; Morenoet, A. Characterization of bacterial contaminants of boar semen: Identification by MALDI-TOF mass spectrometry and antimicrobi susceptibility profiling. *J. Appl. Anim. Res.* **2020**, *48*, 559–565. [CrossRef]
30. Kaewchomphunuch, T.; Charoenpichitnunt, T.; Thongbaiyai, V.; Ngamwongsatit, N.; Kaeoket, K. Cell-free culture supernatan of *Lactobacillus* spp. and *Pediococcus* spp. inhibit growth of pathogenic *Escherichia coli* isolated from pigs in Thailand. *BMC V Res.* **2022**, *18*, 60. [CrossRef] [PubMed]
31. Ngo, C.; Suwimonteerabutr, J.; Prapasarakul, N.; Morrell, M.J.; Tummaruk, P. Bacteriospermia and its antimicrobial resistan in relation to boar sperm quality during short-term storage with or without antibiotics in a tropical environment. *Porc. Heal Manag.* **2023**, *9*, 21. [CrossRef]
32. Grahofer, A.; Björkman, S.; Peltoniemi, O. Diagnosis of endometritis and cystitis in sows: Use of biomarkers. *J. Anim. Sci.* **2020**, (Suppl. S1), S107–S116. [CrossRef]
33. de Winter, P.; Verdoncka, M.; de Kruif, A.; Devriese, L.; Haesebrouck, F. Bacterial endometritis and vaginal discharge in the so Prevalence of different bacterial species and experimental reproduction of the syndrome. *Anim. Reprod. Sci.* **1995**, *37*, 325–3 [CrossRef]
34. Pluchino, N.; Freschi, L.; Wenger, J.M.; Streuli, I. Innovations in classical hormonal targets for endometriosis. *Expert Rev. Cl Pharmacol.* **2016**, *9*, 317–327. [CrossRef]
35. Maroto Martín, L.O.; Muñoz, E.C.; De Cupere, F.; Van Driessche, E.; Echemendia-Blanco, D.; Rodríguez, J.M.; Beeckmans, Bacterial contamination of boar semen affects the litter size. *Anim. Reprod. Sci.* **2010**, *120*, 95–104. [CrossRef]
36. Burch, D.G.S.; Sperling, D. Amoxicillin-current use in swine medicine. *J. Vet. Pharmacol. Ther.* **2018**, *41*, 356–368. [CrossRef]
37. Speck, S.; Courtiol, A.; Junkes, C.; Dathe, M.; Muller, K.; Schulze, M. Cationic Synthetic Peptides: Assessment of their antimicrobi potency in liquid preserved boar semen. *PLoS ONE* **2014**, *9*, e105949. [CrossRef]
38. Ciornei, Ş.; Drugociu, D.; Ciornei, L.M.; Mareş, M.; Roşca, P. Total aseptization of boar semen, to increase the biosecurity reproduction in swine. *Molecules* **2021**, *26*, 6183. [CrossRef]
39. Bussalleu, E.; Yeste, M.; Sepúlveda, L.; Torner, E.; Pinart, E.; Bonet, S. Effects of different concentrations of enterotoxigenic an verotoxigenic *E. coli* on boar sperm quality. *Anim. Reprod Sci.* **2011**, *127*, 176–182. [CrossRef]
40. Sepúlveda, L.; Bussalleu, E.; Yeste, M.; Bonet, S. Effect of *Pseudomonas aeruginosa* on sperm capacitation and protein phosphoryl tion of boar spermatozoa. *Theriogenology* **2016**, *85*, 1421–1431. [CrossRef]
41. Sepúlveda, L.; Bussalleu, E.; Yeste, M.; Torner, E.; Bonet, S. How do different concentrations of *Clostridium perfringens* affect th quality of extended boar spermatozoa? *Anim. Reprod. Sci.* **2013**, *140*, 83–91. [CrossRef]
42. Bonet, S.; Delgado-Bermúdez, A.; Yeste, M.; Pinart, E. Study of boar sperm interaction with *Escherichia coli* and *Clostridiu perfringens* in refrigerated semen. *Anim. Reprod. Sci.* **2018**, *197*, 134–144. [CrossRef]
43. Li, Z.; Hu, Y.H.; Yang, Y.Y.; Lu, Z.Q.; Wang, Y.Z. Antimicrobial resistance in livestock: Antimicrobial peptides provide a ne solution for a growing challenge. *Anim. Front.* **2018**, *8*, 21–29. [CrossRef]
44. Talapko, J.; Meštrović, T.; Juzbašić, M.; Tomas, M.; Erić, S.; Horvat Aleksijević, L.; Bekić, S.; Schwarz, D.; Matić, S.; Neuber M.; et al. Antimicrobial peptides-mechanisms of action, antimicrobial effects and clinical applications. *Antibiotics* **2022**, *11*, 141 [CrossRef]
45. Fazly Bazzaz, B.S.; Seyedi, S.; Hoseini Goki, N.; Khameneh, B. Human antimicrobial peptides: Spectrum, mode of action an resistance mechanisms. *Int. J. Pept. Res.* **2021**, *27*, 801–816. [CrossRef]
46. Magdanz, V.; Gebauer, J.; Sharan, P.; Eltoukhy, S.; Voigt, D.; Simmchen, J. Sperm-particle interactions and their prospects fc charge mapping. *Adv. Biosyst.* **2019**, *3*, e1900061. [CrossRef]
47. Keeratikunakorn, K.; Aunpad, R.; Ngamwongsatit, N.; Kaeoket, K. The effect of antimicrobial peptide (PA-13) on *Escherichia co* carrying antibiotic-resistant genes isolated from boar semen. *Antibiotics* **2024**, *13*, 138. [CrossRef]
48. Pushpanathan, M.; Gunasekaran, P.; Rajendhran, J. Antimicrobial peptides: Versatile biological properties. *Int. J. Pept.* **2013**, *201* 675391. [CrossRef] [PubMed]
49. Bechinger, B.; Gorr, S.U. Antimicrobial peptides: Mechanisms of action and resistance. *J. Dent. Res.* **2017**, *96*, 254–260. [CrossRe [PubMed]
50. Aranha, C.; Gupta, S.; Reddy, K.V. Contraceptive efficacy of antimicrobial peptide Nisin: In vitro and in vivo studies. *Contraceptio* **2004**, *69*, 333–338. [CrossRef] [PubMed]
51. Sancho, S.; Briz, M.; Yeste, M.; Bonet, S.; Bussalleu, E. Effects of the antimicrobial peptide protegrine 1 on sperm viability an bacterial load of boar seminal doses. *Reprod. Domest. Anim.* **2017**, *52* (Suppl. S4), 69–71. [CrossRef]
52. Shaoyong, W.; Li, Q.; Ren, Z.Q.; Wei, C.S.; Chu, G.Y.; Dong, W.Z.; Yang, G.S.; Pang, W.J. Evaluation of ε-polylysine as antimicrobia alternative for liquid-stored boar semen. *Theriogenology* **2019**, *130*, 146–156. [CrossRef]
53. Santos, C.; Rodrigues, G.R.; Lima, L.F.; dos Reis, M.C.G.; Cunha, N.B.; Dias, S.C.; Franco, O.L. Advances and perspectives fo antimicrobial peptide and combinatory therapies. *Front. bioeng. biotechnol.* **2022**, *1*, 1051456.
54. Chanapiwat, P.; Buranasinsup, S.; Kaeoket, K. Transformation of a short-term boar semen extender into a long-term boar seme extender by using penicillamine. *Czech J. Anim. Sci.* **2022**, *67*, 407–415. [CrossRef]

- Waterhouse, K.E.; De Angelis, P.M.; Haugan, T.; Paulenz, H.; Hofmo, P.O.; Farstad, W. Effects of in vitro storage time and semen-extender on membrane quality of boar sperm assessed by flow cytometry. *Theriogenology* **2004**, *62*, 1638–1651. [CrossRef]
- Balogun, K.B.; Stewart, K.R. Effects of air exposure and agitation on quality of stored boar semen samples. *Reprod. Domest. Anim.* **2021**, *56*, 1200–1208. [CrossRef]
- Fernandez-Fuertes, B.; Narciandi, F.; O'Farrelly, C.; Kelly, A.K.; Fair, S.; Meade, K.G.; Lonergan, P. Cauda epididymis-specific Beta-defensin 126 promotes sperm motility but not fertilizing ability in cattle. *Biol. Reprod.* **2016**, *95*, 122. [CrossRef] [PubMed]
- Koketsu, Y.; Tani, S.; Iida, R. Factors for improving reproductive performance of sows and herd productivity in commercial breeding herds. *Porc. Health. Manag.* **2017**, *3*, 1. [CrossRef]
- Lee, S.; Yoo, I.; Han, J.; Ka, H. Antimicrobial peptides cathelicidin, PMAP23, and PMAP37: Expression in the endometrium throughout the estrous cycle and at the maternal-conceptus interface during pregnancy and regulation by steroid hormones and calcitriol in pigs. *Theriogenology* **2021**, *160*, 1–9. [CrossRef] [PubMed]
- Lee, S.; Yoo, I.; Cheon, Y.; Hong, M.; Jeon, B.Y.; Ka, H. Antimicrobial peptides β defensin family: Expression and regulation in the endometrium during the estrous cycle and pregnancy in pigs. *Dev. Comp. Immunol.* **2023**, *139*, 104596. [CrossRef] [PubMed]
- Wongchai, M.; Wongkaewkhiaw, S.; Kanthawong, S.; Roytrakul, S.; Aunpad, R. Dual-function antimicrobial-antibiofilm peptide hybrid to tackle biofilm-forming *Staphylococcus epidermidis*. *Ann. clin. microbiol. antimicrob.* **2024**, *23*, 44. [CrossRef]
- Klubthawee, N.; Adisakwattana, P.; Hanpithakpong, W.; Somsri, S.; Aunpad, R. A novel, rationally designed, hybrid antimicrobial peptide, inspired by cathelicidin and aurein, exhibits membrane-active mechanisms against *Pseudomonas aeruginosa*. *Sci. Rep.* **2020**, *10*, 9117. [CrossRef] [PubMed]
- Kaeoket, K.; Chanapiwat, P. The beneficial effect of resveratrol on the quality of frozen-thawed boar sperm. *Animals* **2023**, *13*, 2829. [CrossRef]
- Henning, H.; Luther, A.M.; Höfner-Schmiing, L.; Waberski, D. Compensability of an enhanced incidence of spermatozoa with cytoplasmic droplets in boar semen for use in artificial insemination: A single cell approach. *Sci. Rep.* **2022**, *12*, 21833. [CrossRef]
- Chanapiwat, P.; Kaeoket, K. L-cysteine prolonged fresh boar semen qualities, but not for docosahexaenoic acid. *Czech J. Anim. Sci.* **2021**, *66*, 21–28. [CrossRef]
- Huo, L.J.; Ma, X.H.; Yang, Z.M. Assessment of sperm viability, mitochondrial activity, capacitation and acrosome intactness in extended boar semen during long-term storage. *Theriogenology* **2002**, *58*, 1349–1360. [CrossRef] [PubMed]
- Kaeoket, K.; Tantasuparuk, W.; Kunavongkrit, A. The effect of post-ovulatory insemination on the subsequent embryonic loss, oestrous cycle length and vaginal discharge in sows. *Reprod. Domest. Anim.* **2005**, *40*, 492–494. [CrossRef] [PubMed]

antibiotics

MDP

Article

The Effects of Different Antimicrobial Peptides (A-11 and AP1 on Isolated Bacteria from Fresh Boar Semen and Semen Qualit during Storage at 18 °C

Krittika Keeratikunakorn [1], Panida Chanapiwat [1], Ratchaneewan Aunpad [2], Natharin Ngamwongsatit [1,3] and Kampon Kaeoket [1,*]

1 Department of Clinical Sciences and Public Health, Faculty of Veterinary Science, Mahidol University, 999 Phuttamonthon 4 Rd., Salaya, Phuttamonthon, Nakhon Pathom 73170, Thailand; krittika.ker@student.mahidol.edu (K.K.); panida.chn@mahidol.edu (P.C.); natharin.nga@mahidol.edu (N.

2 Graduate Program in Biomedical Sciences, Faculty of Allied Health Sciences, Thammasat University, Rangsit Campus, Klongluang, Pathum Thani 12120, Thailand; aunpad@gmail.com

3 Laboratory of Bacteria, Veterinary Diagnostic Center, Faculty of Veterinary Science, Mahidol University, 999 Phuttamonthon 4 Rd., Salaya, Phuttamonthon, Nakhon Pathom 73170, Thailand

* Correspondence: kampon.kae@mahidol.edu

Citation: Keeratikunakorn, K.; Chanapiwat, P.; Aunpad, R.; Ngamwongsatit, N.; Kaeoket, K. The Effects of Different Antimicrobial Peptides (A-11 and AP19) on Isolated Bacteria from Fresh Boar Semen and Semen Quality during Storage at 18 °C. *Antibiotics* **2024**, *13*, 489. https://doi.org/10.3390/antibiotics13060489

Academic Editors: Marisa Di Pietro and Piyush Baindara

Received: 4 May 2024
Revised: 22 May 2024
Accepted: 23 May 2024
Published: 24 May 2024

Correction Statement: This article has been republished with a minor change. The change does not affect the scientific content of the article and further details are available within the backmatter of the website version of this article.

Abstract: Antibiotic resistance (AMR) is a major public health concern. Antimicrobial peptid (AMPs) could be an alternative to conventional antibiotics. The purpose of this research was investigate the antimicrobial ability of the synthetic AMPs (i.e., A-11 and AP19) on the most frequen isolated bacteria in boar semen and their effect on extended boar semen quality during storage. V tested the antimicrobial effect of A-11 and AP19 at different concentrations and compared them wi gentamicin for inhibiting the growth of *E. coli*, *Pseudomonas aeruginosa* and *Proteus mirabilis* that we isolated from fresh boar semen. In order to evaluate the effect of AMP on semen qualities on da 0, 1, 3, and 5 after storage at 18 °C, seven fresh boar semen samples were collected, diluted wi semen extender with antibiotic (i.e., gentamicin at 200 µg/mL, positive control) or without (negati control), and semen extender contained only A-11 or AP19 at different concentrations (i.e., 62.5 31.25, and 15.625 µg/mL). The total bacterial count was also measured at 0, 24, 36, 48, and 72 h aft storage. Comparable to gentamicin, both A-11 and AP19 inhibited the growth of *E. coli*, *Pseudomon aeruginosa*, and *Proteus mirabilis* at 62.50, 31.25, and 15.625 µg/mL, respectively. Comparing the tot bacterial count at 0, 24, 36, 48 and 72 h after storage, the lowest total bacterial concentration was four in the positive control group ($p < 0.05$), and an inferior total bacterial concentration was found in th treatment groups than in the negative control. On day 1, there is a lower percentage of all sper parameters in the AP19 group at a concentration of 62.50 µg/mL compared with the other grou On day 3, the highest percentage of all sperm parameters was found in the positive control and A- at a concentration of 31.25 µg/mL compared with the other groups. The AP19 group at 62.5 µg/m constantly yielded inferior sperm parameters. On day 5, only A-11 at a concentration of 15.625 µg/m showed a total motility higher than 70%, which is comparable to the positive control. A-11 and AP showed antimicrobial activity against *E. coli*, *Pseudomonas aeruginosa* and *Proteus mirabilis* isolate from boar semen. Considering their effect on semen quality during storage, these antimicrobi peptides are an alternative to conventional antibiotics used in boar semen extenders. Neverthele the utilization of these particular antimicrobial peptides relied on the concentration and duratio of storage.

Keywords: antimicrobial peptides; boar semen; semen quality

1. Introduction

Artificial insemination (AI) has been used as assisted reproductive technology in th pig industry for many years [1]. Liquid boar semen preservation is commonly used fo

AI, and the semen extender must be provided for preserving semen [2,3]. The purpose of the semen extender is to protect sperm from cold shock, maintain pH and osmotic pressure, and inhibit bacterial growth, with the goal of preserving the longevity and quality of sperm [3]. Both Gram-positive and Gram-negative bacteria, such as *Streptococcus* spp., *Staphylococcus* spp., *E. coli*, *Klebsiella* spp., *Aeromonas* spp., *Pseudomonas* spp., *Proteus* spp., and *Providencia* spp. have been frequently found in fresh boar semen [4–7]. These abundances of bacteria are resident in boar's skin, hair, and preputial diverticulum and contaminated into fresh boar semen during semen collection [8]. Bacteria contamination has several adverse impacts on the performance and quality of sperm as well as sow reproductive health [8]. In practice, many antibiotics are mixed into the semen extender to inhibit bacterial growth and limit the deleterious effect of this contamination [3,9,10]. Gentamicin, neomycin, streptomycin, and other antibiotics are commonly supplemented in boar semen extenders [3,11,12]. Further, more than one antibiotic is mixed with the boar semen extender, for example, gentamicin and polymyxin B or gentamicin and florfenicol, in order to inhibit both Gram-positive and Gram-negative bacteria [3,13]. Recently, it has been reported that the bacteria isolated from boar semen carried antibiotic resistance genes such as *mcr-3* and *int1* [7,14]. In addition, most bacteria from boar semen are prone to resistance to gentamicin and penicillin [15]. Antibiotic resistance is a worldwide problem owing to the overuse of unnecessary antibiotics in animals and humans, as well as the slow development of novel antibiotic discoveries [14].

Many studies have been performed on substitute strategies that can lower the usage of antibiotics in pig farms, including reducing or replacement the antibiotic supplementation in boar semen extenders. Antimicrobial peptides (AMPs) have been determined to be an alternative antimicrobial agent of interest, in which it showed compromised results for inhibiting *Escherichia coli* isolated from boar semen that carry antibiotic resistance genes [16]. To date, it has been documented that altogether 3257 AMPs were added to the Antimicrobial Peptide Database (APD) [17]. Most AMPs have been discovered and identified as antimicrobial agents, and can be applied for the treatment of antibiotic-resistant bacteria [16,18]. These include proline-rich antimicrobial peptides (PrAMPs), tryptophan- and arginine-rich antimicrobial peptides, histidine-rich antimicrobial peptides, and glycine-rich antimicrobial peptides [19,20]. The differences in the charge between the membranes of animals and bacteria can enable AMPs to become active through direct and rapid binding to the outer bacterial cell wall, such as lipopolysaccharide (LPS) in Gram-negative bacteria or teichoic acid in Gram-positive bacteria [14,21–23]. Additionally, the outermost surface of bacterial cells contains lipopolysaccharides, or teichoic acid [21,24,25]. The positive charge of AMPs strongly interacts with the negative charge there, but it has a weak interaction with the positively charged animal membrane [20–24]. More significantly, the key characteristic of AMPs is their capacity to kill bacteria without damaging the host cell [26]. Therefore, AMPs is an interesting choice to reduce or replace antibiotic usage in boar semen extender. A-11 and AP19 are two novel AMPs, when used in high concentrations, are not damaging animal cells and inhibiting the growth of both Gram-positive and Gram-negative bacteria, including *Salmonella enterica* serovar Typhimurium and *Acinetobacter baumannii* [27,28]. However, the application of these two peptides on the inhibition of bacteria isolated from boar semen has not been reported.

The purpose of this study was to determine the antimicrobial ability of A-11 and AP19 whether to inhibit the growth of most frequently found bacteria (i.e., *E. coli*, *Pseudomonas aeruginosa* and *Proteus mirabilis*) in boar semen and, subsequently, their effect on boar semen quality while being used as a replacement of antibiotics in boar semen extender.

2. Results

2.1. Bacterial Survival Assay

The growth curves of *E. coli*, *Pseudomonas aeruginosa*, and *Proteus mirabilis* are shown in Figures 1–3. Similar to gentamicin, A-11 and AP19 showed their ability to inhibit the growth of *E. coli* (Figure 1). After 12 h of growth, the OD_{600} values of *E. coli* in the gentamicin group increased, while A-11 (Figure 1A) and AP19 (Figure 1B) plateaue In addition, the OD_{600} value of *E. coli* at 14 h in the A-11 at 15.625 µg/mL was slight increased (Figure 1A). *Pseudomonas aeruginosa* was inhibited by gentamicin and AP1 (Figure 2B) throughout the investigation period. However, A-11 at 15.625 µg/mL wa able to inhibit *Pseudomonas aeruginosa* only for 10 h (Figure 2A). Similar to *E. coli* an *Pseudomonas aeruginosa*, *Proteus mirabilis* was inhibited by gentamicin and AP19 througho the investigation period (Figure 3B). However, A-11 at 15.625 µg/mL was able to inhib *Proteus mirabilis* only for 16 h (Figure 3A). The OD_{600} value of *Proteus mirabilis* at 21 h in tl A-11 at 31.25 µg/mL was also slightly increased (Figure 3A).

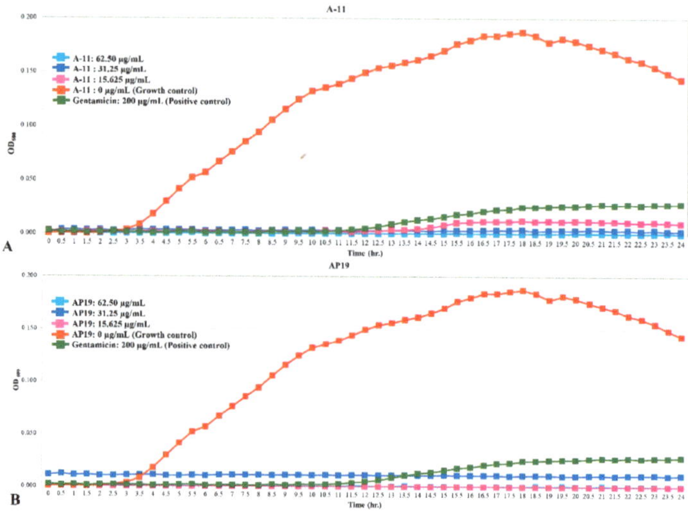

Figure 1. The growth curve of *E. coli* incubated with different concentrations of A-11 (**A**) and AP19 (**B**

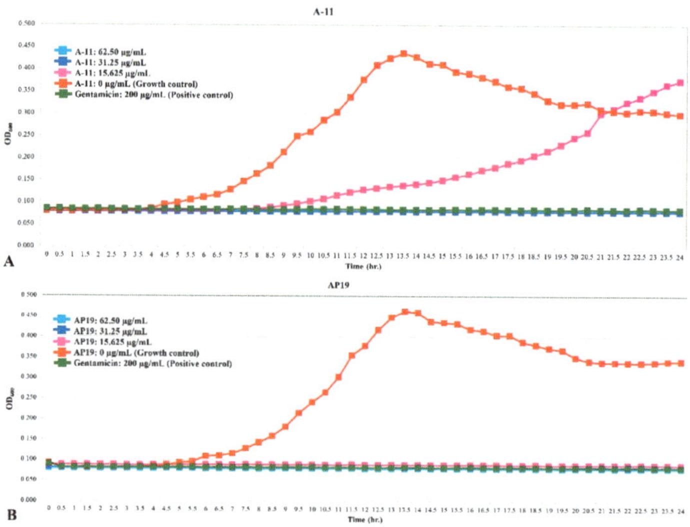

Figure 2. The growth curve of *Pseudomonas aeruginosa* incubated with different concentrations of A-1 (**A**) and AP19 (**B**).

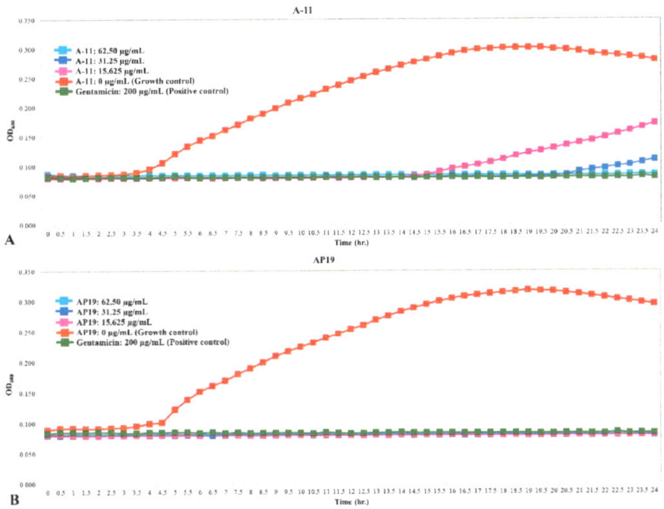

Figure 3. The growth curve of *Proteus mirabilis* incubated with different concentrations of A-11 (**A**) and AP19 (**B**).

2.2. Sperm Quality Parameters Analysis

The sperm quality of fresh boar semen samples is presented in Table 1. On day 1 (Table 2), the sperm quality parameters remained normal, and there was no significant difference in all sperm parameters among the 8 groups. However, there were inferior values for all sperm parameters, particularly viability and MMP, in AP19 at a concentration of 62.50 µg/mL than in other groups. On day 3 (Table 3), no significant difference was observed when compared all sperm parameters in AP19 at a concentration of 31.25 µg/mL with the positive control group. In addition, there was a significantly lower percentage for all sperm quality parameters of AP19 at a concentration of 62.50 µg/mL ($p < 0.05$). The A-11 and AP19 at the same concentration of 15.625 µg/mL yielded an acceptable percentage of more than 70 in terms of total motility, viability and intact acrosome. On day 5 (Table 4), only total motility, viability and intact acrosome parameters in the positive control, negative control and AP19 at a concentration of 31.25 µg/mL remained higher than 70%.

Table 1. Descriptive statistics for sperm parameters measurements of fresh boar semen (n = 7).

Parameters	Mean ± S.D.	Range
Concentration ($\times 10^6$ sperm/mL)	381.9 ± 72.4	292–495
Osmolality (mOsm/kg)	308.80 ± 3.69	305–316
Total motility (%)	90.3 ± 2.22	87.4–94.1
Progressive motility (%)	84.5 ± 3.2	79.7–88.3
Sperm viability (%)	88.2 ± 2.7	85–91
Intact acrosome (%)	88.2 ± 4.5	84–95
MMP (%)	86.1 ± 1.7	83–88
Total bacterial count ($\log_{10}$; CFU/mL)	2.36 ± 0.51	1.74–3.04

MMP: sperm with high mitochondrial membrane potential.

Table 2. Means ± SEM of semen quality parameters on day 1 after storage at 18 °C (n = 7).

Sperm Parameters	Groups							
	BTS	BTS + ABO	A-11 62.50 μg/mL	A-11 31.25 μg/mL	A-11 15.625 μg/mL	AP19 62.5 μg/mL	AP19 31.25 μg/mL	AP19 15.625 μg/mL
MOT (%)	86.1 ± 1.9	87.6 ± 2.3	87.0 ± 1.9	87.0 ± 1.1	85.3 ± 1.7	82.1 ± 3.1	85.1 ± 3.0	87.2 ± 1.7
PMOT (%)	79.7 ± 2.3	81.1 ± 3.1	80.0 ± 2.2	80.2 ± 1.8	77.8 ± 1.9	72.3 ± 4.2	78.5 ± 4.2	80.7 ± 2.0
VCL (μm/s)	98.3 ± 16.5	89.7 ± 4.3	90.8 ± 10.5	88.7 ± 4.9	84.2 ± 3.8	76.0 ± 5.1	82.7 ± 6.9	89.6 ± 3.8
VSL (μm/s)	25.3 ± 2.6	26.1 ± 2.0	25.9 ± 2.1	26.6 ± 2.9	23.5 ± 2.6	22.2 ± 3.7	26.1 ± 3.3	26.9 ± 1.8
VAP (μm/s)	37.6 ± 1.2	36.4 ± 1.9	35.9 ± 2.2	36.1 ± 3.2	32.2 ± 2.6	30.3 ± 2.3	34.4 ± 3.7	36.9 ± 1.9
ALH (μm)	0.91 ± 0.04	0.94 ± 0.06	0.92 ± 0.05	0.90 ± 0.03	0.87 ± 0.03	0.80 ± 0.05	0.85 ± 0.07	0.92 ± 0.04
BCF (Hz)	16.4 ± 1.2	16.9 ± 1.3	17.6 ± 0.9	17.5 ± 1.2	16.7 ± 1.1	15.9 ± 1.3	16.8 ± 1.3	17.4 ± 1.2
STR (%)	70.1 ± 3.8	71.1 ± 3.3	71.3 ± 2.7	72.7 ± 3.2	71.7 ± 2.9	72.4 ± 3.2	69.9 ± 3.9	72.5 ± 2.1
LIN (%)	29.4 ± 2.8	29.4 ± 2.5	28.6 ± 1.8	29.6 ± 2.3	27.6 ± 2.5	28.5 ± 2.7	32.8 ± 3.8	29.8 ± 1.6
Viability (%)	84.9 ± 1.5 ^{a,b}	86.4 ± 1.0 ^{b}	81.5 ± 0.8 ^{a,b}	85.6 ± 1.8 ^{a,b}	83.5 ± 0.7 ^{a,b}	80.7 ± 2.0 ^{a}	82.6 ± 2.2 ^{a,b}	83.2 ± 1.2 ^{a,b}
Intact acrosome (%)	81.3 ± 1.6	81.9 ± 1.6	80.1 ± 1.8	80.2 ± 1.8	78.1 ± 1.8	77.0 ± 2.2	79.1 ± 2.0	79.9 ± 1.1
MMP (%)	78.2 ± 1.0 ^{b}	78.6 ± 5.2.1 ^{b}	78.3 ± 1.7 ^{b}	80.0 ± 1.3 ^{b}	76.6 ± 1.6 ^{a,b}	72.4 ± 2.3 ^{a}	76.9 ± 2.3 ^{a,b}	77.9 ± 1.1 ^{a,b}

Values in each row marked with different superscript letters differ significantly (p-value < 0.05). ABO: antibiotic (gentamicin 200 μg/mL); BTS: Beltsville Thawing Solution; MOT: total motility; PMOT: progressive motility; VCL: curvilinear velocity; VSL: velocity straight line; VAP: average pathway velocity; ALH: amplitude of lateral head displacement; BCF: beat cross frequency, straightness; STR: straightness; LIN: linearity; MMP: sperm with high mitochondrial membrane potential.

Table 3. Means ± SEM of semen quality parameters on day 3 after storage at 18 °C (n = 7).

Sperm Parameters	Groups							
	BTS	BTS + ABO	A-11 62.50 μg/mL	A-11 31.25 μg/mL	A-11 15.625 μg/mL	AP19 62.5 μg/mL	AP19 31.25 μg/mL	AP19 15.625 μg/mL
MOT (%)	80.0 ± 2.2 ^{b}	80.0 ± 2.2 ^{b}	76.6 ± 4.0 ^{b}	82.0 ± 2.0 ^{b}	78.9 ± 4.0 ^{b}	60.9 ± 8.2 ^{a}	76.5 ± 4.0 ^{b}	79.1 ± 3.0 ^{b}
PMOT (%)	68.6 ± 2.9 ^{b}	72.0 ± 3.7 ^{b}	65.9 ± 5.5 ^{b}	71.8 ± 2.9 ^{b}	68.0 ± 4.8 ^{b}	48.8 ± 9.4 ^{a}	64.7 ± 4.7 ^{b}	67.8 ± 4.3 ^{b}
VCL (μm/s)	78.3 ± 3.1 ^{a}	83.7 ± 3.9 ^{a}	72.8 ± 6.7 ^{a,b}	78.1 ± 5.3 ^{a}	72.0 ± 7.2 ^{a,b}	57.4 ± 10.8 ^{b}	74.1 ± 8.3 ^{a,b}	77.5 ± 5.5 ^{a}
VSL (μm/s)	21.2 ± 1.6	21.8 ± 2.1	20.6 ± 2.3	21.4 ± 2.3	23.5 ± 2.6	15.3 ± 3.4	19.5 ± 3.5	20.6 ± 2.2
VAP (μm/s)	30.7 ± 2.1	30.7 ± 2.1	28.8 ± 2.9	30.1 ± 3.0	32.2 ± 2.6	21.5 ± 4.6	28.0 ± 4.4	29.4 ± 2.7
ALH (μm)	0.87 ± 0.03 ^{a,b}	0.90 ± 0.04 ^{a}	0.80 ± 0.07 ^{a,b}	0.82 ± 0.04 ^{a,b}	0.87 ± 0.03 ^{a,b}	0.65 ± 0.09 ^{b}	0.78 ± 0.06 ^{a,b}	0.82 ± 0.05 ^{a,b}
BCF (Hz)	13.7 ± 0.9 ^{a,b}	15.2 ± 0.9 ^{a}	13.2 ± 1.4 ^{a,b}	15.8 ± 1.2 ^{a}	16.7 ± 1.1 ^{a}	10.9 ± 2.0 ^{b}	14.4 ± 1.5 ^{a,b}	14.3 ± 1.2 ^{a,b}

Table 3. *Cont.*

Sperm Parameters	Groups							
	BTS	BTS + ABO	A-11 62.50 μg/mL	A-11 31.25 μg/mL	A-11 15.625 μg/mL	AP19 62.5 μg/mL	AP19 31.25 μg/mL	AP19 15.625 μg/mL
STR (%)	69.0 ± 1.7	68.0 ± 1.8	71.6 ± 1.8	70.7 ± 1.4	71.7 ± 2.9	69.3 ± 2.2	68.4 ± 2.3	69.5 ± 1.8
LIN (%)	26.9 ± 1.1	25.7 ± 1.4	28.1 ± 1.3	27.0 ± 1.4	27.6 ± 2.5	25.7 ± 1.8	25.1 ± 1.8	26.3 ± 1.7
Viability (%)	82.4 ± 0.7 ^{b}	82.4 ± 0.8 ^{b}	76.5 ± 2.7 ^{a,b}	81.0 ± 0.6 ^{b}	78.6 ± 2.7 ^{b}	69.8 ± 5.0 ^{a}	76.7 ± 2.5 ^{a,b}	79.4 ± 1.9 ^{b}
Intact acrosome (%)	77.9 ± 1.7	78.0 ± 1.4	76.8 ± 1.4	77.0 ± 1.4	76.6 ± 1.1	73.6 ± 2.0	76.1 ± 1.3	75.4 ± 1.1
MMP (%)	69.4 ± 1.3 ^{b}	70.5 ± 1.5 ^{b}	66.2 ± 2.3 ^{b}	70.3 ± 2.0 ^{b}	68.3 ± 3.0 ^{b}	53.5 ± 6.3 ^{a}	67.4 ± 2.0 ^{b}	68.3 ± 2.1 ^{b}

Values in each row marked with different superscript letters differ significantly (p-value < 0.05). ABO: antibiotic (gentamicin 200 μg/mL); BTS: Beltsville Thawing Solution; MOT: total motility; PMOT: progressive motility; VCL: curvilinear velocity; VSL: velocity straight line; VAP: average pathway velocity; ALH: amplitude of lateral head displacement; BCF: beat cross frequency, straightness; STR: straightness; LIN: linearity; MMP: sperm with high mitochondrial membrane potential.

Table 4. Means ± SEM of semen quality parameters on day 5 after storage at 18 °C (n = 7).

Sperm Parameters	Groups							
	BTS	BTS + ABO	A-11 62.50 μg/mL	A-11 31.25 μg/mL	A-11 15.625 μg/mL	AP19 62.5 μg/mL	AP19 31.25 μg/mL	AP19 15.625 μg/mL
MOT (%)	74.5 ± 4.6 ^{b}	77.4 ± 5.9 ^{b}	63.8 ± 7.5 ^{a,b}	67.1 ± 7.8 ^{a,b}	70.2 ± 5.9 ^{a,b}	48 ± 9.3 ^{a}	55.6 ± 8.0 ^{a,b}	60.7 ± 8.4 ^{a,b}
PMOT (%)	58.3 ± 7.0 ^{a,b}	64.0 ± 7.1 ^{a,b}	51.2 ± 7.5 ^{a,b}	54.4 ± 9.2 ^{a,b}	58.9 ± 7.7 ^{a,b}	36.8 ± 8.0 ^{a}	42.9 ± 8.1 ^{a,b}	47.5 ± 9.0 ^{a,b}
VCL (μm/s)	71.6 ± 9.4 ^{a}	81.3 ± 13.0 ^{a}	62.2 ± 9.5 ^{a,b}	63.9 ± 12.0 ^{a,b}	67.4 ± 10.3 ^{a,b}	43.6 ± 8.4 ^{b}	55.2 ± 9.8 ^{a,b}	57.7 ± 9.9 ^{a,b}
VSL (μm/s)	18.6 ± 3.8 ^{a,b}	23.3 ± 5.4 ^{a}	18.1 ± 3.6 ^{a,b}	17.6 ± 4.5 ^{a,b}	22.1 ± 4.3 ^{a}	11.6 ± 2.7 ^{b}	15.6 ± 3.6 ^{a,b}	15.5 ± 3.7 ^{a,b}
VAP (μm/s)	26.3 ± 4.8 ^{a,b}	33.0 ± 6.6 ^{a}	25.2 ± 4.6 ^{a,b}	25.0 ± 5.7 ^{a,b}	29.4 ± 5.4 ^{a,b}	17.3 ± 3.8 ^{b}	22.5 ± 4.9 ^{a,b}	23.1 ± 5.0 ^{a,b}
ALH (μm)	0.78 ± 0.09 ^{a,b}	0.86 ± 0.11 ^{a}	0.73 ± 0.09 ^{a,b}	0.73 ± 0.11 ^{a,b}	0.76 ± 0.11 ^{a,b}	0.55 ± 0.09 ^{b}	0.67 ± 0.10 ^{a,b}	0.71 ± 0.10 ^{a,b}
BCF (Hz)	11.9 ± 0.9 ^{a,b}	14.0 ± 1.3 ^{a}	9.5 ± 1.2 ^{b,c}	10.0 ± 1.6 ^{b,c}	12.7 ± 1.7 ^{a,b}	6.8 ± 1.4 ^{c}	8.3 ± 1.2 ^{b,c}	9.3 ± 1.6 ^{b,c}
STR (%)	69.6 ± 1.8	69.0 ± 3.2	71.0 ± 2.2	68.6 ± 2.2	73.5 ± 2.4	63.7 ± 3.0	68.0 ± 2.7	65.4 ± 3.2
LIN (%)	24.7 ± 1.9	27.1 ± 2.1	28.3 ± 1.7	25.6 ± 1.7	31.3 ± 3.4	24.1 ± 2.2	26.1 ± 2.2	24.8 ± 2.4
Viability (%)	77.6 ± 2.1 ^{a,b}	77.8 ± 2.3 ^{b}	68.0 ± 5.3 ^{a,b}	75.0 ± 2.9 ^{b}	73.9 ± 3.6 ^{b}	58.1 ± 7.3 ^{a}	65.0 ± 5.8 ^{a,b}	68.4 ± 6.0 ^{a,b}
Intact acrosome (%)	75.3 ± 1.9 ^{b}	75.0 ± 2.4 ^{b}	69.5 ± 2.3 ^{a,b}	70.7 ± 2.1 ^{a,b}	74.7 ± 1.8 ^{b}	65.3 ± 3.3 ^{a}	69.4 ± 3.8 ^{a,b}	72.0 ± 2.4 ^{a,b}
MMP (%)	58.7 ± 4.5 ^{a,b}	64.5 ± 2.8 ^{b}	57.0 ± 5.1 ^{a,b}	61.3 ± 3.9 ^{b}	64.3 ± 3.4 ^{b}	43.0 ± 6.7 ^{a}	51.0 ± 6.1 ^{a,b}	54.3 ± 5.8 ^{a,b}

Values in each row marked with different superscript letters differ significantly (p-value < 0.05). ABO: antibiotic (gentamicin 200 μg/mL); BTS: Beltsville Thawing Solution; MOT: total motility; PMOT: progressive motility; VCL: curvilinear velocity; VSL: velocity straight line; VAP: average pathway velocity; ALH: amplitude of lateral head displacement; BCF: beat cross frequency, straightness; STR: straightness; LIN: linearity; MMP: sperm with high mitochondrial membrane potential.

2.3. Total Bacterial Concentration

The mean total bacterial concentration of fresh boar semen was log2.36 ± 0.5 CFU/m (ranged log1.74 to log 3.04 CFU/mL) (Table 1). The total bacterial concentration of d luted semen at 0, 24, 36, 48, and 72 h after storage at 18 °C is shown in Table 5. The tot bacterial concentration increased as the incubation period was prolonged. At 0 h aft incubation, the highest total bacterial concentration was found in the negative contr group (log1.32 CFU/mL, BTS without antibiotic) when compared with other groups. Co paring among the AMP groups, AP19 at a concentration of 31.25 µg/mL showed a high bacterial concentration than the other groups (i.e., log0.78 CFU/mL). At 24 h after storag the total bacterial concentration of the positive control group (BTS with gentamicin) w still absent, while the concentration in other groups continued increasing. However, th negative control group (BTS without antibiotic) had the highest total bacterial concentratio (log2.38 CFU/mL), compared with the AMPs and positive control groups. At 36 h aft storage, the highest total bacteria concentration was found in the negative control grou (log3.11 CFU/mL, BTS without antibiotic) when compared with other groups. Howeve there was no significant difference in the bacterial concentration among the AMP grou which varied from log2.47 to 2.78 CFU/mL. At 48 h after storage, the lowest bacterial co centration among the AMP groups was found in AP19 at a concentration of 15.625 µg/m (log2.86 CFU/mL), while the bacterial concentrations in the negative and positive contr were log3.71 CFU/mL and absent, respectively. At 72 h after storage, the bacterial conce tration among the AMP groups varied from log4.54 to 4.96 CFU/mL, while the highe bacterial concentration was still found in the negative control group (i.e., log5.31 CFU/mI

Table 5. Total bacteria count (means ± SD) from boar semen samples (n = 7) after incubated at 18 °C

Groups	Concentrations (µg/mL)	Total Bacteria Concentration (log; CFU/mL)				
		Incubation Time				
		0 h	24 h	36 h	48 h	72 h
BTS	-	1.32 ± 0.27 ^{a}	2.38 ± 0.37 ^{a}	3.11 ± 0.79 ^{a}	3.71 ± 0.96 ^{a}	5.31 ± 1.44 ^{a}
BTS + ABO	-	0.00 ± 0.00 ^{b}	0.00 ± 0.00 ^{b}	0.00 ± 0.00 ^{b}	0.00 ± 0.00 ^{b}	0.00 ± 0.00 ^{b}
A-11	62.50	0.24 ± 0.16 ^{b,c}	1.72 ± 0.54 ^{a}	2.51 ± 0.91 ^{a}	3.50 ± 1.34 ^{a}	4.59 ± 1.58 ^{a}
A-11	31.25	0.39 ± 0.19 ^{b,c}	1.28 ± 0.53 ^{a,b}	2.78 ± 0.76 ^{a}	3.66 ± 1.23 ^{a}	4.78 ± 1.44 ^{a}
A-11	15.625	0.41 ± 0.20 ^{b,c}	1.32 ± 0.63 ^{a,b}	2.65 ± 0.85 ^{a}	3.57 ± 1.27 ^{a}	4.67 ± 1.50 ^{a}
AP19	62.50	0.49 ± 0.18 ^{b,c}	1.14 ± 0.54 ^{a,b}	2.59 ± 0.92 ^{a}	3.68 ± 1.17 ^{a}	4.96 ± 1.44 ^{a}
AP19	31.25	0.78 ± 0.16 ^{c}	1.08 ± 0.51 ^{a,b}	2.57 ± 0.70 ^{a}	3.63 ± 0.99 ^{a}	4.95 ± 1.22 ^{a}
AP19	15.625	0.34 ± 0.17 ^{b,c}	1.16 ± 0.55 ^{a,b}	2.47 ± 0.90 ^{a}	2.86 ± 1.35 ^{a}	4.54 ± 1.44 ^{a}

Values in each column marked with different superscript letters differ significantly ($p < 0.05$). ABO: antibiot (gentamicin 200 µg/mL); BTS: Beltsville Thawing Solution.

3. Discussion

The concentration of AMPs (A-11 and AP19) for this study came from the MIC valu (62.50–15.625 µg/mL) for inhibiting Gram-negative bacteria in the previous studies [27,28 and was further approved for inhibiting the most Gram-negative bacteria observed in fres semen by comparing with 200 µg/mL of gentamicin [7], which is the common antibioti mixed in boar semen extenders [13]. The results of the bacterial survival assay clearl showed the inhibitory effect of AMPs on bacterial growth in each stage of bacterial growt curve. The current findings regarding the total bacterial count clearly demonstrate tha the A-11 and AP19 peptides have the ability to inhibit bacterial growth for a minimum of 36 h when stored at 18 °C. During this period, the total bacterial concentration in al treatment groups remained below log2.80 CFU/mL (ranging from 2.47 to 2.78), in contras to the log3.11 CFU/mL observed in the negative control. Ciornei et al. [29] determined tha the normal range for the overall bacterial concentration in fresh boar semen is betwee 22.40 and 188.20 × 10^3 CFU/mL (equivalent to log4.35–5.27) for optimal reproductiv outcomes in pig farming. According to reports, there was a 6.4% reduction in sperm viability for every $\log_{10}$ increase in total bacterial concentration [30]. Furthermore, if boar semen was found to have a contamination level of *E. coli* exceeding 3.5×10^3 CFU/mL (log3.54), it led to a reduction in litter size and consequently had a negative impact on reproductive performance in pig farms [31]. The primary cause of sow endometritis or post-mating vaginal discharge is typically the presence of *E. coli* contamination in boar semen prior to artificial insemination [4,32,33]. This condition, known as acute endometritis, has the potential to progress into chronic endometritis, which can then have a negative impact on the reproductive performance of pigs [31]. In addition to *E. coli*, recent reports have indicated that *Pseudomonas aeruginosa* and *Proteus mirabilis* are the most common bacteria found in fresh boar semen [7]. In this study, it is noteworthy that A-11 and AP19 effectively inhibited the growth of contaminating bacteria in semen samples, regardless of the concentration of antimicrobial peptides used. Importantly, this inhibition did not have any adverse effects on the quality of the semen.

This study used a short-term semen extender (BTS), which has the ability to preserve semen quality of less than or equal to three days after dilution [34]. For the reasons mentioned, this study observed the sperm quality at days 0, 1, 3, and 5 during storage to ensure that the BTS still maintained sperm quality as claim by the manufacturer. While the bacteria growth during storage was rapid growth and significant growth after 72 h of storage, as a result, the total bacteria concentration was measured at 0, 24, 36, 48, and 72 h of storage [4,8,35]. In practice for the pig farms, it is also worth noting that they usually used extended boar semen within 24 h after storage. Consequently, the present experimental design was correspondent to those clinical practices. When examining semen qualities, specifically total motility and progressive motility, after being stored at 18 °C from days 0 to 5 in all groups, it was found that the total bacterial count increased over time. However, the negative impact on semen qualities was only observed when the semen extender was supplemented with a high concentration of A-11 and AP19 (62.50 µg/mL). The observed effect was evident on day 3 for AP19 and on day 5 for A-11. The semen extender used in this study is BTS based and specifically designed for short-term preservation of boar semen, with a recommended storage period of 3 days. After evaluating the semen quality on day 3 following storage, it was found that only A-11 at a concentration of 31.25 µg/mL produced semen quality similar to that of the positive control group. The present results of A-11 clearly showed that there is no sign of toxicity to sperm cells for all concentrations. This is in agreement with the hemolytic activity examination of A-11, which discovered that A-11 did not cause damage to red blood cells at concentrations between 0.98 and 250 µg/mL [27]. The underlying mechanism might be that in the outer membrane of animal cell (i.e., sperm cell) constituent of neutral components, subsequently the positively charge of AMP were not interaction with this cell [14,22].

Collectively, the antimicrobial peptides employed in this investigation demonstrate the capacity to impede bacterial proliferation within the initial 36 h period and sustain the quality of boar semen for a duration of 3 days. This phenomenon can be attributed to the interaction between positively charged antimicrobial peptides and the negatively charged teichoic acid or lipopolysaccharides present on the outermost membrane of bacterial cells [21,22,24]. The negative charge of the animal cell membrane is situated internally and in close proximity to the cytoplasm and in the outer membrane of were expressed neutral components. Consequently, the positively charged antimicrobial peptides do not interact with this cell [14,22,27]. Prior research has shown that the rupture of the *E. coli* membrane is triggered by the difference in the charge between animal and bacterial cell membranes. This allows active AMPs to exclusively bind to the bacterial membrane, leading to membrane dysfunction. This dysfunction is caused by the induction of membrane curvature, the formation of membrane pores, and ultimately the lysis of the bacterial cell [14,17,22–24]. At the optimal concentrations, AMPs caused damage to bacterial cell membranes. However, at lower concentrations, they moved into the cytoplasm and engaged in electrostatic interactions with bacterial DNA or ribosomes [36–38]. As stated by Schulze et al. [39], a high concentration of AMPs can have a detrimental effect on spermatozoa, which aligns with the findings of this study. The two AMPs examined in th study exhibited contrasting outcomes in terms of their ability to inhibit bacterial grow and preserve semen quality. These differences can potentially be attributed to their varyi hydrophobicity levels (A-11 = 44% and AP19 = 47%), which may also contribute to the toxicity towards sperm cells. Hydrophobicity plays a role in the effectiveness and specifici of AMPs in interacting with the target cell. This character facilitates the incorporation water-soluble AMPs into the lipid bilayer of the membrane. The activity and selectivity a substance are determined by its hydrophobicity. A high level of hydrophobicity can l harmful to the animal cell membrane and reduce antimicrobial activity [40,41]. In order prevent the use of excessively high concentrations of AMPs, it has been shown that combi ing antimicrobial peptides with antibiotics can reduce the negative effects on boar sper For instance, in liquid-stored boar semen, a combination of 0.16 g/L epsilon-polylysi (ε-PL) and 0.125 g/L gentamicin resulted in similar sperm quality compared to usi 0.25 g/L gentamicin alone. Studies have reported using a combination of two distin AMPs or a combination of an AMP and antibiotics to address the issue of multidru resistant bacteria [42]. However, it is crucial to note that one of the key features of th AMPs utilized in boar semen extenders is its ability to prevent bacterial growth witho damaging spermatozoa [39,43,44]. Additional research is required to examine the impa of A-11 and AP19 on farm fertility, specifically in relation to post-mating vaginal dischar pregnancy rate, farrowing rate, and litter size, before introducing these peptides into th pig industry.

4. Materials and Methods

4.1. Synthesis of Peptides and Their Physical-Chemical Analysis

The AMPs in this study were synthesized, determined for physicochemical prope ties (Table 6) and validated for inhibiting *Pseudomonas aeruginosa* isolated from huma clinical cases by Klubthawee et al. [45]. In brief, after being created utilizing solid-pha methods and 9-fuorenylmethoxycarbonyl (Fmoc) chemistry, the A-11 and AP19 peptid were synthesized by solid-phase techniques and purified as trifluoroacetate salts by HPL (ChinaPeptides, Shanghai, China). 19F nuclear magnetic resonance (NMR) revealed th there was less than 1.7% (wt/wt) of residual TFA present. The TAMRA-labeled antim crobial peptide was created via dehydration condensation, and TAMRA was bound antimicrobial peptide via an amide bond at the N-terminus. Reversed-phase HPLC analys revealed that more than 98% of the peptides were purified. Electrospray Ionization Ma Spectrometry (ESI-MS) was used to identify the peptides [45].

Table 6. Physicochemical properties of the A-11 and AP19 peptides.

Peptide	Amino Acid Sequence	Number of Amino Acids	Molecular Weight (g/mol)	Net Charge	Percentage of Hydrophobic Residues
A-11	WVKKVARKVVKIGRKVAR	18	2121.66	+8	44%
AP19	RLFRRVKKVAGKIAKRIWK	19	2353.94	+9	47%

4.2. Bacterial Strains and Culture Conditions

The bacteria in the present study were obtained from our previous report by Kee atikunakorn et al. [7] in which three species of the most frequently found bacteria in fres boar semen including *E. coli*, *Pseudomonas aeruginosa*, and *Proteus mirabilis* were isolate and kept in a culture collection at the Laboratory of Bacteria, Veterinary Diagnostic Ce ter, Faculty of Veterinary Science, Mahidol University (Salaya, Phuttamonthon, Nakho Pathom, Thailand). The bacteria *E. coli*, *Pseudomonas aeruginosa*, and *Proteus mirabilis* we all grown in a brain heart infusion (BHI, Difco, Reno, NV, USA) medium and incubate for 16–18 h at 37 °C. Pre-culture was performed by inoculating BHI broth with a sing isolated colony and then shaking it at 200 rpm for 16–18 h at 37 °C. Before being used, a 1% concentration of the pre-culture was added to the BHI broth and kept to grow at 37 °C.

4.3. Bacterial Survival Assay

The bacteria were cultivated in BHI broth before being moved to a normal saline solution (0.85% NaCl) to achieve the 0.5 McFarland standard (10^8 CFU/mL). A 500 µL bacterial suspension diluted to 10^6 CFU/mL in Mueller–Hinton broth (Difco™, Reno, NV, USA), was used in each well of the triplicate experiments, which used 48-well plates. This was mixed with 500 µL of appropriate antimicrobial peptide dilutions at the doses of 62.50, 31.25. 15.625, 0 µg/mL (growth control). A positive control, gentamicin 200 µg/mL was used. The OD_{600} values were measured every hour for a 24 h period at 37 °C using a microplate spectrophotometer (BMG LABTECH, SPECTROstar Nano, Ortenberg, Germany), and a growth curve was created [17].

4.4. Boar Semen Collection and Preparation

A semen sample was collected from each of the seven mature Duroc boars, theirs ages ranged from 1.5 to 3 years. Boar semen was collected using the gloved-hand technique [46]. Only the sperm-rich fractions of the semen were collected after it was filtered via gauze. The sperm motility, concentration, percentage of viability, intact acrosome, sperm with high mitochondrial membrane potential, osmolality and total bacterial concentration of the fresh semen were measured after collection [42]. Only semen ejaculates with a progressive motility of more than 70% and a concentration more than 100×10^6 spermatozoa/mL were included in the experiment [47].

As shown in Table 7, the fresh boar semen was divided into 8 groups and diluted with different semen extenders as follows: Beltsville Thawing Solution with 200 µg/mL of gentamicin (BTS; Minitube, Tiefenbach, Germany), BTS without antibiotic (Minitube, Tiefenbach, Germany), and BTS without antibiotic plus various concentrations of A-11 and AP19. The sperm concentration was adjusted to 3.0×10^9 sperm/100 mL. The diluted semen samples were stored in a digitally controlled refrigerator at 18 °C until evaluation. The total bacterial concentration was assessed at 0, 24, 36, 48, and 72 h after storage. The quality of sperm was evaluated on days 1, 3, and 5 after storage.

Table 7. The table shows the experimental and control groups by antimicrobial peptide type and concentration.

Group	Antimicrobial Peptides	Concentration (µg/mL)
1	BTS without gentamicin (negative control)	-
2	BTS with gentamicin 200 µg/mL (positive control)	-
3	A-11	62.50
4	A-11	31.25
5	A-11	15.625
6	AP19	62.50
7	AP19	31.25
8	AP19	15.625

4.5. Sperm Parameters Analysis

4.5.1. Total Motility and Progressive Motility

The sperm motility was analyzed by computer-assisted sperm motility analysis (CASA) (AndroVision®, Minitube, Tiefenbach, Germany). In summary, 3 µL of expanded semen was inserted into the counting chamber (Leja®, IMV Technologies, L'Aigle, Basse-Normandie, France) in which the temperature of glass slide and stage were set at 37 °C. The data were then recorded right away using the CASA. Each sample has five fields that are evaluated, and each analysis counts at least 600 cells. The percentage of motile sperm, progressive motile sperm, and motility patterns, including curvilinear velocity (VCL, µm/s), average pathway velocity (VAP, mm/s), straight-line velocity (VSL, mm/s), beat cross frequency straightness (BCF, Hz), amplitude of lateral head displacement (ALH, mn straightness (STR; VSL/VAP, %), and linearity (LIN; VSL/VCL, %) were expressed in tl analysis results [46].

4.5.2. Sperm Acrosomal Integrity

To assess acrosomal integrity, fluorescein isothiocyanate-labeled peanut (*Arachis h pogaea*) agglutinin (FITC-PNA) with staining was employed. A total of 10 μL of the seme sample was incubated with 10 μL of EthD-1 at 37 °C for 15 min A glass slide was covere with the mixture in 5 μL, air-dried, and fixed in 95% ethanol for 30 sec. Spread all throug the slide, 50 μL of diluted FITC-PNA (diluted with PBS 1:10 v/v) was incubated at 4 °C i a moist chamber for 30 min. Thereafter, the slide was washed with cold PBS and given air dry. Under a fluorescence microscope, 200 sperm were examined and separated int two groups: those with intact and damaged acrosomes [34,46,47].

4.5.3. Sperm Viability

SYBR-14 (L7011; Live/Dead™ Sperm viability kit, Invitrogen, Waltham, MA, US and Ethidiumhomodimer-1 (EthD-1, E1169, Invitrogen, Waltham, MA, USA) staining we performed to assess the sperm viability. The mixture contained 10 μL of semen sampl 2.7 μL of SYBR-14 (0.54 μM in DMSO), and 10 μL of EthD-1 (1.17 μM in PBS). The mixtu was incubated at 37 °C for 20 min. After incubation, 5 μL of the processed sample wa pipetted to the glass slide, and the coverslip was placed over it. A fluorescent microscop with a 1000× magnification was used to examine 200 sperm, which were then separate into live and dead sperm [46].

4.5.4. Sperm with High Mitochondrial Membrane Potential (MMP)

The membrane potential of the mitochondria was assessed by a staining approach wit fluorochrome5,5′,6,6′-tetrachloro-1,1′,3,3′-tetraethylbenzimidazolyl-carbocyanine iodid (JC-1; T3168, Invitrogen, Waltham, MA, USA). The mixture contained 50 μL of dilute semen samples, 3 μL of a 2.4 mM propidium iodide (PI) solution, and 3 μL of a 1.53 m JC-1 solution in DMSO and was then incubated at 37 °C for 10 min. Two hundred liv sperm (PI-negative) were examined using a 400× magnification fluorescent microscope an identified as having high mitochondrial membrane potential (yellow-orange fluorescenc and low mitochondrial membrane potential (green fluorescence) [46].

4.6. Total Bacterial Concentration

The spread plate technique was used to determine the total bacterial content followin to the incubation of the boar semen samples at 18 °C. Using a ten-fold dilution metho semen samples were diluted with 0.85% NaCl. Each dilution of a semen sample was spare 100 μL, and these were then cultured at 37 °C on plate count agar (PCA, DifcoTM, Ren NV, USA). After incubation, the colonies were enumerated and converted to CFU/mL a 48 h [48].

4.7. Statistical Analysis

The statistical analysis was performed by using PASW Statistics for Windows, versio 18.0 (SPSS Inc., Chicago, IL, USA). The normal distribution test of the data was evalu ated using the Shapiro–Wilk test. The total bacterial concentration was presented as th mean ± SD and semen parameters data were presented as the mean ± SEM. The bacteria concentration and sperm parameters data analysis were performed by using one-wa analysis of variance (ANOVA) and compared means by using Duncan's test. $p < 0.05$ wa considered statistically significant.

5. Conclusions

The antimicrobial peptides A-11 and AP19 demonstrated the capacity to inhibit th growth of *E. coli*, *Pseudomonas aeruginosa*, and *Proteus mirabilis* that were obtained from fres boar semen and extended boar semen stored at 18 °C. The potential of these peptides as an alternative to antibiotics in boar semen extenders is being remarked. Nevertheless, the utilization of these specific antimicrobial peptides relied on the concentration and duration of storage.

Author Contributions: K.K. (Krittika Keeratikunakorn) performed the experiments, collected and analyzed the data, and wrote the first manuscript. R.A. conducted antimicrobial peptide synthesis, designed and supervised this study and edited the manuscript. P.C. performed the semen quality analysis, collected data, and conducted analysis. N.N. designed and supervised this study, conceived the idea, investigated this study, analyzed data, and edited the manuscript. K.K. (Kampon Kaeoket) conceptualized and supervised this study, performed clinical trials, edited the manuscript, conducted project administration and secured funding. All authors have read and agreed to the published version of the manuscript.

Funding: This project is financially supported by the National Research Council of Thailand (NRCT) and Mahidol University (NRCT5-RSA63015-05), and the National Science, Research and Innovation Fund (NSRF) via the Program Management Unit for Competitiveness Enhancement (grant number C10F650150) and Pornchai Intertrade Co., Ltd.

Institutional Review Board Statement: This study was conducted in compliance with the ARRIVE guidelines. The research ethics were approved by the Faculty of Veterinary Science, Mahidol University-Institute Animal Care and Use Committee (FVS-MU-IACUC-Protocol No. MUVS-2021-10-41), Animal use license No. U1-01281-2558.

Informed Consent Statement: Not applicable.

Data Availability Statement: Data are contained within the article.

Acknowledgments: We really appreciated to Semen Laboratory, Veterinary Diagnostic Center, Faculty of Veterinary Science, Mahidol University and other supportive staff for providing us materials and kind assistance.

Conflicts of Interest: The authors declare no competing interests. The authors declare that this study received funding from Pornchai Intertrade Co., Ltd. The funder was not involved in the study design, collection, analysis, interpretation of data, the writing of this article or the decision to submit it for publication.

References

1. Waberski, D.; Luther, A.-M.; Grünther, B.; Jäkel, H.; Henning, H.; Vogel, C.; Peralta, W.; Weitze, K.F. Sperm function in vitro and fertility after antibiotic-free, hypothermic storage of liquid preserved boar semen. *Sci. Rep.* **2019**, *9*, 14748. [CrossRef] [PubMed]
2. Pezo, F.; Romero, F.; Zambrano, F.; Sanchez, R.S. Preservation of boar semen: An update. *Reprod. Domest. Anim.* **2019**, *54*, 423–434. [CrossRef] [PubMed]
3. Bustani, G.S.; Baiee, F.H. Semen extenders: An evaluative overview of preservative mechanisms of semen and semen extenders. *Vet. World.* **2021**, *4*, 1220–1233. [CrossRef] [PubMed]
4. Gączarzewicz, D.; Udała, J.; Piasecka, M.; Błaszczyk, B.; Stankiewicz, T. Bacterial contamination of boar semen and its relationship to sperm quality preserved in commercial extender containing gentamicin sulfate. *Pol. J. Vet. Sci.* **2016**, *19*, 451–459. [CrossRef] [PubMed]
5. Lopez Rodriguez, A.; Van Soom, A.; Arsenakis, I.; Maes, D. Boar management and semen handling factors affect the quality of boar extended semen. *Porc. Health Manag.* **2017**, *3*, 15. [CrossRef]
6. Bennemann, P.E.; Machado, S.A.; Girardini, L.K.; Sonalio, K.; Tonin, A.A. Bacterial contaminants and antimicrobial susceptibility profile of boar semen in Southern Brazil studs. *Rev. Mvz. Cordoba.* **2018**, *23*, 6637–6648.
7. Keeratikunakorn, K.; Kaewchomphunuch, T.; Kaeoket, K.; Ngamwongsatit, N. Antimicrobial activity of cell free supernatants from probiotics inhibits against pathogenic bacteria isolated from fresh boar semen. *Sci. Rep.* **2023**, *13*, 5995. [CrossRef] [PubMed]
8. Contreras, M.J.; Núñez-Montero, K.; Bruna, P.; García, M.; Leal, K.; Barrientos, L.; Weber, H. Bacteria and boar semen storage: Progress and challenges. *Antibiotics* **2022**, *11*, 1796. [CrossRef] [PubMed]
9. Schulze, M.; Grobbel, M.; Riesenbeck, A.; Brüning, S.; Schaefer, J.; Jung, M.; Grossfeld, R. Dose rates of antimicrobial substances in boar semen preservation-time to establish new protocols. *Reprod. Domest. Anim.* **2017**, *52*, 397–402. [CrossRef]
10. Vickram, A.S.; Kuldeep Dhama, K.D.; Archana, K.; Parameswari, R.; Pathy, M.R.; Iqbal, H.M.N.; Sridharan, T.B. Antimicrobial peptides in semen extenders: A valuable replacement option for antibiotics in cryopreservation—A prospective review. *J. Exp. Biol. Agric. Sci.* **2017**, *5*, 578–588.
11. Gadea, J. Review: Semen extenders used in the artificial insemination of swine. *Span. J. Agric. Res.* **2003**, *1*, 17–27. [CrossRef]

12. Santos, C.S.; Silva, A.R. Current and alternative trends in antibacterial agents used in mammalian semen technology. *An Reprod.* **2020**, *17*, e20190111. [CrossRef] [PubMed]
13. Bryła, M.; Trzcińska, M. Quality and fertilizing capacity of boar spermatozoa during liquid storage in extender supplement with different antibiotics. *Anim. Reprod. Sci.* **2015**, *163*, 157–163. [CrossRef] [PubMed]
14. Kumar, R.; Ali, S.A.; Singh, S.K.; Bhushan, V.; Mathur, M.; Jamwal, S.; Mohanty, A.K.; Kaushik, K.J.; Kumar, S. Antimicrob peptides in farm animals: An updated review on its diversity, function, modes of action and therapeutic prospects. *Vet. Sci.* **20** *7*, 206. [CrossRef] [PubMed]
15. Costinar, L.; Herman, V.; Pitoiu, E.; Iancu, I.; Degi, J.; Hulea, A.; Pascu, C. Boar semen contamination: Identification of Gra negative bacteria and antimicrobial resistance profile. *Animals* **2021**, *12*, 43. [CrossRef] [PubMed]
16. Keeratikunakorn, K.; Aunpad, R.; Ngamwongsatit, N.; Kaeoket, K. The effect of antimicrobial peptide (PA-13) on *Escherichia c* carrying antibiotic-resistant genes isolated from boar semen. *Antibiotics* **2024**, *13*, 138. [CrossRef] [PubMed]
17. Wang, G.; Zietz, C.M.; Mudgapalli, A.; Wang, S.; Wang, Z. The evolution of the antimicrobial peptide database over 18 yea Milestones and new features. *Protein Sci.* **2022**, *31*, 92–106. [CrossRef] [PubMed]
18. Xuan, J.; Feng, W.; Wang, J.; Wang, R.; Zhang, B.; Bo, L.; Chen, Z.-S.; Yang, H.; Sun, L. Antimicrobial peptides for combati drug-resistant bacterial infections. *Drug. Resist. Updat.* **2023**, *68*, 100954. [CrossRef] [PubMed]
19. Huan, Y.; Kong, Q.; Mou, H.; Yi, H. Antimicrobial peptides: Classification, design, application and research progress in multip fields. *Front. Microbiol.* **2020**, *11*, 582779. [CrossRef]
20. Zhang, Q.Y.; Yan, Z.B.; Meng, Y.M.; Hong, Y.X.; Shao, G.; Ma, J.J.; Fu, C.Y. Antimicrobial peptides: Mechanism of action, activi and clinical potential. *Mil. Med. Res.* **2021**, *8*, 48. [CrossRef]
21. Fazly Bazzaz, B.S.; Seyedi, S.; Hoseini Goki, N.; Khameneh, B. Human antimicrobial peptides: Spectrum, mode of action ar resistance mechanisms. *Int. J. Pept. Res.* **2021**, *27*, 801–816. [CrossRef]
22. Mahlapuu, M.; Håkansson, J.; Ringstad, L.; Björn, C. Antimicrobial peptides: An emerging category of therapeutic agents. *Fro Cell. Infect. Microbiol.* **2016**, *6*, 194. [CrossRef] [PubMed]
23. Islam, M.M.; Asif, F.; Zaman, S.U.; Arnab, M.K.H.; Rahman, M.M.; Hasan, M. Effect of charge on the antimicrobial activity alpha-helical amphibian antimicrobial peptide. *Curr. Res. Microb. Sci.* **2023**, *4*, 100182. [CrossRef] [PubMed]
24. Luo, Y.; Song, Y. Mechanism of Antimicrobial Peptides: Antimicrobial, Anti-Inflammatory and Antibiofilm Activities. *Int. J. M Sci.* **2021**, *22*, 11401. [CrossRef]
25. Talapko, J.; Meštrović, T.; Juzbašić, M.; Tomas, M.; Erić, S.; Horvat Aleksijević, L.; Bekić, S.; Schwarz, D.; Matić, S.; Neuberg, M et al. Antimicrobial Peptides-Mechanisms of Action, Antimicrobial Effects and Clinical Applications. *Antibiotics* **2022**, *11*, 14 [CrossRef] [PubMed]
26. Wang, G.; Mechesso, A.F. Realistic and critical review of the state of systemic antimicrobial peptides. *ADMET DMPK* **2022**, *1* 91–105. [CrossRef]
27. Sengkhui, S.; Klubthawee, N.; Aunpad, R. A novel designed membrane-active peptide for the control of foodborne *Salmonel enterica* serovar Typhimurium. *Sci. Rep.* **2023**, *13*, 3507. [CrossRef]
28. Jariyarattanarach, P.; Klubthawee, N.; Wongchai, M.; Roytrakul, S.; Aunpad, R. Novel D-form of hybrid peptide (D-AP19) rapid kills *Acinetobacter baumannii* while tolerating proteolytic enzymes. *Sci. Rep.* **2022**, *12*, 15852. [CrossRef]
29. Ciornei, Ş.; Drugociu, D.; Ciornei, L.M.; Mareş, M.; Roşca, P. Total aseptization of boar semen, to increase the biosecurity reproduction in swine. *Molecules* **2021**, *26*, 6183. [CrossRef]
30. Ngo, C.; Suwimonteerabutr, J.; Prapasarakul, N.; Morrell, M.J.; Tummaruk, P. Bacteriospermia and its antimicrobial resistanc in relation to boar sperm quality during short-term storage with or without antibiotics in a tropical environment. *Porc. Heal Manag.* **2023**, *9*, 21. [CrossRef]
31. Maroto Martín, L.O.; Muñoz, E.C.; De Cupere, F.; Van Driessche, E.; Echemendia-Blanco, D.; Rodríguez, J.M.M.; Beeckmans, Bacterial contamination of boar semen affects the litter size. *Anim. Reprod. Sci.* **2020**, *120*, 95–104. [CrossRef] [PubMed]
32. de Winter, P.; Verdoncka, M.; de Kruif, A.; Devriese, L.; Haesebrouck, F. Bacterial endometritis and vaginal discharge in the sov Prevalence of different bacterial species and experimental reproduction of the syndrome. *Anim. Reprod. Sci.* **1995**, *37*, 325–33 [CrossRef]
33. Kaeoket, K.; Tantasuparuk, W.; Kunavongkrit, A. The effect of post-ovulatory insemination on the subsequent embryonic los oestrous cycle length and vaginal discharge in sows. *Reprod. Domest. Anim.* **2005**, *40*, 492–494. [CrossRef] [PubMed]
34. Chanapiwat, P.; Buranasinsup, S.; Kaeoket, K. Transformation of a short-term boar semen extender into a long-term boar seme extender by using penicillamine. *Czech J. Anim. Sci.* **2022**, *67*, 407–415. [CrossRef]
35. Balogun, K.B.; Stewart, K.R. Effects of air exposure and agitation on quality of stored boar semen samples. *Reprod. Domest. Anir* **2021**, *56*, 1200–1208. [CrossRef] [PubMed]
36. Gottschalk, S.; Gottlieb, C.T.; Vestergaard, M.; Hansen, P.R.; Gram, L.; Ingmer, H.; Thomsen, L.E. Amphibian antimicrobi peptide fallaxin analogue FL9 affects virulence gene expression and DNA replication in *Staphylococcus aureus*. *J. Clin. Microbic* **2015**, *64*, 1504–1513. [CrossRef] [PubMed]
37. Polikanov, Y.S.; Aleksashin, N.A.; Beckert, B.; Wilson, D.N. The mechanisms of action of ribosome-targeting peptide antibiotic *Front. Mol. Biosci.* **2018**, *5*, 48. [CrossRef]
38. Vasilchenko, A.S.; Rogozhin, E.A. Sub-inhibitory effects of antimicrobial peptides. *Front. Microbiol.* **2019**, *10*, 1160. [CrossRef]

9. Schulze, M.; Grobbel, M.; Müller, K.; Junkes, C.; Dathe, M.; Rüdiger, K.; Jung, M. Challenges and limits using antimicrobial peptides in boar semen preservation. *Reprod. Domest. Anim.* **2015**, *50*, 5–10. [CrossRef]
0. Kumar, P.; Kizhakkedathu, J.N.; Straus, S.K. Antimicrobial Peptides: Diversity, mechanism of action and strategies to improve the activity and biocompatibility in vivo. *Biomolecules* **2018**, *8*, 4. [CrossRef]
1. Meier, S.; Ridgway, Z.M.; Picciano, A.L.; Caputo, G.A. Impacts of hydrophobic mismatch on antimicrobial peptide efficacy and bilayer permeabilization. *Antibiotics* **2023**, *12*, 1624. [CrossRef]
2. Shaoyong, W.K.; Li, Q.; Ren, Z.Q.; Xiao, J.Y.; Diao, Z.X.; Yang, G.S.; Pang, W. Effects of kojic acid on boar sperm quality and anti-bacterial activity during liquid preservation at 17 C. *Theriogenology* **2019**, *140*, 124–135. [CrossRef] [PubMed]
3. Schulze, M.; Dathe, M.; Waberski, D.; Muller, K. Liquid storage of boar semen: Current and future perspectives on the use of cationic antimicrobial peptides to replace antibiotics in semen extenders. *Theriogenology* **2016**, *85*, 39–46. [CrossRef]
4. Li, Z.; Hu, Y.H.; Yang, Y.Y.; Lu, Z.Q.; Wang, Y.Z. Antimicrobial resistance in livestock: Antimicrobial peptides provide a new solution for a growing challenge. *Anim. Front.* **2018**, *8*, 21–29. [CrossRef] [PubMed]
5. Klubthawee, N.; Adisakwattana, P.; Hanpithakpong, W.; Somsri, S.; Aunpad, R. A novel, rationally designed, hybrid antimicrobial peptide, inspired by cathelicidin and aurein, exhibits membrane-active mechanisms against *Pseudomonas aeruginosa*. *Sci. Rep.* **2020**, *10*, 9117. [CrossRef] [PubMed]
6. Kaeoket, K.; Chanapiwat, P. The beneficial effect of resveratrol on the quality of frozen-thawed boar sperm. *Animals* **2023**, *13*, 2829. [CrossRef] [PubMed]
7. Chanapiwat, P.; Kaeoket, K. L-cysteine prolonged fresh boar semen qualities, but not for docosahexaenoic acid. *Czech J. Anim. Sci.* **2021**, *66*, 21–28. [CrossRef]
8. Zeyad, A.; Hamad, M.; Amor, H.; Hammadeh, M.E. Relationships between bacteriospermia, DNA integrity, nuclear protamine alteration, sperm quality and ICSI outcome. *Reprod. Biol.* **2018**, *18*, 115–121. [CrossRef]

antibiotics

Article

Anti-Microbial Activities of Mussel-Derived Recombinant Proteins against Gram-Negative Bacteria

Dong Yun Kim [1], You Bin Oh [2], Je Seon Park [2], Yu-Hong Min [3] and Min Chul Park [1,*]

[1] College of Pharmacy, Inje Institute of Pharmaceutical Sciences and Research, Inje University, Gimhae 50832, Republic of Korea; kdy9903@gmail.com
[2] Department of Pharmaceutical Engineering, Inje University, Gimhae 50832, Republic of Korea; dhdbqls1208@gmail.com (Y.B.O.); dolphin3922@naver.com (J.S.P.)
[3] College of Health and Welfare, Daegu Hanny University, Gyeongsan 38610, Republic of Korea; yhmin@dhu.ac.kr
* Correspondence: mcpark@inje.ac.kr

Abstract: Many anti-microbial peptides (AMPs) and pro-apoptotic peptides are considered as nov anti-microbial agents, distinguished by their different characteristics. Nevertheless, AMPs exhib certain limitations, including poor stability and potential toxicity, which hinder their suitability f applications in pharmaceutics and medical devices. In this study, we used recombinant muss adhesive protein (MAP) as a robust scaffold to overcome these limitations associated with AMI Mussel adhesive protein fused with functional peptides (MAP-FPs) was used to evaluate an microbial activities, minimal inhibitory concentration (MIC), and time-kill kinetics (TKK) assay against six of bacteria strains. MAP and MAP-FPs were proved to have an anti-microbial effe with MIC of 4 or 8 μM against only Gram-negative bacteria strains. All tested MAP-FPs killed fo different Gram-negative bacteria strains within 180 min. Especially, MAP-FP-2 and -5 killed thre Gram-negative bacteria strain, including *E. coli*, *S. typhimurium*, and *K. pneumoniae*, within 10 min. cytotoxicity study using Vero and HEK293T cells indicated the safety of MAP and MAP-FP-2 and - Thermal stability of MAP-FP-2 was also validated by HPLC analysis at an accelerated condition f 4 weeks. This study identified that MAP-FPs have novel anti-microbial activity, inhibiting the growt and rapidly killing Gram-negative bacteria strains with high thermal stability and safety.

Keywords: recombinant anti-microbial protein; mussel adhesive protein; Gram-negative bacteri thermal stability

Citation: Kim, D.Y.; Oh, Y.B.; Park, J.S.; Min, Y.-H.; Park, M.C. Anti-Microbial Activities of Mussel-Derived Recombinant Proteins against Gram-Negative Bacteria. *Antibiotics* **2024**, *13*, 239. https://doi.org/10.3390/antibiotics13030239

Academic Editors: Piyush Baindara and Marisa Di Pietro

Received: 5 February 2024
Revised: 29 February 2024
Accepted: 3 March 2024
Published: 5 March 2024

1. Introduction

Infectious diseases, such as SARS-CoV-2, among other viruses, have gained majo attention owing to large-scale pandemics, which can severely affect human quality of lif and be exacerbated by antibiotic resistance [1,2]. A great need exists for effective antiv rals and antimicrobials to impede the transmission of infectious diseases [2,3]. Antivira and antimicrobial agents are commonly applied on safeguarding materials ranging fron consumer goods to medical devices to mitigate the risk of public outbreaks. Althoug many substances exhibit robust antiviral or antimicrobial activities, a prevailing concern i their unfavorable therapeutic index and resistance [4,5]. Throughout prokaryotes, plant and vertebrates, antimicrobial peptides (AMPs) are important biomolecules participatin in the innate immune response and host defense against infections [6–8]. AMPs (typica length of <200 amino acid residues) are secreted by host cells and act as bacteriostati or bactericidal agents [9]. They often exhibit a distinctive composition featuring cationi and hydrophobic residues that represent amphiphilic structures. Based on their compos tional and structural characteristics, AMPs interact with and disrupt the bacterial plasm membranes bearing negative charges in their outer layers [10,11]. Given their diversit and potential as antimicrobial agents, active research is being conducted on the use o

AMPs in the therapeutic, medical, and consumer industries. Some AMPs with strong and broad-spectrum antimicrobial activities have been developed for use as antimicrobial substances and antibiotics. FDA-approved AMP-derived antibiotics are used to treat various diseases, depending on their specific characteristics and antimicrobial spectrum. For example, daptomycin, a cyclic lipopeptide sourced from *Streptomyces roseosporus*, is used to treat *Staphylococcus aureus*-associated skin infections [12,13]. Vancomycin, a glycopeptide originating from *Amycolatopsis orientalis*, plays a pivotal role in combatting life-threatening Gram-positive bacterial infections, such as methicillin-resistant *S. aureus* (MRSA) [14]. Nonetheless, the practical application of AMPs is limited by their half-life, cytotoxicity, and production costs [15,16]. Representative AMP-derived antibiotics, including daptomycin, vancomycin, and telavancin, are susceptible to rapid structural degradation owing to instability and high cytotoxicity when administered at high doses, which is often required in clinical treatment. These limitations can be overcome by using recombinant proteins derived from natural materials as scaffolds.

Mussels have a remarkable ability to adhere to surfaces with great stability, even under harsh conditions, such as high pressures and salinity. MAPs are characterized by good biocompatibility and biodegradability, making them useful scaffolds for recombinant proteins [17–20]. Among the recombinant MAP scaffold proteins, fp-151 (derived from *Mytilus edulis*) is composed of six mussel foot protein 1 decapeptide repeats (fp1) at each mussel terminus of foot protein 5 (fp5) and commonly employed for its high yield, easy purification, and cost-effective manufacturing [21]. Extracellular-matrix-derived peptide-fused MAPs can further be used to regulate cell attachment and growth [22,23]. However, the antimicrobial activities of AMP-fused MAPs have not yet been examined. In this study, we constructed AMP-fused MAPs and determined their antimicrobial performance against various bacteria in vitro.

2. Results

2.1. Recombinant MAP-Fused Functional Peptides (MAP-FPs)

Many different proteins, which are derived from living organisms such as metabolites and components of some species, have been tried as biocompatible materials to apply pharmaceutics and medical devices. Although many AMPs were developed in a few decades, there were some limitations, such as low stability and cytotoxicity. To overcome the limitations of AMPs, MAP-derived recombinant protein system was used to give stability and safety to AMPs. Since recombinant mussel adhesive proteins, MAP fp-151, have been developed as biomaterial, having stability and cost-effective manufacturing was used as a scaffold for recombinant AMPs. MAP fp-151 was composed as fp1-fp5-fp1, derived from *Mytilus edulis* [23]. To make MAP-fused functional peptide for anti-microbial application, a number of antibiotic peptides originating from AMPs and proapoptotic peptides were fused to C-terminal of MAP fp-151 (Figure 1a). Anti-microbial-activity-associated peptides, including AKRHHGYKRKFH from anti-microbial histatin 5-derived P-113 (MAP-FP-1) [24], LKKLAKLALAF from anti-cancer peptide (MAP-FP-2) [25], THRPPMWSPVWP from transferrin receptor binding peptide (MAP-FP-3) [26,27], ILRWPWWPWRRK from anti-microbial omiganan (MAP-FP-4) [28,29], and KLAKLAKKLAKLAK from proapoptotic/anti-microbial KLAK peptide (MAP-FP-5) [30,31], were selected for MAP-fused functional peptides (MAP-FPs) (Table 1). Recombinant MAP-FPs were overexpressed and purified as previously described [21]. Since overexpressed recombinant MAP-FPs were aggregated as an inclusion body in *E. coli*, they lost anti-microbial activity and could be purified. Purified recombinant MAP-FPs showed >80% purity, as determined by SDS-PAGE (Figure 1b).

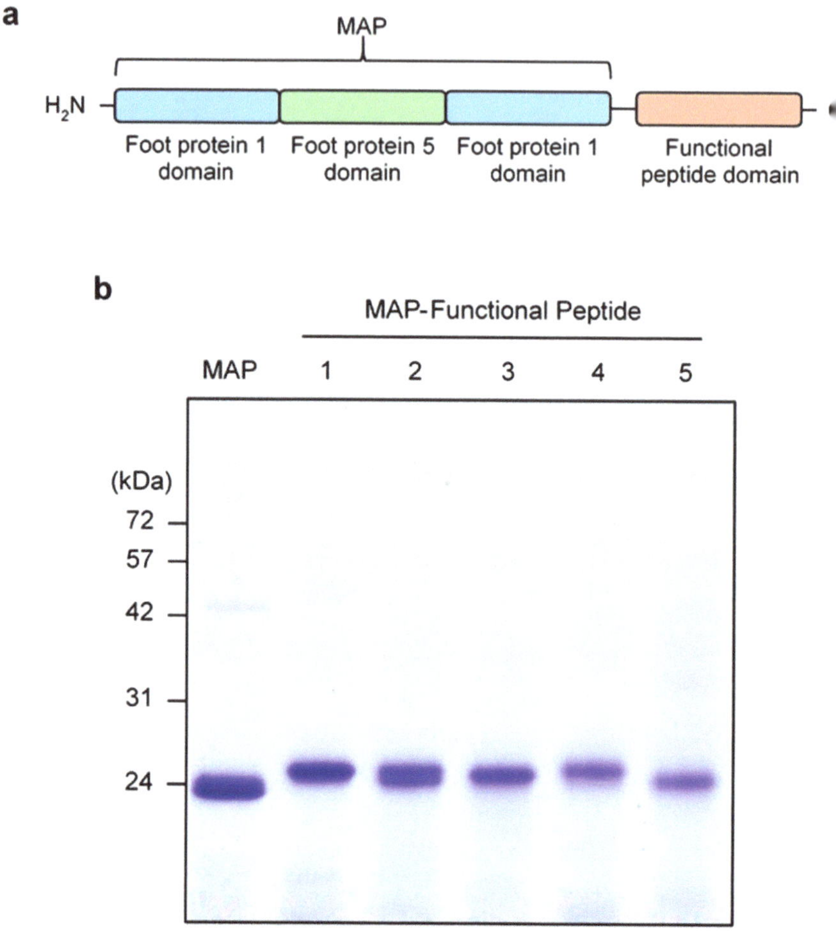

Figure 1. Schematic representation for MAP-FPs. (**a**) Schematic diagram of construction for MA FPs. (**b**) Electrophoresis analysis of purified recombinant MAP-FPs for purity. Purified recombina MAP-FPs (5 µg) were loaded with SDS-PAGE and separated using electrophoresis. After electrophor sis, SDS-PAGE was stained with Coomassie blue dye.

Table 1. Chemical properties of the MAP-fused functional peptides (MAP-FPs).

Peptide	Sequence	Molecular Weight	Hydrophobic Residues (%)	Net Charge at pH 7.0
MAP-FP-1	AKRHHGYKRKFH	1564.82	42	5.27
MAP-FP-2	LKKLAKLALAF	1215.59	27	3.00
MAP-FP-3	THRPPMWSPVWP	1490.75	17	1.09
MAP-FP-4	ILRWPWWPWRRK	1780.16	33	4.00
MAP-FP-5	KLAKLAKKLAKLAK	1524.01	43	6.00
Magainin I	GIGKFLHSAGKFGKAFVGEIMKS	2409.88	30	3.09

2.2. *Determination of Minimal Inhibitory Concentrations (MICs) of MAP-FPs*

MICs of MAP-FPs and Magainin I were determined against standard laboratory m crobial strains, including four Gram-negative and two Gram-positive strains, using th broth microdilution method. The growth inhibitory activities of MAP-FPs were assesse at concentrations of 0.25–8 µM against Gram-negative bacteria, including *Escherichia co*

Salmonella typhimurium, *Klebsiella pneumoniae*, and *Citrobacter freundii*, and Gram-positive bacteria, including *Staphylococcus aureus* and *Bacillus cereus* (Table 2). MAP-FPs and Magainin I at 4–8 μM inhibited the growth of Gram-negative strains. However, neither the MAP-FPs nor Magainin I inhibited the growth of Gram-positive strains. Interestingly, MAP, used as the backbone and adhesive domain for the surface coating, also inhibited the growth of the Gram-negative *E. coli* and *S. typhimurium* strains. Compared to the other MAP-FPs, MAP-FP-2, -4, and -5 showed a two-fold higher growth inhibitory potency against *E. coli* and *S. typhimurium*. These results demonstrate that 4–8 μM of MAP and MAP-FPs could inhibit the growth of susceptible Gram-negative bacteria.

Table 2. Minimum inhibitory concentration of MAP-FPs against Gram-negative and -positive bacteria. Microbials (1.5 × 10^5 CFU/mL), including *E. coli*, *S. typhimurium*, *K. pneumoniae*, *C. freundii*, *S. aureus*, and *B. cereus*, were incubated with MAP-FPs and Magainin I (concentration range 0.25~8 μM) in CAMHB. After 18 h of incubation, the growth inhibition of microbials was measured using CCK-8 reagent. A volume of 10 μL of CCK-8 reagent was added to each well and incubated plate for 90 min at 37 °C. Absorbance of 450 nm wavelength was determined using microplate reader. Data represent the result of the experiment performed in triplicate. n.d. means not determined. Data are presented as the mean ± standard deviation.

Peptide	Gram-Negative								Gram-Positive			
	E. coli		*S. typhimurium*		*K. pneumoniae*		*C. freundii*		*S. aureus*		*B. cereus*	
	MIC (μM)	*p* Value	MIC (μM)	*p* Value	MIC (μM)	*p* Value	MIC (μM)	*p* Value	MIC (μM)	*p* Value	MIC (μM)	*p* Value
MAP	8	<0.0001	8	<0.0001	n. d.	-	n. d.	-	n. d.	-	n. d.	-
MAP-FP-1	8	<0.0001	8	<0.0001	8	<0.0001	8	0.0001	n. d.	-	n. d.	-
MAP-FP-2	4	<0.0001	4	<0.0001	8	0.0001	8	0.0011	n. d.	-	n. d.	-
MAP-FP-3	8	<0.0001	8	<0.0001	8	0.0091	n. d.	-	n. d.	-	n. d.	-
MAP-FP-4	4	<0.0001	4	<0.0001	8	<0.0001	8	<0.0001	n. d.	-	n. d.	-
MAP-FP-5	4	<0.0001	4	0.0001	8	<0.0001	8	<0.0001	n. d.	-	n. d.	-
Magainin	4	<0.0001	4	<0.0001	4	<0.0001	8	<0.0001	n. d.	-	n. d.	-

2.3. Time-Dependent Microbicidal Activity of MAP-FPs

To check the time-dependent microbicidal activity of MAP-FPs, TKK assay was performed on six microbial strains. TKK of MAP-FPs was investigated using the CFU counting method after indicated times, 0, 10, 30, 60, and 180 min with 1 × MIC of MAP-FPs. Since some MICs have not yet been determined, for all MAP-FPs against Gram-positive bacteria and MAP against *K. pneumoniae* and *C. freundii*, 8 μM was used as 1 × MIC of MAP-FPs and Magainin I. Time-kill kinetics of 1 × MIC MAP and MAP-FPs showed all four Gram-negative bacteria were killed within 180 min, but Magainin I did not (Figure 2a–d). *S. typhimurium* and *K. pneumoniae* were killed within 60 min with MAP and MAP-FPs (Figure 2b,c). Similar to the findings of a previous MIC study, MAP-FP-2 and -5 showed the fastest antimicrobial effects and could kill *E. coli*, *S. typhimurium*, and *K. pneumoniae* within 30 min (Figure 2a–c). Although none of the 1 × MIC MAP-FPs killed *S. aureus* (Figure 2e), MAP-FP-5 killed *B. cereus* within 180 min (Figure 2f). A 1/4 × MIC of MAP-FPs killed all Gram-negative bacteria except *C. freundii* within 180 min (Figure S1a–f). Compared with the TKK of 1 × MIC results, it was observed that even smaller amounts of MAP-FPs also killed susceptible bacteria within 180 min. These results demonstrate the direct microbicidal effects of MAP and MAP-FPs against susceptible Gram-negative bacteria.

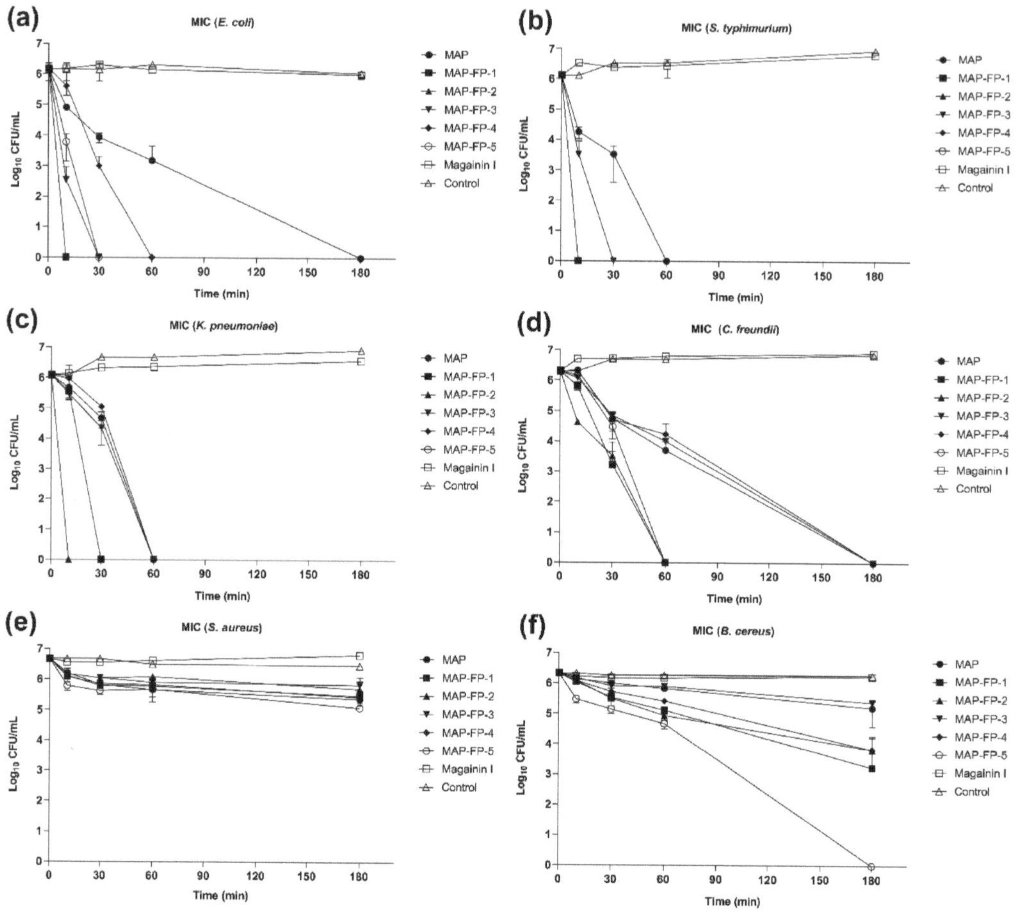

Figure 2. Time-killing kinetics of MAP-FPs against Gram-negative and -positive bacteria. Micr bials (1.0 × 10^7 CFU/mL), including *E. coli* (**a**), *S. typhimurium* (**b**), *K. pneumoniae* (**c**), *C. freundii* (**d** *S. aureus* (**e**), and *B. cereus* (**f**), were treated with 1 × MIC of MAP-FPs and Magainin I in PBS. 1 × MI stands for 4 μM. After indicated times (10, 30, 60, and 180 min), bacteria were harvested and the spread onto an agar plate. After 18 h of incubation, the CFU was calculated. All data represent th mean value ± standard deviation of three independent experiments.

2.4. Cytotoxicity of the MAP-FPs

To investigate the cytotoxicity of MAP-FPs, Vero, monkey kidney epithelial cells, an HEK293T, human kidney epithelial cells, cell lines were used. Two normal kidney epithelia cells were treated with MAP-FPs in a dose-dependent manner. After 24 h of incubation, cel viability was measured to determine the 50% cytotoxic concentration (CC_{50}) of MAP-FP to normal cells. Although MAP showed good safety even at 12 μM, MAP-FP-1, -4, and - derived from P-113, omiganan, and KLA proapoptotic peptide, showed a CC_{50} of 8.3, 5.8 and 7.3 μM, respectively, in only the Vero cell line. Although some MAP-FPs were toxic a high concentrations, MAP-FP-2 and -3 demonstrated good safety at 12 μM (Table 3 an Figure S2a–f).

Table 3. Cytotoxicity of the MAP-FPs in mammalian cells. Vero and HEK293T cells were used for cytotoxicity of MAP-FPs. Cells (5000 cells/well) were seeded and then treated with different concentration of MAP-FPs for 24 h. After 24 h of incubation, cytotoxicity was determined by CCK-8 solution. A volume of 10 μL of CCK-8 reagent was added to each well and incubated plate for 90 min at 37 °C, 5% CO_2. Data represent the result of the experiment performed in triplicate.

Peptide	CC_{50} (μM)	
	Vero	HEK293T
MAP	>12	>12
MAP-FP-1	8.3	>12
MAP-FP-2	>12	>12
MAP-FP-3	>12	>12
MAP-FP-4	5.8	>12
MAP-FP-5	7.3	>12

2.5. Thermal Stability of the MAP-FPs

Since MAP-FP-2 exhibited anti-microbial activity without cytotoxicity in normal mammalian cells among MAP-FPs, thermal stability of MAP and MAP-FP-2 was analyzed using RP-HPLC. For thermal stability, MAP and MAP-FP-2 were incubated at 42 °C for 14 and 28 days, and then stability was determined by RP-HPLC. MAPs, incubated at 42 °C, showed 100.02 ± 0.19% and 96.36 ± 0.3% stabilities after 14 and 28 days, respectively. Stabilities of MAP-FP-2 were exhibited as 90.89 ± 1.44% and 91.33 ± 2.6% after 14 and 28 days incubation at 42 °C (Table 4). Although MAP-FP-2 showed a 9% decrease in stability compared to MAP, its stability did not decrease in a time-dependent manner, and its antimicrobial activity was sustained (Table 4). Based on these results, MAP and MAP-FP-2 exhibit strong thermal stability.

Table 4. Thermal stability analysis of MAP and MAP-FP-2. MAP and MAP-FP-2, dissolved in deionized water (8 μM), were incubated at 42 °C for 14 and 28 days. After incubation, MAP and MAP-FP-2 were analyzed using RT-HPLC. To check anti-microbial activity, the MIC of incubated MAP and MAP-FP-2 against *E. coli* was determined by the microdilution method.

Day of Storage	MAP		MAP-FP-2	
	Recovery ± RSD	MIC (μM)	Recovery ± RSD	MIC (μM)
0	100.00 ± 0.78	8	100.00 ± 3.61	4
14	100.02 ± 0.19	8	90.89 ± 1.44	4
28	96.36 ± 0.30	8	91.33 ± 2.60	4

3. Discussion

There is an urgent global need for new strategies and drugs to treat anti-microbial-resistant bacterial infections. While the presence of antibiotic-resistant strains does not render all existing antibiotics completely useless, it significantly reduces the survival rates of individuals infected with antibiotic-resistant strains. To eliminate antibiotic-resistant bacteria, novel anti-microbial agents should be developed. The traditional antibiotic development pipeline has been unable to address the clinical issues associated with antimicrobial resistance. Thus, the development of novel classes of antibiotics with new targets or modes of action is necessary. New anti-microbial alternatives such as AMP are especially promising because of their potent and broad anti-microbial activity with their slow rate inducing resistance in bacteria [32,33]. In this study, AMPs exhibiting antimicrobial, anticancer, and proapoptotic activities were fused with a biocompatible MAP scaffold to produce a novel antimicrobial MAP-FP, serving as a functional peptide domain. The manufactured MAP and MAP-FPs exhibited strong antimicrobial activities against specific Gram-negative bacteria. Some antibiotics, including aztreonam, cephalosporin, and polymyxin B, also exhibit antimicrobial activities against Gram-negative bacteria. In particular, similar to MAP-FPs, aztreonam did not show anti-microbial activity against Gram-positive bacteria, but on against Gram-negative bacteria. Similar to Gram-negative-specific antibiotic mechanism it disrupted the Gram-negative cell wall by binding enriched lipopolysaccharide (LPS) synthesis enzyme; MAP-FPs are also expected to have a Gram-negative cell wall targeti mechanism [34–36].

MAP-FPs not only exhibit unique anti-microbial activity against Gram-negative bac ria but also have favorable therapeutic index and physicochemical properties, such as hi thermal stability and adhesiveness. Unlike synthetic adhesives, MAPs possess distincti properties; mussel filaments (byssus) can attach to inorganic surfaces in the sea witho degradation or alteration [37]. This high stability is also reflected in MAP-FPs, whi remained structurally stable at 42 °C over 4 weeks, far surpassing the rapid degradatic of peptides under similar conditions. This highlights the potential of MAPs as scaffo materials for attaching to inorganic surfaces and incorporating with specific peptides as functional peptide domain for various applications. Based on these physicochemical pro erties of MAP-FPs, MAP-FPs possess promise as anti-microbial coating materials again Gram-negative bacteria like as *E. coli*, *S. typhimurium*, and *P. aeruginosa* in the medic devices and consumer goods. Especially, *E. coli* and *P. aeruginosa* are representative bacter to make biofilm on the surface of medical devices and consumer goods [38,39]. Biofilm having resistance to anti-microbial agents and host immunity, are the product of microbi cells sticking to each other and adhesive surfaces. Because biofilm is not easily remove by common sterilization and biofilm formation takes some time, biomedical devices nee to be coated with an anti-microbial and stable agent to suppress biofilm formation [4 Since not only MAP-FPs have stable and adhesive properties based on MAP but also ha anti-microbial activity based on anti-microbial peptides, MAP-FPs could be good canc dates for anti-microbial/biofilm coating agents. Further studies are needed to validate t anti-biofilm activity and adhesiveness to elucidate the practical applications in consum products and medical devices.

4. Materials and Methods

4.1. Preparation of Recombinant Proteins

MAP fp-151, cloned into pET-22b(+) vector (Novagen, Darmstadt, Germany), w used as the backbone vector [2]. cDNAs encoding anti-microbial peptides or cell lyt peptides were subcloned into C-terminal of MAP fp-151 at sites *Hind*III and *Xho*I restrictio enzymes and expressed in *E. coli* Rosetta (DE3) (Novagen, Darmstadt, Germany) by IPT (Sigma-Aldrich, St. Louis, MO, USA) induction. MAP-fused functional peptides (MA FPs) were purified as previously described [21]. Briefly, induced *E. coli* was lysed usin a high-pressure homogenizer. MAP-FPs were extracted using 25% (v/v) acetic acid fro the inclusion body of lysates. Extracts were dialyzed in phosphate-buffered saline (PB containing 20% glycerol using dialysis tubing (Sigma-Aldrich, St. Louis, MO, USA). Purifie MAP-FPs were freeze-dried and then stored at −80 °C.

4.2. Microorganisms

The microorganisms used in this study were *Escherichia coli* (E. coli, ATCC, 25922 *Salmonella typhimurium* (*S. typhimurium*, ATCC, 13311), *Klebsiella pneumoniae* (*K. pneumonia* ATCC, 10031), *Citrobacter freundii* (*C. freundii*, ATCC, 6750), *Staphylococcus aureus* (*S. aure* ATCC, 29213), and *Bacillus cereus* (*B. cereus*, ATCC, 27348), obtained from the American Typ Culture Collection (ATCC, Manassas, VA, USA). The strains were maintained at −80 ° with broth containing 20% glycerol. The broth that was used for the storage of *E. co* *S. typhimurium*, *K. pneumoniae*, *C. freundii*, and *B. cereus* in cryogenics was nutrient broth (M Cell, Seoul, Republic of Korea), while Tryptic soy broth (MB Cell, Seoul, Republic of Kore was used for *S. aureus*.

4.3. Minimal Inhibitory Concentrations (MICs) Determinations

The minimal inhibitory concentrations (MICs) of the MAP-FPs were determined using the broth microdilution method in Cation-adjusted Mueller Hinton Broth (CAMHB, MB Cell, Seoul, Republic of Korea) against all six strains following the guidelines of the Clinical and Laboratory Standard Institute (CLSI). The M7-A9 protocol was used for MIC assay. Briefly, *E. coli*, *S. typhimurium*, *K. pneumoniae*, *C. freundii*, and *B. cereus* were incubated in nutrient broth, and *S. aureus* was incubated in Tryptic soy broth overnight at 37 °C. Then, six strains of bacteria were grown overnight at 37 °C, 150 RPM in the CAMHB and were diluted with CAMHB to achieve 0.5 McFarland turbidity standard (1.5×10^8 CFU/mL, OD_{600}: 0.07~0.08) using a microplate reader. Next, a total of 100 μL of the 0.5 McFarland solution was added to 9900 μL of CAMHB (1/100 dilution, 1.5×10^6 CFU/mL). To make final inoculum as 1.5×10^5 CFU/mL, 10 μL of bacterial inoculum containing 1.5×10^6 CFU/mL cells was added into the plate (SPL, Gyeonggi-do, Republic of Korea) wells containing 90 μL of serially diluted peptide samples (final concentrations ranging from 0.25 to 8 μM) and incubated at 37 °C for 18 h. The inhibition of microbial growth was determined by cell counting kit-8 (CCK-8) reagent (Dojindo, Kumamoto, Japan). Incubated broth of each well was diluted 10-fold with PBS using a multi-channel pipette. A volume of 10 μL of CCK-8 reagent was added to each well and incubated for 90 min. The absorbance at 450 nm wavelength was measured using the microplate reader (BioTek Synergy, Winooski, VT, USA, S1LFA). All measurements were performed in triplicate.

4.4. Time-Kill Kinetics (TKK) Assay

The time-kill kinetics (TKK) assay of the MAP-FPs was studied against six different microbial strains. A volume of 670 μL of the 0.5 McFarland turbidity standard solution, having microbials of 1.5×10^8 CFU/mL, was added to 9330 μL of PBS without bacteria to adjust the inoculum concentration, having 1.0×10^7 CFU/mL. Adjusted microbials (1.0×10^7 CFU/mL) were incubated with PBS containing 1 × MIC (4 μM: MAP-FP-2, -4, -5, Magainin I against *E. coli*, *S. typhimurium*; Magainin I against *K. pneumoniae*/8 μM: MAP, MAP-FP-1, -3 against *E. coli*, *S. typhimurium*; MAP, MAP-FPs against *K. pneumoniae*; MAP, MAP-FPs, Magainin I against *C. freundii*, *S. aureus*, and *B. cereus*) and 1/4 × MIC (1 μM: MAP-FP-2, -4, -5, Magainin I against *E. coli*, *S. typhimurium*; Magainin I against *K. pneumoniae*/2 μM: MAP, MAP-FP-1, -3 against *E. coli*, *S. typhimurium*; MAP, MAP-FPs against *K. pneumoniae*; MAP, MAP-FPs, Magainin I against *C. freundii*, *S. aureus*, and *B. cereus*) of MAP and MAP-FPs. Then, incubation of PBS containing MAP-FPs and microbials was performed under the conditions of 37 °C, 150 RPM. After 0, 10, 30, 60, and 180 min, incubated microbials were harvested, serially diluted, and then 100 μL of serially diluted sample containing microbials were spread onto a nutrient agar plate. Agar plates were then further incubated at 37 °C for 24 h. CFUs were determined by counting colonies manually. Non-treated cultured bacteria were used as a control.

4.5. Cytotoxicity

Vero (KCLB, 10081) and HEK293T (ATCC, CRL-3216) were obtained from the Korean Cell Line Bank (KCLB, Seoul, Republic of Korea) and ATCC (Manassas, VA, USA), respectively. Vero and HEK293T cells were cultured at 37 °C, 5% of CO_2 with Roswell Park Memorial Institute (RPMI)-1640 and Dulbecco's Modified Eagle's Medium (DMEM) High Glucose (Hyclone, Logan, UT, USA) containing 10% fetal bovine serum (FBS, Hyclone, Logan, UT, USA), 100 Unit/mL penicillin, 100 μg/mL streptomycin, respectively (Hyclone, Logan, UT, USA). Vero and HEK293T (5×10^3/well) cells were seeded in 96-well cell culture plates (SPL, Gyeonggi-do, Republic of Korea) with 10% FBS medium. After 15 h, cells were exposed to media only or media containing various concentration of MAP and MAP-FPs for 24 h. Cytotoxicity was measured using CCK-8 solution following manufacturer's instructions (Dojindo, Kumamoto, Japan). Briefly, 10 μL of CCK-8 solution was added to cultured media and incubated for 90 min to detect absorbance. The absorbance was measured at 450 nm wavelength using a microplate reader (BioTek Synergy, Winoos VT, USA). CC_{50} values were calculated using Prism (GraphPad, version 9.1.1).

4.6. Thermal Stability Analysis

MAP and MAP-FPs, dissolved in deionized water, were aliquoted and incubated a 42 °C for 2 and 4 weeks (final concentration was 8 μM). After incubation, thermal stabili of MAP-FPs was analyzed by reverse-phase high-performance liquid chromatograph (RP-HPLC) instrument. RP-HPLC was performed on an Agilent 1100 series HPLC syste coupled with a diode array detector (Agilent, Santa Clara, CA, USA, G1315A). A Ph nomenex gemini C18 column (250 × 4.6 mm, 5 μm; Phenomenex, 00G-4435-E0, Torranc CA, USA) protected by a C18 security guard column (SecurityGuard; Phenomenex, To rance, CA, USA) was used for chromatographic separation at 40 °C. Chromatograph separation was performed in gradient mode with a flow rate of 1 mL/min. Gradient elutio consisting of an organic phase (A, 0.1% trifluoroacetic acid (TFA) in acetonitrile (ACN)) an an aqueous phase (B, 0.1% TFA in water), was changed from 0:100 (A:B, v/v) to 100:0 (A: v/v) in 50 min. The sample injection volume was 10 μL, and protein identification wa made by UV detection at 280 nm. The data were analyzed by ChemStation B.04.03 (Agiler Santa Clara, CA, USA). All measurements were performed in triplicate. Stability wa expressed as the percentage recovery which was calculated from the following equation

$$\text{Recovery (\%)} = \text{purity of analyzed sample / purity of analyzed control (day 0)} \times 100$$

4.7. Statistical Analysis

Statistical analysis was performed using an unpaired, two-tailed, Student's t-tes with GraphPad Prism 9.1.1 (GraphPad, San Diego, CA, USA). The data were presented a mean ± standard deviation. $p < 0.05$ was considered statistically significant.

Supplementary Materials: The following supporting information can be downloaded at: https:/ www.mdpi.com/article/10.3390/antibiotics13030239/s1, Figure S1: Time killing kinetics of MAP-FP 1/4 × MIC against Gram-negative and positive bacteria; Figure S2: Cytotoxicity of the MAP-FPs i Vero and HEK293T cells.

Author Contributions: M.C.P. designed and supervised the experiment; D.Y.K., Y.B.O. and J.S. performed the experiment; D.Y.K., Y.-H.M. and M.C.P. analyzed the data; D.Y.K., M.C.P. and Y.-H.M wrote the paper; D.Y.K., Y.B.O., J.S.P., Y.-H.M. and M.C.P. discussed the results and commented. A authors have read and agreed to the published version of the manuscript.

Funding: This work was supported by grant from Inje University, 2023 (2023-0024).

Institutional Review Board Statement: Not applicable.

Informed Consent Statement: Not applicable.

Data Availability Statement: Data are contained within the article and Supplementary Materials.

Conflicts of Interest: The authors declare no conflicts of interest.

References

1. Wu, F.; Zhao, S.; Yu, B.; Chen, Y.M.; Wang, W.; Song, Z.G.; Hu, Y.; Tao, Z.W.; Tian, J.H.; Pei, Y.Y.; et al. A new coronaviru associated with human respiratory disease in China. *Nature* **2020**, *579*, 265–269. [CrossRef] [PubMed]
2. Tacconelli, E.; Carrara, E.; Savoldi, A.; Harbarth, S.; Mendelson, M.; Monnet, D.L.; Pulcini, C.; Kahlmeter, G.; Kluytmans, J Carmeli, Y.; et al. Discovery, research, and development of new antibiotics: The WHO priority list of antibiotic-resistant bacteri and tuberculosis. *Lancet Infect. Dis.* **2018**, *18*, 318–327. [CrossRef] [PubMed]
3. Balasubramaniam, B.; Prateek, R.S.; Saraf, M.; Kar, P.; Singh, S.P.; Thakur, V.K.; Singh, A.; Gupta, R.K. Antibacterial and Antivira Functional Materials: Chemistry and Biological Activity toward Tackling COVID-19-like Pandemics. *ACS Pharmacol. Transl. Sc* **2020**, *4*, 8–54. [CrossRef] [PubMed]
4. Huemer, M.; Mairpady, S.S.; Brugger, S.D.; Zinkernagel, A.S. Antibiotic resistance and persistence-Implications for human healt and treatment perspectives. *EMBO Rep.* **2020**, *21*, e51034. [CrossRef] [PubMed]
5. Filippone, E.J.; Kraft, W.K.; Farber, J.L. The Nephrotoxicity of Vancomycin. *Clin. Pharmacol. Ther.* **2017**, *102*, 459–469. [CrossRef

Bulet, P.; Hetru, C.; Dimarcq, J.L.; Hoffmann, D. Antimicrobial peptides in insects; structure and function. *Dev. Comp. Immunol.* **1999**, *23*, 329–344. [CrossRef] [PubMed]

Steinstraesser, L.; Kraneburg, U.; Jacobsen, F.; Al-Benna, S. Host defense peptides and their antimicrobial-immunomodulatory duality. *Immunobiology* **2011**, *216*, 322–333. [CrossRef] [PubMed]

Hancock, R.E.; Nijnik, A.; Philpott, D.J. Modulating immunity as a therapy for bacterial infections. *Nat. Rev. Microbiol.* **2012**, *10*, 243–254. [CrossRef] [PubMed]

Pasupuleti, M.; Schmidtchen, A.; Malmsten, M. Antimicrobial peptides: Key components of the innate immune system. *Crit. Rev. Biotechnol.* **2012**, *32*, 143–171. [CrossRef]

Aisenbrey, C.; Marquette, A.; Bechinger, B. The Mechanisms of Action of Cationic Antimicrobial Peptides Refined by Novel Concepts from Biophysical Investigations. *Adv. Exp. Med. Biol.* **2019**, *1117*, 33–64.

Torrent, M.; Navarro, S.; Moussaoui, M.; Nogués, M.V.; Boix, E. Eosinophil cationic protein high-affinity binding to bacteria-wall lipopolysaccharides and peptidoglycans. *Biochemistry* **2008**, *47*, 3544–3555. [CrossRef] [PubMed]

2. Bin, H.A.; Jiang, X.; Bergen, P.J.; Zhu, Y. Antimicrobial Peptides: An Update on Classifications and Databases. *Int. J. Mol. Sci.* **2021**, *22*, 11691.

3. Tally, F.P.; DeBruin, M.F. Development of daptomycin for gram-positive infections. *J. Antimicrob. Chemother.* **2000**, *46*, 523–526. [CrossRef] [PubMed]

4. Jones, R.N. Microbiological features of vancomycin in the 21st century: Minimum inhibitory concentration creep, bactericidal/static activity, and applied breakpoints to predict clinical outcomes or detect resistant strains. *Clin. Infect. Dis. Off. Publ. Infect. Dis. Soc. Am.* **2006**, *42* (Suppl. S1), S13–S24. [CrossRef]

5. Rai, A.; Ferrão, R.; Palma, P.; Patricio, T.; Parreira, P.; Anes, E.; Tonda-Turo, C.; Martins, M.C.L.; Alves, N.; Ferreira, L. Antimicrobial peptide-based materials: Opportunities and challenges. *J. Mater. Chem. B* **2022**, *10*, 2384–2429. [CrossRef]

6. Rodríguez, A.A.; Otero-González, A.; Ghattas, M.; Ständker, L. Discovery, Optimization, and Clinical Application of Natural Antimicrobial Peptides. *Biomedicines* **2021**, *9*, 1381. [CrossRef]

7. Choi, Y.S.; Kang, D.G.; Lim, S.; Yang, Y.J.; Kim, C.S.; Cha, H.J. Recombinant mussel adhesive protein fp-5 (MAP fp-5) as a bulk bioadhesive and surface coating material. *Biofouling* **2011**, *27*, 729–737. [CrossRef]

8. Kaushik, N.K.; Kaushik, N.; Pardeshi, S.; Sharma, J.G.; Lee, S.H.; Choi, E.H. Biomedical and Clinical Importance of Mussel-Inspired Polymers and Materials. *Mar. Drugs* **2015**, *13*, 6792–6817. [CrossRef]

9. Choi, B.H.; Cheong, H.; Jo, Y.K.; Bahn, S.Y.; Seo, J.H.; Cha, H.J. Highly purified mussel adhesive protein to secure biosafety for in vivo applications. *Microb. Cell Factories* **2014**, *13*, 52. [CrossRef]

0. Cha, H.J.; Hwang, D.S.; Lim, S.; White, J.D.; Matos-Perez, C.A.; Wilker, J.J. Bulk adhesive strength of recombinant hybrid mussel adhesive protein. *Biofouling* **2009**, *25*, 99–107. [CrossRef]

1. Hwang, D.S.; Gim, Y.; Yoo, H.J.; Cha, H.J. Practical recombinant hybrid mussel bioadhesive fp-151. *Biomaterials* **2007**, *28*, 3560–3568. [CrossRef] [PubMed]

2. Hwang, D.S.; Sim, S.B.; Cha, H.J. Cell adhesion biomaterial based on mussel adhesive protein fused with RGD peptide. *Biomaterials* **2007**, *28*, 4039–4046. [CrossRef] [PubMed]

3. Kim, B.J.; Cheong, H.; Choi, E.S.; Yun, S.H.; Choi, B.H.; Park, K.S.; Kim, I.S.; Park, D.H.; Cha, H.J. Accelerated skin wound healing using electrospun nanofibrous mats blended with mussel adhesive protein and polycaprolactone. *J. Biomed. Mater. Res. Part A* **2017**, *105*, 218–225. [CrossRef]

4. Rothstein, D.M.; Spacciapoli, P.; Tran, L.T.; Xu, T.; Roberts, F.D.; Dalla, S.M.; Buxton, D.K.; Oppenheim, F.G.; Friden, P. Anticandida activity is retained in P-113, a 12-amino-acid fragment of histatin 5. *Antimicrob. Agents Chemother.* **2001**, *45*, 1367–1373. [CrossRef] [PubMed]

5. Dharmar, M.; Shanmugam, G.; Menakha, S.; Jagatheesh, K.; Sangeetha, M.; Thangavel, M.; Rajendran, V.; Krishnan, M.; Rengasamy, B.; Namasivayam, E. Molecular insights of newly identified potential peptide inhibitors of hypoxia inducible factor 1α causing breast cancer. *J. Mol. Struct.* **2019**, *1177*, 558–563.

6. Lee, J.H.; Engler, J.A.; Collawn, J.F.; Moore, B.A. Receptor mediated uptake of peptides that bind the human transferrin receptor. *Eur. J. Biochem.* **2001**, *268*, 2004–2012. [CrossRef] [PubMed]

7. Bruhn, K.W.; Spellberg, B. Transferrin-mediated iron sequestration as a novel therapy for bacterial and fungal infections. *Curr. Opin. Microbiol.* **2015**, *27*, 57–61. [CrossRef]

8. Czechowicz, P.; Jaśkiewicz, M.; Neubauer, D.; Gościniak, G.; Kamysz, W. Anticandidal Activity of Omiganan and Its Retro Analog Alone and in Combination with Fluconazole. *Probiotics Antimicrob. Proteins* **2021**, *13*, 1173–1182. [CrossRef]

9. Melo, M.N.; Castanho, M.A. Omiganan interaction with bacterial membranes and cell wall models. Assigning a biological role to saturation. *Biochim. Biophys. Acta* **2007**, *1768*, 1277–1290. [CrossRef]

0. Obeid, M. Anticancer activity of targeted proapoptotic peptides and chemotherapy is highly improved by targeted cell surface calreticulin-inducer peptides. *Mol. Cancer Ther.* **2009**, *8*, 2693–2707. [CrossRef]

1. Lin, L.; Chi, J.; Yan, Y.; Luo, R.; Feng, X.; Zheng, Y.; Xian, D.; Li, X.; Quan, G.; Liu, D.; et al. Membrane-disruptive peptides/peptidomimetics-based therapeutics: Promising systems to combat bacteria and cancer in the drug-resistant era. *Acta Pharm. Sin. B* **2021**, *11*, 2609–2644. [CrossRef] [PubMed]

2. Browne, K.; Chakraborty, S.; Chen, R.; Willcox, M.D.; Black, D.S.; Walsh, W.R.; Kumar, N. A New Era of Antibiotics: The Clinical Potential of Antimicrobial Peptides. *Int. J. Mol. Sci.* **2020**, *21*, 7047. [CrossRef]

33. Matsuzaki, K. Control of cell selectivity of antimicrobial peptides. *Biochim. Biophys. Acta* **2009**, *1788*, 1687–1692. [CrossRef]
34. Sykes, R.B.; Bonner, D.P. Aztreonam: The first monobactam. *Am. J. Med.* **1985**, *78*, 2–10. [CrossRef]
35. Yotsuji, A.; Mitsuyama, J.; Hori, R.; Yasuda, T.; Saikawa, I.; Inoue, M.; Mitsuhashi, S. Mechanism of action of cephalosporins a resistance caused by decreased affinity for penicillin-binding proteins in *Bacteroides fragilis*. *Antimicrob. Agents Chemother.* **1988**, 1848–1853. [CrossRef] [PubMed]
36. Domingues, M.M.; Inácio, R.G.; Raimundo, J.M.; Martins, M.; Castanho, M.A.; Santos, N.C. Biophysical characterization polymyxin B interaction with LPS aggregates and membrane model systems. *Biopolymers* **2012**, *98*, 338–344. [CrossRef] [PubMe
37. Lee, B.P.; Messersmith, P.B.; Israelachvili, J.N.; Waite, J.H. Mussel-Inspired Adhesives and Coatings. *Annu. Rev. Mater. Res.* **20** *41*, 99–132. [CrossRef] [PubMed]
38. Tagliaferri, T.L.; Jansen, M.; Horz, H.P. Fighting Pathogenic Bacteria on Two Fronts: Phages and Antibiotics as Combined Strate *Front. Cell. Infect. Microbiol.* **2019**, *9*, 22. [CrossRef]
39. Römling, U.; Balsalobre, C. Biofilm infections, their resilience to therapy and innovative treatment strategies. *J. Intern. Med.* **20** *272*, 541–561. [CrossRef]
40. Salwiczek, M.; Qu, Y.; Gardiner, J.; Strugnell, R.A.; Lithgow, T.; McLean, K.M.; Thissen, H. Emerging rules for effecti antimicrobial coatings. *Trends Biotechnol.* **2014**, *32*, 82–90. [CrossRef]

antibiotics

MDPI

Review

Gut-Antimicrobial Peptides: Synergistic Co-Evolution with Antibiotics to Combat Multi-Antibiotic Resistance

Piyush Baindara [1,*] and Santi M. Mandal [2]

[1] Radiation Oncology, NextGen Precision Health, School of Medicine, University of Missouri, Columbia, MO 65211, USA

[2] Department of Biotechnology, Indian Institute of Technology Kharagpur, Kharagpur 721302, India; mandalsm@gmail.com

* Correspondence: piyush.baindara@gmail.com

Abstract: Due to huge diversity and dynamic competition, the human gut microbiome produces a diverse array of antimicrobial peptides (AMPs) that play an important role in human health. The gut microbiome has an important role in maintaining gut homeostasis by the AMPs and by interacting with other human organs via established connections such as the gut–lung, and gut–brain axis. Additionally, gut AMPs play a synergistic role with other gut microbiota and antimicrobials to maintain gut homeostasis by fighting against multi-antibiotic resistance (MAR) bacteria. Further, conventional antibiotics intake creates a synergistic evolutionary pressure for gut AMPs, where antibiotics and gut AMPs fight synergistically against MAR. Overall, gut AMPs are evolving under a complex and highly synergistic co-evolutionary pressure created by the various interactions between gut microbiota, gut AMPs, and antibiotics; however, the complete mechanism is not well understood. The current review explores the synergistic action of gut AMPs and antibiotics along with possibilities to fight against MAR bacteria.

Keywords: gut microbiota; gut peptides; multi-antibiotic resistance; co-evolution

Citation: Baindara, P.; Mandal, S.M. Gut-Antimicrobial Peptides: Synergistic Co-Evolution with Antibiotics to Combat Multi-Antibiotic Resistance. *Antibiotics* **2023**, *12*, 1732. https://doi.org/10.3390/antibiotics12121732

Academic Editor: Paul B. Savage

Received: 29 November 2023
Revised: 11 December 2023
Accepted: 13 December 2023
Published: 14 December 2023

1. Introduction

The rapid emergence of MAR and bacterial infections are global health concerns that urgently need to be addressed. The unavailability of new antibiotics and failure of available therapeutic strategies due to resistance development results in severe health complications and a sharp rise in deaths throughout the world [1,2]. In light of these facts, there is an urgent need for new antimicrobials and the development of new antimicrobial therapeutic strategies with effective outcomes to win the battle against MAR. AMPs are one of the promising options to fight against MAR due to their ubiquitous availability and diverse activity spectrum [3,4]. Additionally, the amenability of AMPs to bioengineering and drug repurposing may also play an important role in the development of new strategies to treat MAR [5–7]. Interestingly, the human gut is a complex environment where the cohabitation of pathogens with a beneficial gut microbiome and host appeases the synergistic co-evolution and action of gut AMPs and antibiotics. AMPs are also known to have multiple antimicrobial properties within a single peptide including membrane permeabilization and inhibition of both transcription and translation [8]. In the complex environment of the gut, high antimicrobial strength and complexity are observed in the tightly synchronized secretion of AMPs enriched with interdependent properties [9]. Host defense peptide-producing cells in the gut also take advantage of this synergistic action of gut AMPs in specific combinations that result in higher efficiency against pathogens at low concentrations. Similarly, the synergistic action of gut AMPs is observed with conventional antibiotics and could be used to develop new therapeutics against MAR. Interestingly, because of their known benefits, AMP-based drugs are now under consideration by the European Medicines Agency (EMA) and the Food and Drug Administration (FDA) [10,11].

Although gut AMPs have already been discussed extensively as a potential alternative fighting against MAR, new strategies are required to control the development and evolutic of rapid resistance [12–14]. Also, it is important to understand the synergistic and rap evolutionary process of gut AMPs with antibiotics. Here in the present review, we discu the use of gut microbiota-produced AMPs in conjugation with conventional antibioti their synergistic co-evolution, and their action in controlling MAR.

2. Influence of Antibiotics on Gut Microbiota, Susceptibility to Infections, and Resistance Development

The most frequent and significant factor altering the normal gut microbiome cor position and function is the use of antibiotics; however, many other factors that mig impair the beneficial gut microbiota include mental and physical stress, radiation treatme altered gut peristalsis, gastrointestinal infections, and dietary changes [15]. Antibioti have a major impact on changing the gut microbiota, resulting in decreased bacterial c versity and increased numbers of some taxa [16]. This change in gut microbiome furth results in the altered production of AMPs produced by gut microbiota and their associate functions impacting host immunity. Additionally, antibiotics' activity spectrum, mode action, potency, pharmacokinetics, dosage, and length of administration are also maj factors that influence the gut AMPs and microbiome [17]; however, the presence of pree isting antimicrobial resistance genes in an individual's microbiome is another concernin factor. Changes in the variety of gut bacteria can result in *Clostridium difficile* infectio which is naturally resistant to many antibiotics [18]. Other unintended consequences antibiotic use on gut microbiota include selection for a reservoir of bacterial antibiot resistance genes, and progression of horizontal gene transfer between bacterial strains th affects the expression, production, and regulation of gut AMPs further leading to immur dysregulation and antibiotic resistance development [16].

Antibiotics affect the local gut immune system by changing the composition of the gu resident microbiota and their metabolites, specifically AMPs. It has been shown that pos antibiotic treatment, the small intestine showed lower IL-17 and INF-γ production, whi the colon showed decreased numbers of T_{reg} cells. This suggests that antibiotics induc altered host–microbiota interactions that cause immune imbalance [19]. Additionally, th gut microbiota stimulates mucin production, whereas antibiotics cause the weakening the mucus barrier, making the body more vulnerable to bacterial invasion and subseque infections [20]. Intestinal infections may be brought on by newly acquired pathogens or b the overgrowth and pathogenic potential of opportunistic microorganisms due to changes i the bacterial populations that ordinarily inhabit the gut lumen. Numerous studies on infan receiving antibiotics, particularly preterm ones, have been conducted. The normal bacteri microbiota of infants is changed by treatment with different antibiotics, such as cephalexi gentamicin, vancomycin, and erythromycin, by increasing the percentage of potentiall pathogenic *Enterobacteriaceae* and decreasing the number of bacteria like *Bifidobacteriacea Bacilli*, and *Lactobacillus* which are part of the healthy microbiota [21]. Overall, antibiotic displayed a significant role in the modulation of gut microbiota that leads to infectio susceptibility at one end and resistance development as another counterpart.

3. Interplay of Gut Microbiota with Gut AMPs

Gut microbiota plays an essential role in the regulation of the host defense system b maintaining gut homeostasis. Bacteroidetes, Firmicutes, Actinobacteria, and Proteobacteri form the majority of the gut microbiome [22]. The diverse array of gut AMPs produced b gut microbiota plays an important role in various functional activities in the gut such a immunomodulatory activities and protection against pathogens by disrupting bacterial ce membranes and halting the RNA and DNA synthesis of metabolism [23]. Bacteriocins ar the major bacterially produced gut AMPs and efficiently compete with other microbes i the gut. However, much of the gut microbiome's diversity is still unknown; a study of som isolated microbes and metagenomic analysis suggested that there are many unreveale classes of antibiotics and AMP-producing microbes present in the gut that are as yet unknown [24]. Gut microbiota-derived AMPs have been reported to protect against various disease-causing pathogens in the human gut (Table 1). A bacteriocin, Abp118, produced by a gut microbe *Lactobacillus salivarius* UCC118 in the gut is confirmed to protect against the foodborne pathogen *Listeria monocytogenes*. It has been confirmed that mutant *L. salivarius* UCC118, expressing the cognate Abp118 immunity protein AbpIM, failed to protect against *L. monocytogenes* infections in mice [25]. Another bacteriocin, thuricin CD, produced by *Bacillus thuringiensis* DPC 6431 has been shown to have efficient killing potential against disease-causing clinical isolates of *C. difficile* without any antagonistic effect on commensal gut microbiota [26]. Bacteriocin encoded by pheromone-responsive plasmids is common in enterococcus strains residing in the gut which are reported as gut commensals as well as for casing hospital-acquired infections [27]. Bacteriocin 21, produced by conjugative plasmid pPD1 of *Enterococcus faecalis*, is demonstrated to protect against vancomycin-resistant enterococci without affecting the other commensal microbiota in the gut. Interestingly, *E. faecalis* containing pPD1 plasmid outcompetes and replaced other *E. faecalis* lacking pPD1. This suggests that gut bacteriocin can also regulate the niche in the gut and can be used as potential therapeutic peptides able to target MAR bacteria specifically [28]. Another study showed that microcins produced by a probiotic bacterium *Escherichia coli* Nissle 1917 (EcN) can regulate inter- and intra-species competition among the *Enterobacteriaceae* and other related pathogens in the inflamed gut environment and are suggested as potential narrow-spectrum therapeutic agents against enteric pathogens [29].

Table 1. Gut microbiota-produced AMPs and involvement in the treatment of different diseases.

Gut AMPs	Producing Bacteria	Targeted Pathogens or Diseases	References
Bacteriocin Abp118	*L. salivarius*	Listeriosis	[30]
Bacteriocin OR-7	*L. salivarius* NRRLB	*Campylobacter jejuni*	[31]
Bactofencin A	*L. salivarius*	Antilisterial, antistaphylococcal	[32]
Lactocin AL705	*L. curvatus*	Listeriosis	[33]
Lactocin 160	*L. rhamnosus*	*Escherichia coli* *Bordetella pertussis*	[34]
Lacticin3147	*Lactococcus lactis* DPC3147	*C. difficile*-associated diarrhea (CDAD)	[35]
Garvicin ML	*L. garvieae*	*Streptococcus pneumonia*	[36]
Nisin Z	*L. lactis*	Immunomodulatory effect	[37]
Nisin F	*L. lactis*	Respiratory infection	[38]
Nisin	*L. lactis*	Meningitis, sepsis, pneumonia	[39]
Nisin Z	*L. lactis*	Enteric pathogens	[37]
Nisin A	*L. lactis*	Colorectal cancer	[40]
Pediocin PA1	*Pediococcus acidilactici*	Listeriosis	[41]
Pediocin AcH	*P. acidilactici*	Enteric pathogens	[37]
Enterocin CRL35	*Enterococcus mundtii* RL35	Listeriosis	[42]
Avicin	*E. avium*	Listeriosis	[43]
Enterocin P	*E. faecium* P13	Enteric pathogens	[44]
Piscicolin 126, carnobacteriocin	*Carnobacterium maltaromaticum*	Listeriosis	[45]
Kimchichin	*Leuconostoc citreum* GJ7	*Salmonella typhi*	[46]
Erwinaocin NA4	*Erwinia carotovora* NA4	Coliphage	[47]

Gut-epithelium-derived peptides are also reported to have potential antimicrob activities against gut pathogens. In the gastrointestinal tract, enterocytes and Paneth ce are the primary cells responsible for the production of AMPs; however, macrophages, de dritic cells, neutrophils, and lymphocytes present in the lamina propria can also produ AMPs [48,49]. Defensins are the major AMPs secreted within the intestinal mucosa. The and β defensins are abundant AMPs in the gut which are primarily secreted by Paneth an epithelial cells, respectively, in the intestine and the colon [50]. Further, it has been report that the secretion of gut AMPs by Paneth cells is regulated and stimulated by exposu to live pathogens (both Gram-positive and Gram-negative) or bacterial products such lipopolysaccharide, lipoteichoic acid, lipid A, and muramyl dipeptide [51]. In another stu the gut resident *Lactobacillus* population exhibited a correlation with the gene expressi of α defensins, where α defensin gene expression is restored in antibiotic-treated mice *Lactobacillus* administration. Further, it has been confirmed that α defensin gene expressi by Paneth cells is regulated by commensal bacteria via the TLR-MyD88 signaling pathw that provides a deeper understanding of the involvement of gut microbiota and AMPs gut homeostasis [52]. Overexpression of α defensin 5 is found associated with a seve reduction in the colonization of segmented filamentous bacteria that are further linked wi reduced levels of Th17 cells in the lamina propia and suggests the role of α defensins in t regulation of commensal microbiota [53]. Gut epithelial-produced β defensins 2 and 3 we reported to reduce the intestinal damage caused by a gut pathogen *Salmonella typhimuriu* via enhancing the probiotic activity of *Enterococcus faecium* by alteration of cytokine expre sion [54]. Similarly to defensins, cathelicidins are also reported to produce and act again gut pathogens by improving the gut epithelial barrier. In a recent, cathelicidin-WA h been shown to improve host defense and epithelial barrier functions by reducing enteroh morrhagic *Escherichia coli*-induced inflammation and microbiota reduction in the intesti of mice [55]. Cathelicidin-related antimicrobial peptides (CRAMP) were found to prote against an enteric pathogen, *Citrobacter rodentium*, by reducing epithelial cell damage an systemic clearance of infection [56]. In another study, cathelicidins significantly improve the gut barrier against pathogens in mouse colon mucosa where endogenous stimulation administration of cathelicidin is able to clear the infection caused by *Escherichia coli* O157:H and also regulate the gut microbiota balance; this aids mucosal homeostasis [57,58]. Oth major gut peptides are regenerating AMPs (RegAMP) which are soluble lectins and main produced by Paneth cells. A RegAMP, RegIIIγ, is demonstrated to play an important ro in maintaining gut homeostasis by spatial segregation of gut microbiota and host in th intestine [59]. Another study reported that RegIIIγ can protect against *L. monocytogen* infection via MyD88-mediated conditioning of gut epithelium [60]. Further, RegAMP reported to have a role in pathogen clearance that is dependent on the presence of initi healthy gut microbiota and it has been suggested that gut microbiota and gut AMPs a the key factors that regulate the host response during antibiotic treatment [61]. Overa available reports suggest a complex relationship between gut-epithelium-derived AM and gut microbiota; however, further studies are required to explore the regulatory switch that drive the production of gut epithelium AMPs in response to specific gut commensa or pathogens.

4. Synergistic Action of Gut AMPs with Conventional Antibiotics

Due to the rapid emergence of multidrug resistance and the reduced efficacy of co ventional antibiotics, the synergistic action of gut AMPs with antibiotics is explored an suggested as a new approach to control drug-resistant bacteria (Table 2). Interestingl AMPs display multiple mechanisms of action at a time that include membrane pore fo mation, inhibition of cell wall synthesis, biofilm disruption, inhibition of spore formatio and inhibition of protein synthesis and folding, along with inhibition of DNA and RN synthesis [12]. Especially in the complex gut environment with the possibility of numerou unknown interactions, the multiple-mode-of-action scenario of AMPs is intriguing. Th synergistic action of conventional antibiotics with gut AMPs is possibly benefited by e tended pore opening on the target cell membrane, increased membrane permeabilization, and subsequently increased repair time that further results in altered bacterial intracellular functions and overall bactericidal activity (Figure 1). In one of our previous studies, we have shown that laterosporulin10, a defensin-like bacteriocin produced by *Brevibacillus laterosporus* SKDU10, exhibits a synergetic effect with rifampicin against *Mycobacterium tuberculosis* H37Rv. It is confirmed that the addition of 0.25 μM laterosporulin10 results in a four-fold reduction of the rifampicin MIC values against *M. tuberculosis* [62]. Gut AMPs, nisin Z, and pediocin PA-1 including colistin were reported to have potential synergistic effects against MAR *P. fluorescens* when used in combination with antibiotics such as kanamycin, tetracycline, and chloramphenicol [63]. Nisin is also reported to have synergistic antimicrobial action with peptidoglycan-modulating antibiotics and ramoplanin, and exhibits promising activity against methicillin-resistant *S. aureus* (MRSA) and vancomycin-resistant enterococci (VRE). Furthermore, nisin demonstrates improved antibiofilm and antibacterial activity against *E. faecalis* by exhibiting synergistic effects with antibiotics such as penicillin, ciprofloxacin, and chloramphenicol [64]. A different study showed the synergistic effects of nisin with several antibiotics, including penicillin, amoxicillin, tetracycline, streptomycin, and ceftiofur against the swine pathogen *Streptococcus suis* that is known to cause severe infections in pigs [65]. Another study reported the synergic effects of subtilosin with clindamycin and metronidazole when used against *Gardnerella vaginalis*, which causes bacterial vaginosis [66]. In vitro, the activity of various human AMPs, LL-37, HBD1 to HBD3, HNP1, and HD5 have been checked against *C. difficile* in combinations of different antibiotics including tigecycline, moxifloxacin, piperacillin-tazobactam, and meropenem. Interestingly, LL-37 and HBD3 were found to have synergistic action against *C. difficile* with all the tested antibiotics [67]. Cryptdin 2, an AMP produced by Paneth cells, showed a synergistic effect against MAR *S. typhimurium* when used in combination with ampicillin [68]. HNP-1 was also confirmed to exhibit synergistic action with rifampicin and isoniazid against *M. tuberculosis* H37Rv [69]. Further, LL-37-derived membrane active analogs, FK13-a1 and FK13-a7, showed synergistic action against multidrug-resistant *P. aeruginosa* (MDRPA) and methicillin-resistant *S. aureus* (MRSA) when used in combination with chloramphenicol [70]. Next, LL-37 and colistin are reported to have synergistic action against MAR carbapenem-resistant *P. aeruginosa*, *Klebsiella pneumoniae*, and *Acinetobacter baumannii* when used in combination with the antibiotic azithromycin [71]. In a different study, short-cationic AMPs exhibited a synergistic effect with the antibiotics polymyxin B, erythromycin, and tetracycline against MDRPA [72]. A pilot study confirmed the synergistic action of colistin and the antibiotic tobramycin against *P. aeruginosa* [73]. Although multiple reports are available with improved results concerning the synergistic actions of gut AMPs with conventional antibiotics, the specific mechanisms of action need to be studied in further detail. However, in light of the evidence and analyzed results, it is possible to develop synergistic combinations of gut AMPs and antibiotics for the treatment of MAR human pathogens.

Table 2. Gut AMPs in synergy with conventional antibiotics.

Gut AMPs	Antibiotics	Target	References
Nisin	Ramoplanin	MRSA	[74]
	Polymyxin E Clarithromycin	*P. aeruginosa*	[75]
	Amoxicillin Penicillin Streptomycin Ceftiofur Tetracycline	*S. suis*	[65]

Table 2. *Cont.*

Gut AMPs	Antibiotics	Target	References
Nisin Z	Ampicillin Chloramphenicol Kanamycin Lincomycin Penicillin G Rifampicin Streptomycin Tetracycline Vancomycin	*P. fluorescens* LRC-R73 and its Penicillin-resistant/Streptomycin-resistant/Lincomycin-resistant/Rifampicin-resistant variant	[63]
Lacticin 3147	Polymyxin B	*S. aureus* 5247	[76]
Actagardine	Ramoplanin Metronidazole Vancomycin	*C. difficile*	[77]
Thuricin CD	Ramoplanin	*C. difficile*	[77]
	Vancomycin	*C. difficile*	[77]
Subtilosin A	Clindamycin phosphate Metronidazole	*G. vaginalis*	[66]
	Lauramide arginate Ester poly-lysine	*G. vaginalis*	[66]
PsVP-10	Chlorhexidine	*S. mutans* *S. sobrinus*	[78]
Plantaricin E, F, J, K	Several antibiotics	*C. albicans*	[79]
Colistin	Tobramycin	*P. aeruginosa*	[73]
Cryptdin 2	Ampicillin	*S. typhimurium*	[68]
Laterosporulin10	Rifampicin	*M. tuberculosis* H37Rv	[62]
Colistin	Azithromycin	*A. baumannii* *K. pneumoniae* *P. aeruginosa*	[71]
LL-37	Azithromycin	*A. baumannii* *K. pneumoniae* *P. aeruginosa*	[71]
Human defensin 5 (HD5)	Meropenem	*C. difficile*	[67]
Human neutrophil peptide-1 (HNP1)	Rifampicin	*M. tuberculosis* H37Rv	[69]
Human β-defensin 3 (HBD3)	Meropenem Moxifloxacin Piperacillin-Tazobactam Tigecycline	*C. difficile*	[67]

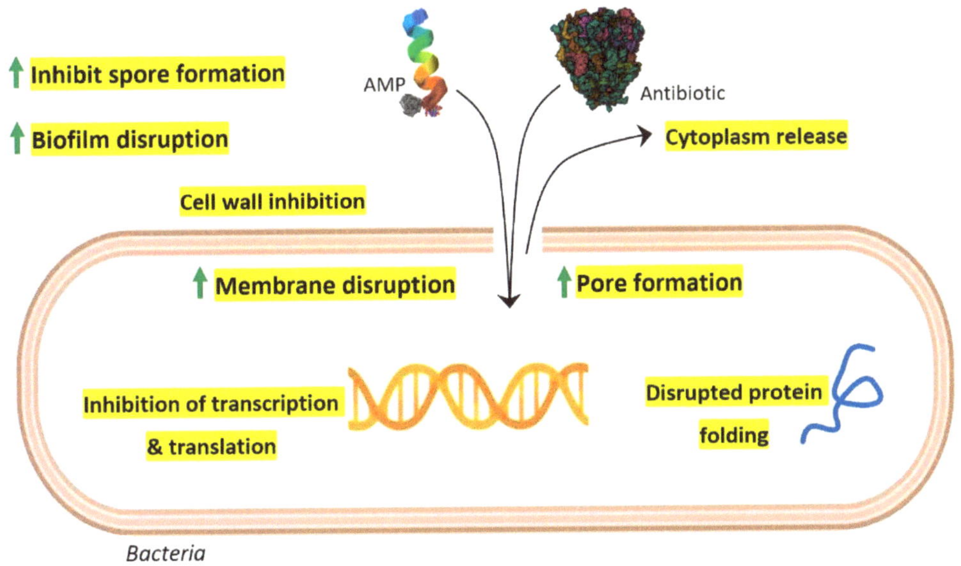

Figure 1. Possible model for synergistic antimicrobial activity of gut AMPs with conventional antibiotics. As per their membrane-acting properties, continuous pore formation and increased membrane permeabilization by AMPs allow more influx of antibiotics and AMPs which results in efficient bactericidal activity along with improved targeting of intracellular components such as transcription, protein synthesis machinery, and protein folding. Gut AMPs might also facilitate enhanced biofilm disruption and inhibition of spore formation when used in combination with antibiotics.

5. Gut AMPs, Conventional Antibiotics, and Evolution of Resistance Development

A major global public health concern is bacterial resistance to small-molecule antibiotics that are already on the market. The global spread of antibiotic-resistant bacteria has created the possibility of a post-antibiotic age in which ordinary illnesses and small wounds could develop into potentially fatal conditions. Such resistance has resulted in the creation of multidrug-resistant bacteria over the past few decades, which can both endanger healthy people and cause serious infections in immunocompromised patients. For instance, hospital-acquired infections with ampicillin-resistant *E. coli*, vancomycin-resistant *E. faecalis*, and methicillin-resistant *Staphylococcus aureus* (MRSA) have all increased in frequency [80,81]. Thus, there is an urgent need for novel antimicrobial strategies given the rising threat of MAR bacteria.

Antibiotic misuse has contributed to the emergence of MAR organisms. MAR infections are a leading source of morbidity and mortality worldwide [57]. Microbes can create and use defense and resistance mechanisms against the substances used to eradicate them in a complex environment such as the human gut, which is the home of over 100 trillion bacteria. Interestingly, not only the external antibiotics but also the antimicrobial substances produced by competitors present a challenge for the gut resident bacterial community to survive. Further, in the case of dysbiosis, an additional competition force exists between beneficial and harmful gut microbiota. AMPs are one such strategy used by bacteria (beneficial or harmful) to kill their competitors present in the surrounding complex environment. In addition to all of this, host-gut-derived AMPs are also present in the gut under the regulatory pressure of foreign AMPs, antibiotics, and the presence of their producers. Overall, there are multiple dynamic interactions present in the complex environment of the gut between various gut AMPs and antibiotics, whether internal or external. Together, these dynamic interactions and regulatory pressures create an evolutionary force under which microorganisms acquire a fair chance to evolve survival strategies and eventually develop antibiotic resistance (Figure 2). A pool of resistant genes belonging to several classes antibiotics has been identified in a recent metagenomic study of gut resistome conduct across different continents [82].

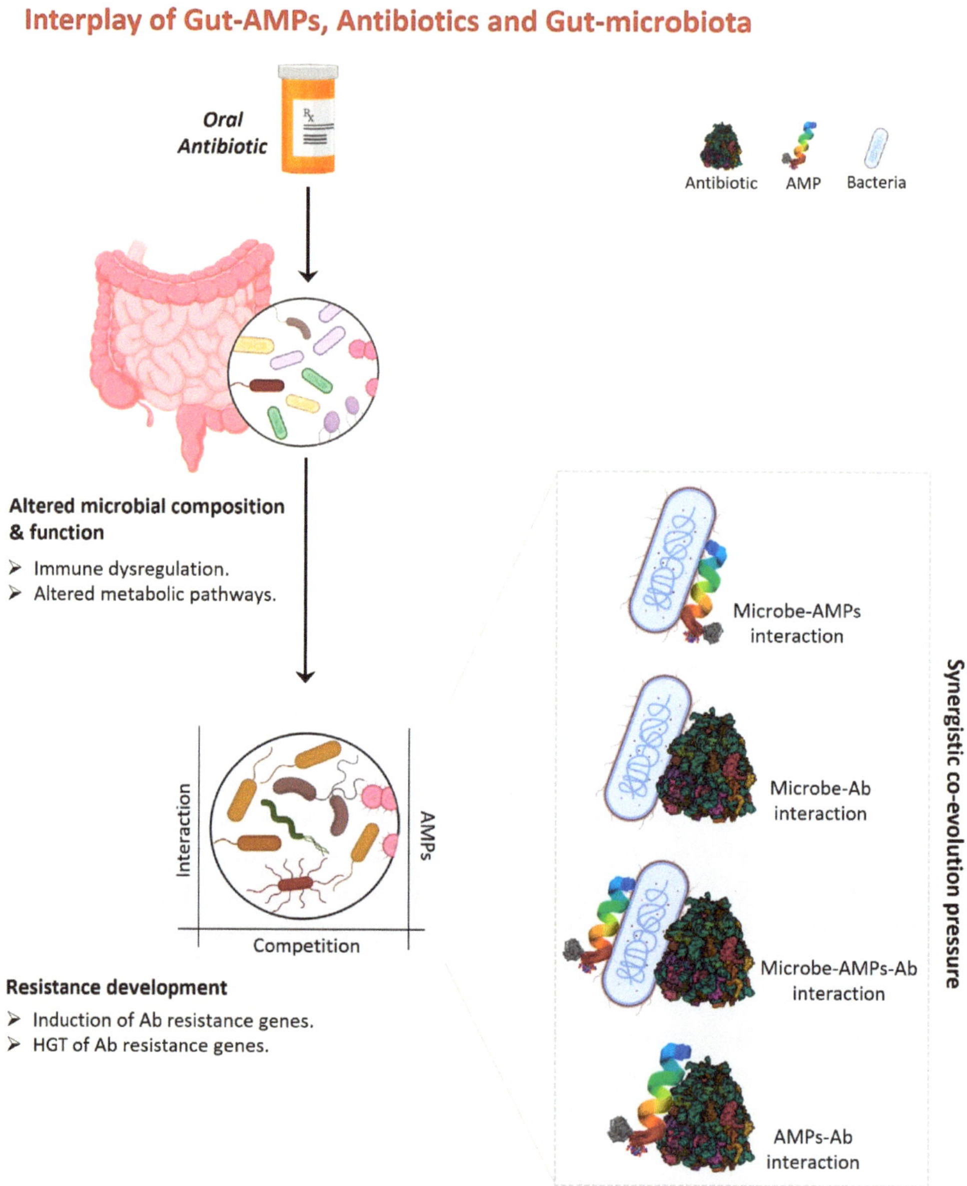

Figure 2. The interplay of gut AMPs, antibiotics, and gut microbiota is driven by various interactior among them. These interactions develop a synergistic co-evolutionary pressure under which gu AMPs are co-evolved to fight against MAR.

Natural gut resident bacteria, bacteria with acquired resistance genes, and acquired bacteria with resistance genes that do not typically colonize the gut are all included in the gut resistance reservoir [83]. Although it is uncommon, it is conceivable for resistance genes or virulence features to be transferred between pathogenic and non-pathogenic gut resident bacteria. The interesting question of how the resident gut bacteria and gut AMPs have maintained their efficiency through evolutionary timeframes is prompted by the growing issue of antibiotic-resistant bacteria. This question may have a partial explanation in the fact that there is a huge diversity of AMPs in the gut produced by intestinal epithelial cells as well as by healthy gut microbiota, decreasing the chance of combination resistance. Furthermore, since AMPs usually target bacterial cell walls and cell membranes that bacteria typically cannot modify without endangering their fitness, targeting crucial cell walls or cell membrane components likely also adds to the long-term efficiency of gut AMPs.

6. Antimicrobial Stewardship and Modulation of Gut AMPs as a Tool to Fight against Resistance Development

The rationalized use of antibiotics is an important aspect of fighting against antimicrobial resistance by maintaining gut homeostasis and reducing alterations to gut AMPs. The rationalized use of antibiotics includes the choice, dose, and duration of antibiotic therapy. It has been reported in several large meta-analysis studies that using antimicrobial stewardship programs resulted in a reduced number of infections with MAR organisms [84,85]. The type and spectrum of antibiotics employed are critical factors in the development of resistance in the targeted microorganisms. The majority of the commensal population is anaerobic; thus, inappropriate and extended usage of anti-anaerobic antibiotics has been linked to an increased risk of MAR [86,87]. It has been reported that the use of narrow-spectrum antibiotics in place of anti-anaerobic antibiotics is favorable to the human gut microbiota since fewer commensals are impacted [88,89]. Further, the duration of antibiotic therapy has a direct impact on gut microbiota composition as it has been reported that shorter antibiotic courses result in fewer microbial disturbances and quicker gut microbiota restoration [88,90]. Moreover, the right dose of antibiotic is very important as lower doses are linked with less chance of resistance gene development; however, lower doses for extended periods can also cause resistance [91,92].

Interestingly, gut AMPs have great promise as innovative therapeutic antibiotics because they do not easily develop bacterial resistance. The broad development of AMPs as medicines, however, has been hampered by several factors. First, AMPs can have relatively short half-lives because they are extremely sensitive to proteolytic breakdown by microbial and host enzymes. Second, many AMPs are harmful to the membranes of eukaryotic cells and display cytotoxicity.

Protein engineering techniques can be used to improve the bioavailability or efficacy of the AMPs because of their proteinaceous nature. It is possible to generate AMP versions that are resistant to enzymatic digestion. Also, using engineering peptidomimetics, new variants of AMPs could be generated with an altered number of charged amino acid residues with decreased hydrophobicity and cytotoxicity as well [93]. Additionally, the majority of AMPs kill bacteria by direct interaction with bacterial membranes. Interestingly, D-entantiomers of AMPs have longer half-lives and are just as effective at penetrating membranes as their natural L-entantiomers, so they can be used to improve the therapeutic efficacy of AMPs [94]. Further, packaging and delivering natural AMPs or their peptidomimetic analogs via nanoparticles can minimize non-specific cytotoxicity and improve stability with targeted bioavailability [95]. Additionally, novel AMPs should be employed for in vivo screening because the actual gut environment is completely different with the presence of different interactions with other commensals and their secreted AMPs which are already present in the gut. It is worth investigating the efficacy of new AMPs in the real dynamic gut environment against pathogenic bacteria or in conjunction with conventional antibiotics. Further new animal models with a controlled gut environment can be employed to check

AMP efficacy in combination with antibiotics. Additionally, the gut has a diverse microbi ecology that differs for each individual. It is important to fully understand the gut microbi ecology for a detailed understanding of the interaction of gut AMPs with convention antibiotics in the presence of other eukaryotic organisms including viruses, bacteriophag and fungi. The interaction of gut AMPs with these diverse ecological community membe individually or as a whole should be considered to understand their impact on gut AM evolution and resistance development. Next-generation sequencing, transcriptomics, ar gene expression analysis can further elucidate the mechanistic overview of complex g environments that sheds light on unanswered questions and will further help in tl development of a strategy to fight against resistance development.

7. Conclusions and Future Perspective

Combinatory use of AMPs produced by both host and gut microbiota with conve tional antibiotics could result in synergistic actions in different ways. It has been predicte that every species contains a unique set of AMPs that are evolved to defend the host again the microorganisms they might encounter [96]. This phenomenon becomes more comple and functionally specific in the case of the gut. The human gut is inhabited by millior of commensals which constitute the specific set of bacteria for every individual that further affected by dietary habits, environment, and many more factors. Interestingly, the is a highly competitive environment in the gut so gut microbes are known to produc AMPs with various biological activities including immunomodulatory activities. Alon with AMPs produced by gut microbiota, there are multiple host AMPs secreted in the gu in the proximity of gut epithelium and gut microbiota. It is hypothesized that all the gu AMPs synergistically affect each other's functions to drive complex gut functions such regulation of gut homeostasis; however, the mechanisms of this are not fully understoo In addition to fighting against infectious pathogens, gut AMPs play an essential role in th regulation of bacterial symbionts and communities in the gut, thus maintaining a balanc between health and pathogenic microbes [97]. Further, gut microbiota exhibit high intrins resistance to AMPs which suggests that gut AMPs could be a customizable tool to mai tain healthy gut communities [98]. Additionally, recent accumulating pieces of evidenc suggest a functional synergism among the different gut AMPs [99]. The gut synergis may also reduce the chances of resistance evolution. Further, the synergistic mechanism of gut AMPs could be used effectively in combination with conventional antibiotics t combat MAR. Another factor is that the host regulates the gut AMPs synergistically in suc a way that limits the chances of rapid resistance evolution. These synergistic strategie could be further used for the effective translation of AMPs alone or in combination wit conventional antibiotics into therapeutic applications.

The human gut and AMP-producing intestinal epithelium constantly face a challen ing dynamic microbial environment and also produce various antimicrobials for thei survival that eventually affect the overall gut immune response including the efficacy c antibiotics during infection. To meet this challenge of the dynamic microbiome of th gut, epithelial cells also produce a wide variety of AMPs that quickly kill or inactivat bacteria, while a similar action is performed by the gut commensals to maintain the health gut environment which is called homeostasis. However, how the gut immune systen differentiates between the healthy and pathogenic microbiota is still not well understoo and remains a question of further research. On the other hand, in addition to this interna healthy equilibrium within the gut immune system, antibiotic treatment during infectio creates another challenge for gut homeostasis. While both gut epithelium and commensal bear the adverse effects of antibiotics, gut AMPs have enough of a chance to interact wit antibiotics, which affects the treatment efficacy as well (Table 2). However, it is not clea how AMPs interact with antibiotics and what is the response of AMP-producing gut ep ithelium and commensals in this dynamic complex gut environment. The emerging pictur is that epithelial AMPs influence the structure and location of gut commensals in additio to protecting against pathogen colonization and invasion in synergism with the AMP produced by commensals. Overall, gut AMPs are evolved for their antimicrobial action, efficacy, and spectrum under synergistic co-evolution with host immunity and commensals, along with interactions with other AMPs and conventional antibiotics (Figure 2).

Finally, MAR resistance is rapidly growing while the discovery and availability of new antimicrobials are slow which generates an urgent demand for new antimicrobials along with a fully elucidated mechanism of resistance to overcome this crisis. At this point, the dissection of the gut microbiome as an antimicrobial resistance reservoir is much needed. This can be achieved by clinical and translational studies exploring the interaction of gut microbiome and gut AMPs within the gut microbial ecology and with conventional antibiotics. Functional metagenomic studies might be very helpful in identifying the uncultivable gut microbes and their role in resistance development and evolution.

8. Unanswered Questions about Gut Microbiota and Gut AMPs

- What makes the gut microbiome healthy and what are the deciding bio-markers?
- What is the genetic machinery that regulates the production of gut AMPs?
- How do gut AMPs play a role in resistance development?
- Could diet help in the fight against resistance by manipulating gut microbiota? How?
- How do gut AMPs regulate the immune response to fight against resistance?
- How to reconstruct the gut microbiome and gut AMPome to counter antibiotic resistance?

Author Contributions: P.B. conceptualized and wrote the manuscript and prepared the illustrations. P.B. and S.M.M. edited and proofread the manuscript. All authors have read and agreed to the published version of the manuscript.

Funding: This research received no external funding.

Acknowledgments: The authors are thankful to the Department of Radiation Oncology, University of Missouri, Columbia, and the Department of Biotechnology, Indian Institute of Technology, Kharagpur for providing the space and necessary facilities for this work.

Conflicts of Interest: The authors declare no conflict of interest.

eferences

Murray, C.J.; Ikuta, K.S.; Sharara, F.; Swetschinski, L.; Robles Aguilar, G.; Gray, A.; Han, C.; Bisignano, C.; Rao, P.; Wool, E.; et al. Global burden of bacterial antimicrobial resistance in 2019: A systematic analysis. *Lancet* **2022**, *399*, 629–655. [CrossRef] [PubMed]

Ventola, C.L. The antibiotic resistance crisis: Causes and threats. *Pharm. Ther.* **2015**, *40*, 277–283.

Mwangi, J.; Hao, X.; Lai, R.; Zhang, Z.Y. Antimicrobial peptides: New hope in the war against multidrug resistance. *Zool. Res.* **2019**, *40*, 488–505. [CrossRef]

Baindara, P.; Chaudhry, V.; Mittal, G.; Liao, L.M.; Matos, C.O.; Khatri, N.; Franco, O.L.; Patil, P.B.; Korpole, S. Characterization of the antimicrobial peptide penisin, a class Ia novel lantibiotic from *Paenibacillus* sp. strain A3. *Antimicrob. Agents Chemother.* **2016**, *60*, 580–591. [CrossRef] [PubMed]

Josef, J. Drug repurposing to overcome microbial resistance. *Drug Discov. Today* **2022**, *27*, 2028–2041.

Deslouches, B.; Montelaro, R.C.; Urish, K.L.; Di, Y.P. Engineered cationic antimicrobial peptides (eCAPs) to combat multidrug-resistant bacteria. *Pharmaceutics* **2020**, *12*, 501. [CrossRef]

Baindara, P.; Mandal, S.M. Antimicrobial Peptides and Vaccine Development to Control Multi-drug Resistant Bacteria. *Protein Pept. Lett.* **2019**, *26*, 324–331. [CrossRef]

Mathur, H.; Field, D.; Rea, M.C.; Cotter, P.D.; Hill, C.; Ross, R.P. Fighting biofilms with lantibiotics and other groups of bacteriocins. *NPJ Biofilms Microbiomes* **2018**, *4*, 9. [CrossRef]

Wang, S.; Thacker, P.; Watford, M.; Qiao, S. Functions of Antimicrobial Peptides in Gut Homeostasis. *Curr. Protein Pept. Sci.* **2015**, *16*, 582–591. [CrossRef]

0. Lewies, A.; Du Plessis, L.H.; Wentzel, J.F. Antimicrobial Peptides: The Achilles' Heel of Antibiotic Resistance? *Probiotics Antimicrob. Proteins* **2019**, *11*, 370–381. [CrossRef] [PubMed]

1. Costa, F.; Teixeira, C.; Gomes, P.; Martins, M.C.L. Clinical application of AMPs. In *Advances in Experimental Medicine and Biology*; Springer: Berlin/Heidelberg, Germany, 2019; Volume 1117, pp. 281–298.

2. Baindara, P.; Ghosh, A.K.; Mandal, S.M. Coevolution of Resistance Against Antimicrobial Peptides. *Microb. Drug Resist.* **2020**, *26*, 880–899. [CrossRef] [PubMed]

3. El Shazely, B.; Yu, G.; Johnston, P.R.; Rolff, J. Resistance Evolution Against Antimicrobial Peptides in Staphylococcus aureus Alters Pharmacodynamics Beyond the MIC. *Front. Microbiol.* **2020**, *11*, 103. [CrossRef] [PubMed]

14. Dinata, R.; Baindara, P. Laterosporulin25: A probiotically produced, novel defensin-like bacteriocin and its immunoger properties. *Int. Immunopharmacol.* **2023**, *121*, 110500. [CrossRef] [PubMed]
15. Francino, M.P. Antibiotics and the human gut microbiome: Dysbioses and accumulation of resistances. *Front. Microbiol.* **20** *6*, 1543. [CrossRef]
16. Langdon, A.; Crook, N.; Dantas, G. The effects of antibiotics on the microbiome throughout development and alternati approaches for therapeutic modulation. *Genome Med.* **2016**, *8*, 39. [CrossRef]
17. Rafii, F.; Sutherland, J.B.; Cerniglia, C.E. Effects of treatment with antimicrobial agents on the human colonic microflora. *Th Clin. Risk Manag.* **2008**, *4*, 1343–1357. [CrossRef]
18. Rashid, M.U.; Weintraub, A.; Nord, C.E. Effect of new antimicrobial agents on the ecological balance of human microflo *Anaerobe* **2012**, *18*, 249–253. [CrossRef]
19. Ivanov, I.I.; de Frutos, R.L.; Manel, N.; Yoshinaga, K.; Rifkin, D.B.; Sartor, R.B.; Finlay, B.B.; Littman, D.R. Specific Microbio Direct the Differentiation of IL-17-Producing T-Helper Cells in the Mucosa of the Small Intestine. *Cell Host Microbe* **2008**, 337–349. [CrossRef]
20. Wlodarska, M.; Willing, B.; Keeney, K.M.; Menendez, A.; Bergstrom, K.S.; Gill, N.; Russell, S.L.; Vallance, B.A.; Finlay, B Antibiotic treatment alters the colonic mucus layer and predisposes the host to exacerbated Citrobacter rodentium-induced coli *Infect. Immun.* **2011**, *79*, 1536–1545. [CrossRef]
21. Greenwood, C.; Morrow, A.L.; Lagomarcino, A.J.; Altaye, M.; Taft, D.H.; Yu, Z.; Newburg, D.S.; Ward, D.V.; Schibler, K.R. Ear empiric antibiotic use in preterm infants is associated with lower bacterial diversity and higher relative abundance of enterobact *J. Pediatr.* **2014**, *165*, 23–29. [CrossRef]
22. Lee, J.-Y.; Tsolis, R.M.; Bäumler, A.J. The microbiome and gut homeostasis. *Science* **2022**, *377*, eabp9960. [CrossRef]
23. Cotter, P.D.; Ross, R.P.; Hill, C. Bacteriocins-a viable alternative to antibiotics? *Nat. Rev. Microbiol.* **2013**, *11*, 95–105. [CrossRe [PubMed]
24. Donia, M.S.; Cimermancic, P.; Schulze, C.J.; Wieland Brown, L.C.; Martin, J.; Mitreva, M.; Clardy, J.; Linington, R.G.; Fischbac M.A. A systematic analysis of biosynthetic gene clusters in the human microbiome reveals a common family of antibiotics. *C* **2014**, *158*, 1402–1414. [CrossRef] [PubMed]
25. Corr, S.C.; Li, Y.; Riedel, C.U.; O'Toole, P.W.; Hill, C.; Gahan, C.G.M. Bacteriocin production as a mechanism for the antiinfecti activity of Lactobacillus salivarius UCC118. *Proc. Natl. Acad. Sci. USA* **2007**, *104*, 7617–7621. [CrossRef] [PubMed]
26. Rea, M.C.; Sit, C.S.; Clayton, E.; O'Connor, P.M.; Whittal, R.M.; Zheng, J.; Vederas, J.C.; Ross, R.P.; Hill, C. Thuricin CD, posttranslationally modified bacteriocin with a narrow spectrum of activity against Clostridium difficile. *Proc. Natl. Acad. S USA* **2010**, *107*, 9352–9357. [CrossRef]
27. Gilmore, M.S.; Lebreton, F.; van Schaik, W. Genomic transition of enterococci from gut commensals to leading causes multidrug-resistant hospital infection in the antibiotic era. *Curr. Opin. Microbiol.* **2013**, *16*, 10–16. [CrossRef] [PubMed]
28. Kommineni, S.; Bretl, D.J.; Lam, V.; Chakraborty, R.; Hayward, M.; Simpson, P.; Cao, Y.; Bousounis, P.; Kristich, C.J.; Salzma N.H. Bacteriocin production augments niche competition by enterococci in the mammalian gastrointestinal tract. *Nature* **20** *526*, 719–722. [CrossRef]
29. Sassone-Corsi, M.; Nuccio, S.P.; Liu, H.; Hernandez, D.; Vu, C.T.; Takahashi, A.A.; Edwards, R.A.; Raffatellu, M. Microci mediate competition among Enterobacteriaceae in the inflamed gut. *Nature* **2016**, *540*, 280–283. [CrossRef]
30. Riboulet-Bisson, E.; Sturme, M.H.J.; Jeffery, I.B.; O'Donnell, M.M.; Neville, B.A.; Forde, B.M.; Claesson, M.J.; Harris, H.; Gardin G.E.; Casey, P.G.; et al. Effect of lactobacillus salivarius bacteriocin ABP118 on the mouse and pig intestinal microbiota. *PLoS ON* **2012**, *7*, e31113. [CrossRef]
31. Ilinskaya, O.N.; Ulyanova, V.V.; Yarullina, D.R.; Gataullin, I.G. Secretome of Intestinal bacilli: A natural guard against pathologi *Front. Microbiol.* **2017**, *8*, 1666. [CrossRef]
32. O' Connor, P.M.; O' Shea, E.F.; Cotter, P.D.; Hill, C.; Ross, R.P. The potency of the broad spectrum bacteriocin, bactofencin against staphylococci is highly dependent on primary structure, N-terminal charge and disulphide formation. *Sci. Rep.* **20** *8*, 11833. [CrossRef] [PubMed]
33. Huang, F.; Teng, K.; Liu, Y.; Cao, Y.; Wang, T.; Ma, C.; Zhang, J.; Zhong, J. Bacteriocins: Potential for Human Health. *Oxid. Me Cell. Longev.* **2021**, *2021*, 5518825. [CrossRef] [PubMed]
34. Belfiore, C.; Castellano, P.; Vignolo, G. Reduction of Escherichia coli population following treatment with bacteriocins from lact acid bacteria and chelators. *Food Microbiol.* **2007**, *24*, 223–229. [CrossRef] [PubMed]
35. Rea, M.C.; Clayton, E.; O'Connor, P.M.; Shanahan, F.; Kiely, B.; Ross, R.P.; Hill, C. Antimicrobial activity of lacticin 3147 again clinical Clostridium difficile strains. *J. Med. Microbiol.* **2007**, *56*, 940–946. [CrossRef] [PubMed]
36. Borrero, J.; Brede, D.A.; Skaugen, M.; Diep, D.B.; Herranz, C.; Nes, I.F.; Cintas, L.M.; Hernández, P.E. Characterization of garvici ML, a novel circular bacteriocin produced by lactococcus garvieae DCC43, isolated from mallard ducks (*Anas platyrhynchos*). *Ap Environ. Microbiol.* **2011**, *77*, 369–373. [CrossRef]
37. Millette, M.; Cornut, G.; Dupont, C.; Shareck, F.; Archambault, D.; Lacroix, M. Capacity of human nisin- and pediocin-producin lactic acid bacteria to reduce intestinal colonization by vancomycin-resistant enterococci. *Appl. Environ. Microbiol.* **200** *74*, 1997–2003. [CrossRef] [PubMed]
38. De Kwaadsteniet, M.; Doeschate, K.T.; Dicks, L.M.T. Nisin F in the treatment of respiratory tract infections caused by Staphyloco cus aureus. *Lett. Appl. Microbiol.* **2009**, *48*, 65–70. [CrossRef]

9. Goldstein, B.P.; Wei, J.; Greenberg, K.; Novick, R. Activity of nisin against Streptococcus pneumoniae, in vitro, and in a mouse infection model. *J. Antimicrob. Chemother.* **1998**, *42*, 277–278. [CrossRef]
0. Norouzi, Z.; Salimi, A.; Halabian, R.; Fahimi, H. Nisin, a potent bacteriocin and anti-bacterial peptide, attenuates expression of metastatic genes in colorectal cancer cell lines. *Microb. Pathog.* **2018**, *123*, 183–189. [CrossRef]
1. Dabour, N.; Zihler, A.; Kheadr, E.; Lacroix, C.; Fliss, I. In vivo study on the effectiveness of pediocin PA-1 and Pediococcus acidilactici UL5 at inhibiting Listeria monocytogenes. *Int. J. Food Microbiol.* **2009**, *133*, 225–233. [CrossRef]
2. Salvucci, E.; Saavedra, L.; Hebert, E.M.; Haro, C.; Sesma, F. Enterocin CRL35 inhibits Listeria monocytogenes in a murine model. *Foodborne Pathog. Dis.* **2012**, *9*, 68–74. [CrossRef]
3. Birri, D.J.; Brede, D.A.; Forberg, T.; Holo, H.; Nes, I.F. Molecular and genetic characterization of a novel bacteriocin locus in Enterococcus avium isolates from infants. *Appl. Environ. Microbiol.* **2010**, *76*, 483–492. [CrossRef]
4. De Kwaadsteniet, M.; Fraser, T.; Van Reenen, C.A.; Dicks, L.M.T. Bacteriocin T8, a novel class IIa sec-dependent bacteriocin produced by Enterococcus faecium T8, isolated from vaginal secretions of children infected with human immunodeficiency virus. *Appl. Environ. Microbiol.* **2006**, *72*, 4761–4766. [CrossRef] [PubMed]
5. Martin-Visscher, L.A.; van Belkum, M.J.; Garneau-Tsodikova, S.; Whittal, R.M.; Zheng, J.; McMullen, L.M.; Vederas, J.C. Isolation and characterization of carnocyclin a, a novel circular bacteriocin produced by *Carnobacterium maltaromaticum* UAL307. *Appl. Environ. Microbiol.* **2008**, *74*, 4756–4763. [CrossRef] [PubMed]
6. Chang, J.Y.; Chang, H.C. Growth Inhibition of Foodborne Pathogens by Kimchi Prepared with Bacteriocin-Producing Starter Culture. *J. Food Sci.* **2011**, *76*, M72–M78. [CrossRef]
7. Dey, D.; Ema, T.I.; Biswas, P.; Aktar, S.; Islam, S.; Rinik, U.R.; Firoz, M.; Ahmed, S.Z.; Al Azad, S.; Rahman, A.; et al. Antiviral effects of bacteriocin against animal-to-human transmittable mutated SARS-CoV-2: A systematic review. *Front. Agric. Sci. Eng.* **2021**, *8*, 603–622. [CrossRef]
8. Filipp, D.; Brabec, T.; Vobořil, M.; Dobeš, J. Enteric α-defensins on the verge of intestinal immune tolerance and inflammation. *Semin. Cell Dev. Biol.* **2019**, *88*, 138–146. [CrossRef] [PubMed]
9. Phadke, S.M.; Deslouches, B.; Hileman, S.E.; Montelaro, R.C.; Wiesenfeld, H.C.; Mietzner, T.A. Antimicrobial peptides in mucosal secretions: The importance of local secretions in mitigating infection. *J. Nutr.* **2005**, *135*, 1289–1293. [CrossRef]
0. Dutta, P.; Das, S. Mammalian Antimicrobial Peptides: Promising Therapeutic Targets Against Infection and Chronic Inflammation. *Curr. Top. Med. Chem.* **2015**, *16*, 99–129. [CrossRef]
1. Ayabe, T.; Satchell, D.P.; Wilson, C.L.; Parks, W.C.; Selsted, M.E.; Ouellette, A.J. Secretion of microbicidal α-defensins by intestinal Paneth cells in response to bacteria. *Nat. Immunol.* **2000**, *1*, 113–118. [CrossRef]
2. Menendez, A.; Willing, B.P.; Montero, M.; Wlodarska, M.; So, C.C.; Bhinder, G.; Vallance, B.A.; Finlay, B.B. Bacterial stimulation of the TLR-MyD88 pathway modulates the homeostatic expression of ileal paneth cell α-defensins. *J. Innate Immun.* **2013**, *5*, 39–49. [CrossRef] [PubMed]
3. Salzman, N.H.; Hung, K.; Haribhai, D.; Chu, H.; Karlsson-Sjöberg, J.; Amir, E.; Teggatz, P.; Barman, M.; Hayward, M.; Eastwood, D.; et al. Enteric defensins are essential regulators of intestinal microbial ecology. *Nat. Immunol.* **2010**, *11*, 76–83. [CrossRef] [PubMed]
4. Fusco, A.; Savio, V.; Cammarota, M.; Alfano, A.; Schiraldi, C.; Donnarumma, G. Beta-Defensin-2 and Beta-Defensin-3 Reduce Intestinal Damage Caused by Salmonella typhimurium Modulating the Expression of Cytokines and Enhancing the Probiotic Activity of *Enterococcus faecium*. *J. Immunol. Res.* **2017**, *2017*, 6976935. [CrossRef] [PubMed]
5. Yi, H.; Hu, W.; Chen, S.; Lu, Z.; Wang, Y. Cathelicidin-WA Improves Intestinal Epithelial Barrier Function and Enhances Host Defense against Enterohemorrhagic Escherichia coli O157:H7 Infection. *J. Immunol.* **2017**, *198*, 1696–1705. [CrossRef]
6. Iimura, M.; Gallo, R.L.; Hase, K.; Miyamoto, Y.; Eckmann, L.; Kagnoff, M.F. Cathelicidin Mediates Innate Intestinal Defense against Colonization with Epithelial Adherent Bacterial Pathogens. *J. Immunol.* **2005**, *174*, 4901–4907. [CrossRef]
7. Chromek, M.; Arvidsson, I.; Karpman, D. The Antimicrobial Peptide Cathelicidin Protects Mice from Escherichia coli O157:H7-Mediated Disease. *PLoS ONE* **2012**, *7*, e46476. [CrossRef]
8. Yoshimura, T.; McLean, M.H.; Dzutsev, A.K.; Yao, X.; Chen, K.; Huang, J.; Gong, W.; Zhou, J.; Xiang, Y.; Badger, J.H.; et al. The Antimicrobial Peptide CRAMP Is Essential for Colon Homeostasis by Maintaining Microbiota Balance. *J. Immunol.* **2018**, *200*, 2174–2185. [CrossRef]
9. Vaishnava, S.; Yamamoto, M.; Severson, K.M.; Ruhn, K.A.; Yu, X.; Koren, O.; Ley, R.; Wakeland, E.K.; Hooper, L.V. The antibacterial lectin RegIII gamma promotes the spatial segregation of microbiota and host in the intestine. *Science* **2011**, *334*, 255–258. [CrossRef]
0. Brandl, K.; Plitas, G.; Schnabl, B.; DeMatteo, R.P.; Pamer, E.G. MyD88-mediated signals induce the bactericidal lectin RegIIIγ and protect mice against intestinal Listeria monocytogenes infection. *J. Exp. Med.* **2007**, *204*, 1891–1900. [CrossRef]
1. Ju, T.; Shoblak, Y.; Gao, Y.; Yang, K.; Fouhse, J.; Finlay, B.B.; So, Y.W.; Stothard, P.; Willing, B.P. Initial gut microbial composition as a key factor driving host response to antibiotic treatment, as exemplified by the presence or absence of commensal Escherichia coli. *Appl. Environ. Microbiol.* **2017**, *83*, e01107-17. [CrossRef]
2. Baindara, P.; Singh, N.; Ranjan, M.; Nallabelli, N.; Chaudhry, V.; Pathania, G.L.; Sharma, N.; Kumar, A.; Patil, P.B.; Korpole, S. Laterosporulin10: A novel defensin like class iid bacteriocin from brevibacillus sp. strain SKDU10 with inhibitory activity against microbial pathogens. *Microbiology* **2016**, *162*, 1286–1299. [CrossRef]
3. Naghmouchi, K.; Le Lay, C.; Baah, J.; Drider, D. Antibiotic and antimicrobial peptide combinations: Synergistic inhibition of Pseudomonas fluorescens and antibiotic-resistant variants. *Res. Microbiol.* **2012**, *163*, 101–108. [CrossRef]

64. Tong, Z.; Zhang, Y.; Ling, J.; Ma, J.; Huang, L.; Zhang, L. An in vitro study on the effects of nisin on the antibacterial activities 18 antibiotics against *Enterococcus faecalis*. *PLoS ONE* **2014**, *9*, e89209. [CrossRef]
65. Lebel, G.; Piché, F.; Frenette, M.; Gottschalk, M.; Grenier, D. Antimicrobial activity of nisin against the swine pathogen Streptoc cus suis and its synergistic interaction with antibiotics. *Peptides* **2013**, *50*, 19–23. [CrossRef]
66. Cavera, V.L.; Volski, A.; Chikindas, M.L. The Natural Antimicrobial Subtilosin A Synergizes with Lauramide Arginine Eth Ester (LAE), ε-Poly-l-lysine (Polylysine), Clindamycin Phosphate and Metronidazole, Against the Vaginal Pathogen *Gardnere vaginalis*. *Probiotics Antimicrob. Proteins* **2015**, *7*, 164–171. [CrossRef]
67. Nuding, S.; Frasch, T.; Schaller, M.; Stange, E.F.; Zabel, L.T. Synergistic effects of antimicrobial peptides and antibiotics agair clostridium difficile. *Antimicrob. Agents Chemother.* **2014**, *58*, 5719–5725. [CrossRef]
68. Rishi, P.; Preet, S.; Bharrhan, S.; Verma, I. In vitro and in vivo synergistic effects of cryptdin 2 and ampicillin against Salmone *Antimicrob. Agents Chemother.* **2011**, *55*, 4176–4182. [CrossRef]
69. Kalita, A.; Verma, I.; Khuller, G.K. Role of human neutrophil peptide-1 as a possible adjunct to antituberculosis chemotherapy *Infect. Dis.* **2004**, *190*, 1476–1480. [CrossRef]
70. Rajasekaran, G.; Kim, E.Y.; Shin, S.Y. LL-37-derived membrane-active FK-13 analogs possessing cell selectivity, anti-biofil activity and synergy with chloramphenicol and anti-inflammatory activity. *Biochim. Biophys. Acta Biomembr.* **2017**, *1859*, 722–7 [CrossRef]
71. Lin, L.; Nonejuie, P.; Munguia, J.; Hollands, A.; Olson, J.; Dam, Q.; Kumaraswamy, M.; Rivera, H.; Corriden, R.; Rohde, M.; et Azithromycin Synergizes with Cationic Antimicrobial Peptides to Exert Bactericidal and Therapeutic Activity against High Multidrug-Resistant Gram-Negative Bacterial Pathogens. *EBioMedicine* **2015**, *2*, 690–698. [CrossRef]
72. Ruden, S.; Rieder, A.; Chis Ster, I.; Schwartz, T.; Mikut, R.; Hilpert, K. Synergy Pattern of Short Cationic Antimicrobial Peptid against Multidrug-Resistant Pseudomonas aeruginosa. *Front. Microbiol.* **2019**, *10*, 2740. [CrossRef]
73. Herrmann, G.; Yang, L.; Wu, H.; Song, Z.; Wang, H.; Høiby, N.; Ulrich, M.; Molin, S.; Riethmüller, J.; Döring, G. Colist tobramycin combinations are superior to monotherapy concerning the killing of biofilm Pseudomonas aeruginosa. *J. Infect. D* **2010**, *202*, 1585–1592. [CrossRef] [PubMed]
74. Brumfitt, W.; Salton, M.R.J.; Hamilton-Miller, J.M.T. Nisin, alone and combined with peptidoglycan-modulating antibioti Activity against methicillin-resistant *Staphylococcus aureus* and vancomycin-resistant enterococci. *J. Antimicrob. Chemother.* **20** *50*, 731–734. [CrossRef] [PubMed]
75. Giacometti, A.; Cirioni, O.; Barchiesi, F.; Fortuna, M.; Scalise, G. In-vitro activity of cationic peptides alone and in combinati with clinically used antimicrobial agents against *Pseudomonas aeruginosa*. *J. Antimicrob. Chemother.* **1999**, *44*, 641–645. [CrossR [PubMed]
76. Draper, L.A.; Cotter, P.D.; Hill, C.; Ross, R.P. The two peptide lantibiotic lacticin 3147 acts synergistically with polymyxin inhibit Gram negative bacteria. *BMC Microbiol.* **2013**, *13*, 212. [CrossRef] [PubMed]
77. Mathur, H.; O'Connor, P.M.; Hill, C.; Cotter, P.D.; Ross, R.P. Analysis of anti-clostridium difficile activity of thuricin C vancomycin, metronidazole, ramoplanin, and actagardine, both singly and in paired combinations. *Antimicrob. Agents Chemoth* **2013**, *57*, 2882–2886. [CrossRef]
78. Lobos, O.; Padilla, A.; Padilla, C. In vitro antimicrobial effect of bacteriocin PsVP-10 in combination with chlorhexidine ar triclosan against Streptococcus mutans and Streptococcus sobrinus strains. *Arch. Oral Biol.* **2009**, *54*, 230–234. [CrossRef]
79. Sharma, A.; Srivastava, S. Anti-Candida activity of two-peptide bacteriocins, plantaricins (Pln E/F and J/K) and their mode action. *Fungal Biol.* **2014**, *118*, 264–275. [CrossRef]
80. Larsen, J.; Raisen, C.L.; Ba, X.; Sadgrove, N.J.; Padilla-González, G.F.; Simmonds, M.S.J.; Loncaric, I.; Kerschner, H.; Apfalter, Hartl, R.; et al. Emergence of methicillin resistance predates the clinical use of antibiotics. *Nature* **2022**, *602*, 135–141. [CrossRe
81. O'Toole, R.F.; Leong, K.W.C.; Cumming, V.; Van Hal, S.J. Vancomycin-resistant Enterococcus faecium and the emergence of ne sequence types associated with hospital infection. *Res. Microbiol.* **2023**, *174*, 104046. [CrossRef]
82. Forslund, K.; Sunagawa, S.; Kultima, J.R.; Mende, D.R.; Arumugam, M.; Typas, A.; Bork, P. Country-specific antibiotic u practices impact the human gut resistome. *Genome Res.* **2013**, *23*, 1163–1169. [CrossRef] [PubMed]
83. Anthony, W.E.; Burnham, C.A.D.; Dantas, G.; Kwon, J.H. The gut microbiome as a reservoir for antimicrobial resistance. *J. Infe Dis.* **2021**, *223*, S209–S213. [CrossRef] [PubMed]
84. Baur, D.; Gladstone, B.P.; Burkert, F.; Carrara, E.; Foschi, F.; Döbele, S.; Tacconelli, E. Effect of antibiotic stewardship on t incidence of infection and colonisation with antibiotic-resistant bacteria and Clostridium difficile infection: A systematic revie and meta-analysis. *Lancet Infect. Dis.* **2017**, *17*, 990–1001. [CrossRef] [PubMed]
85. Ya, K.Z.; Win, P.T.N.; Bielicki, J.; Lambiris, M.; Fink, G. Association Between Antimicrobial Stewardship Programs and Antibio Use Globally: A Systematic Review and Meta-Analysis. *JAMA Netw. Open* **2023**, *6*, E2253806. [CrossRef]
86. Chanderraj, R.; Baker, J.M.; Kay, S.G.; Brown, C.A.; Hinkle, K.J.; Fergle, D.J.; McDonald, R.A.; Falkowski, N.R.; Metcalf, J.I Kaye, K.S.; et al. In critically ill patients, anti-anaerobic antibiotics increase risk of adverse clinical outcomes. *Eur. Respir. J.* **202** *61*, 2200910. [CrossRef] [PubMed]
87. Bhalla, A.; Pultz, N.J.; Ray, A.J.; Hoyen, C.K.; Eckstein, E.C.; Donskey, C.J. Antianaerobic Antibiotic Therapy Promotes Overgrow of Antibiotic-Resistant, Gram-Negative Bacilli and Vancomycin-Resistant Enterococci in the Stool of Colonized Patients. *Infe Control Hosp. Epidemiol.* **2003**, *24*, 644–649. [CrossRef] [PubMed]

Mitchell, B.G.; Hall, L.; White, N.; Barnett, A.G.; Halton, K.; Paterson, D.L.; Riley, T.V.; Gardner, A.; Page, K.; Farrington, A.; et al. An environmental cleaning bundle and health-care-associated infections in hospitals (REACH): A multicentre, randomised trial. *Lancet Infect. Dis.* **2019**, *19*, 410–418. [CrossRef] [PubMed]

Shahi, F.; Redeker, K.; Chong, J. Rethinking antimicrobial stewardship paradigms in the context of the gut microbiome. *JAC—Antimicrob. Resist.* **2019**, *1*, dlz015. [CrossRef]

Leo, S.; Lazarevic, V.; von Dach, E.; Kaiser, L.; Prendki, V.; Schrenzel, J.; Huttner, B.D.; Huttner, A. Effects of antibiotic duration on the intestinal microbiota and resistome: The PIRATE RESISTANCE project, a cohort study nested within a randomized trial. *EBioMedicine* **2021**, *71*, 103566. [CrossRef]

Zhang, L.; Huang, Y.; Zhou, Y.; Buckley, T.; Wang, H.H. Antibiotic administration routes significantly influence the levels of antibiotic resistance in gut microbiota. *Antimicrob. Agents Chemother.* **2013**, *57*, 3659–3666. [CrossRef]

Wistrand-Yuen, E.; Knopp, M.; Hjort, K.; Koskiniemi, S.; Berg, O.G.; Andersson, D.I. Evolution of high-level resistance during low-level antibiotic exposure. *Nat. Commun.* **2018**, *9*, 1599. [CrossRef]

Mojsoska, B.; Jenssen, H. Peptides and peptidomimetics for antimicrobial drug design. *Pharmaceuticals* **2015**, *8*, 366–415. [CrossRef] [PubMed]

Lu, J.; Xu, H.; Xia, J.; Ma, J.; Xu, J.; Li, Y.; Feng, J. D- and Unnatural Amino Acid Substituted Antimicrobial Peptides with Improved Proteolytic Resistance and Their Proteolytic Degradation Characteristics. *Front. Microbiol.* **2020**, *11*, 563030. [CrossRef] [PubMed]

Fadaka, A.O.; Sibuyi, N.R.S.; Madiehe, A.M.; Meyer, M. Nanotechnology-based delivery systems for antimicrobial peptides. *Pharmaceutics* **2021**, *13*, 1795. [CrossRef] [PubMed]

Zasloff, M. Antimicrobial peptides of multicellular organisms. *Nature* **2002**, *415*, 389–395. [CrossRef]

Bevins, C.L.; Salzman, N.H. Paneth cells, antimicrobial peptides and maintenance of intestinal homeostasis. *Nat. Rev. Microbiol.* **2011**, *9*, 356–368. [CrossRef]

Cullen, T.W.; Schofield, W.B.; Barry, N.A.; Putnam, E.E.; Rundell, E.A.; Trent, M.S.; Degnan, P.H.; Booth, C.J.; Yu, H.; Goodman, A.L. Antimicrobial peptide resistance mediates resilience of prominent gut commensals during inflammation. *Science* **2015**, *347*, 170–175. [CrossRef]

Yan, H.; Hancock, R.E.W. Synergistic interactions between mammalian antimicrobial defense peptides. *Antimicrob. Agents Chemother.* **2001**, *45*, 1558–1560. [CrossRef]

antibiotics

MDP

Article

Impacts of Hydrophobic Mismatch on Antimicrobial Peptide Efficacy and Bilayer Permeabilization

Steven Meier [1], Zachary M. Ridgway [1], Angela L. Picciano [1] and Gregory A. Caputo [1,2,*]

[1] Department of Chemistry & Biochemistry, Rowan University, Glassboro, NJ 08028, USA; angelalpicciano@gmail.com (A.L.P.)
[2] Department of Biological & Biomedical Sciences, Rowan University, Glassboro, NJ 08028, USA
* Correspondence: caputo@rowan.edu

Abstract: Antimicrobial resistance continues to be a major threat to world health, with the continu emergence of resistant bacterial strains. Antimicrobial peptides have emerged as an attractive opti for the development of novel antimicrobial compounds in part due to their ubiquity in nature ar the general lack of resistance development to this class of molecules. In this work, we analyzed t antimicrobial peptide C18G and several truncated forms for efficacy and the underlying mechanis effects of the sequence truncation. The peptides were screened for antimicrobial efficacy agair several standard laboratory strains, and further analyzed using fluorescence spectroscopy to evalua binding to model lipid membranes and bilayer disruption. The results show a clear correlati between the length of the peptide and the antimicrobial efficacy. Furthermore, there is a correlati between peptide length and the hydrophobic thickness of the bilayer, indicating that hydrophob mismatch is likely a contributing factor to the loss of efficacy in shorter peptides.

Keywords: membrane destabilization; AMP; liposome; fluorescence spectroscopy

Citation: Meier, S.; Ridgway, Z.M.; Picciano, A.L.; Caputo, G.A. Impacts of Hydrophobic Mismatch on Antimicrobial Peptide Efficacy and Bilayer Permeabilization. *Antibiotics* **2023**, *12*, 1624. https://doi.org/10.3390/antibiotics12111624

Academic Editors: Piyush Baindara and Marisa Di Pietro

Received: 24 October 2023
Revised: 6 November 2023
Accepted: 8 November 2023
Published: 14 November 2023

1. Introduction

The rapid development of antibiotic resistance has been recognized by the Worl Health Organization and the United States Centers for Disease Control (CDC) as or of the major challenges that global health faces [1]. Beyond the immediate impacts c human health, numerous NGOs and academic researchers have estimated that, by 2050, tl continued growth of antimicrobial resistance may impact annual global GDPs by up to 5' resulting in increased healthcare costs up to USD 1T per year, with more severe impac being found in impoverished and less-developed areas of the globe [2,3].

The resistance phenomenon has been observed in numerous organisms includin both Gram-positive and Gram-negative bacteria, fungi, and viruses. While methicilli resistant *Staphylococcus aureus* (MRSA) and vancomycin-resistant enterococci (VRE) a widely known, there are many other pathogenic organisms which have been clinical isolated displaying antibiotic resistance phenotypes [4]. These developments have le to significant interest in the development of new antibiotics to combat the growing resi tance phenomenon. There are numerous different approaches and classes of molecule being investigated as potential new antimicrobial treatments including traditional sma molecules [5,6], peptides [7–9], peptide and protein mimetics [10–14], hydrogels [15,16], sy thetic polymers [17,18], bacterial communication inhibitors [19,20], metals [21–23], nanopa ticles [24,25], extracts from natural products [26,27], and combinatorial approaches [28–3(

While many different approaches are being explored for novel antimicrobials, antim crobial peptides (AMPs) represent one of the most thoroughly studied and diverse classe of potential leads. AMPs represent a broad class of peptides with wide-ranging source structures, and mechanisms of action. While naturally occurring, and although AMPs ar often found as components of the innate immune system (often referred to as host defens peptides), there have been many modified and synthetic variants investigated. The mos well studied versions of AMPs are those which adopt an amphiphilic, α-helical secondary structure when interacting with bacterial membranes. Examples of this class of peptides include magainin, melittin, LL-37, and cecropins [31,32]. These peptides are typically short (10–30 amino acids in length), are overall net positively charged, have numerous hydrophobic groups, and often form facially amphiphilic structures when in an α-helical conformation. There is significant evidence that these peptides can provoke a disruption of the bacterial membrane; however, there are also some indications that membrane activity may only be one component of a multi-faceted mechanism of action [33–36]. Importantly, AMPs have shown a very low propensity to induce resistance development in bacteria. Despite these benefits, AMPs have found limited success in clinical applications [37].

Among the many naturally derived AMPs that have been investigated, the peptide C18G has proven to be a versatile platform for studying AMPs' mechanism of action. The C18G sequence was originally developed as a modified extension of the C-terminal 13 amino acids of the platelet factor IV protein [38,39]. Subsequently, C18G was found to modulate signaling through several different bacterial two-component sensor systems by the disruption of protein–lipid contacts and that it may be linked to the bacterial "sensing" of AMPs [40–42]. Biophysical studies from our group on the C18G peptide and derivatives have demonstrated the importance of overall hydrophobic character as well as the impact of cationic amino acid side chain length on binding and antimicrobial activity [43–45]. These results demonstrate that C18G can cause the permeabilization of bacterial and model membranes and this activity is linked to the ability of the peptide to bind to and partition into the lipid bilayer.

The study presented here extends on the biophysical characterization of C18G with the goal of further determining the mechanism of membrane disruption. Specifically, a series of truncated peptides was created which are shorter in length than the parent C18G and thus have an overall lower hydrophobic character, lower net charge, and an overall shorter length. Herein, we show that peptide efficacy was directly tied to the length of the peptide sequence. Moreover, the shortest peptides lost the ability to disrupt model and bacterial membranes. By varying the bilayer thickness, we demonstrate that a key component of this phenomenon is linked to hydrophobic mismatch between the peptide and the bilayer.

2. Results

2.1. Peptide Composition

The amino acid sequences and selected physicochemical characteristics of the parent C18G peptide and the truncated versions are shown in Table 1. The full-length peptide, C18G-18, was modified from the original sequence by changing the amino acid at position 10 to a tryptophan residue. This change serves two purposes, the first being the incorporation of the environmentally sensitive Trp residue allowing for interrogation using fluorescence methods. The second reason for the incorporation of Trp at position 10 was to facilitate the synthesis of the series of peptides from a single precursor batch. Since solid-phase peptide synthesis proceeds from the C-terminus to the N-terminus, all three peptides would start with the same synthetic protocol, and subsequently batches of resin can be removed from the synthesis reaction to yield the truncated form while synthesis continues on the remaining resin to create the longer peptides.

Table 1. Peptide sequences and properties.

Peptide	Sequence	Length	MW	Net Charge	GRAVY [a]	Hydrophobicity [b]
C18G-18	ALYKKLLKKWLKSAKKLG-NH2	18	2116.7	+7	−0.45	9.58
C18G-13	LLKKWLKSAKKLG-NH2	13	1512.9	+5	−0.354	7.24
C18G-10	KWLKSAKKLG-NH2	10	1158.5	+4	−0.83	4.97

[a]—Grand average of hydropathicity calculated from reference [46], [b]—Total hydrophobic moment calculated from reference [47]. MW = Molecular weight.

The full-length and truncated peptides vary in net charge, length, and overall h drophobicity, and all three factors are believed to play a role in the mechanism of actic of many AMPs. The peptide length varies from 18 to 10 amino acids, and it has a r sultant effect of varying the molecular weight of the peptides from 2216.7 Da down 1158.5 Da, while the net charge of the peptides at pH 7 decreases from +7 to +4. The overa hydrophobic character of the peptides is also impacted because both hydrophobic ar cationic residues are removed with each truncation. The grand average of hydropathici calculates the overall hydrophobicity of a sequence, and the calculated values remain re tively similar for each of the peptides ranging from −0.354 (most hydrophobic, C18G-1 to −0.83 (least hydrophobic, C18G-10), although in the context of the range of the sca (−4.50 to +4.50), these differences may not be very significant [48]. However, using a mo specialized hydrophobicity scale developed specifically for the partitioning of peptides lipid bilayer interfaces, more significant differences are observed in the properties of th peptides [47]. Helical wheel representations of the peptides can be seen in Supplement Figure S1 [49]. These representations show that all of the truncated versions of the peptid maintain the facial amphiphilicity which is associated with AMP activity.

2.2. Antimicrobial Activity

The overarching goal of this study is to help understand the core physicochemical pro erties that drive antimicrobial activity. Thus, the antimicrobial efficacy of the peptides wa evaluated using the standard broth microdilution assay to evaluate the minimal inhibito concentration (MIC) of the peptides. MIC values represent the lowest concentration of th compound able to prevent growth in an overnight assay. The MIC results are shown i Table 2. The results show that the placement of Trp at position 10 had minimal to no impa on antimicrobial activity. Additionally, the C18G-13 peptide exhibited mixed antimicrobi activity compared to the C18G-18, while the C18G-10 peptide lost all antimicrobial activi against the strains tested over the range of peptide concentrations tested.

Table 2. Minimal inhibitory concentration (μM).

Peptide	*E. coli*	*S. aureus*	*B. subtilis*
C18G	2.5	2.5	2.5
C18G-18	5	2.5	2.5
C18G-13	5	>15	5
C18G-10	>15	>15	>15

2.3. Binding Assays

The first step in the activity of AMPs is the interaction with the bacterial membran This process is driven by both electrostatic and hydrophobic interactions, and it is comp cated by the diversity of molecules presented on bacterial cell surfaces such as polysacch rides, complex lipids, and proteins. Tryptophan fluorescence emission was used to monitc peptide binding to lipid vesicles, as shown in Figure 1A,B. The binding experiments involv the addition of pre-formed lipid vesicles in which the lipid concentration is controlled to sample containing the peptide of interest. In these experiments, we used vesicles containin 100% DOPC lipids, an approximation of mammalian or host cells, and vesicles compose of 3:1 DOPC:DOPG, an approximation of the bacterial cell membrane. Natural membrane contain significant complexity in lipid, sterol, and protein components, but these mod systems are meant to replicate the two major components driving the interaction of AMP with membranes: hydrophobicity and anionic charge. The Trp emission spectrum exhibit a blue shift when the Trp moves from a more aqueous environment (in solution) to a mor non-polar environment (bound to the bilayer surface). This is analyzed by monitoring th barycenter of the emission spectrum. Consistent with previous results on C18G and othe variants, the truncates in this study can interact with liposomes composed of zwitterioni lipids (Figure 1A) and a mixture of zwitterionic and anionic lipids (Figure 1B). There is clear preference for binding to vesicles containing anionic lipids, but all peptides do interac with bilayers in the absence of the anionic lipids. Representative emission spectra can be found in Supplemental Figure S2.

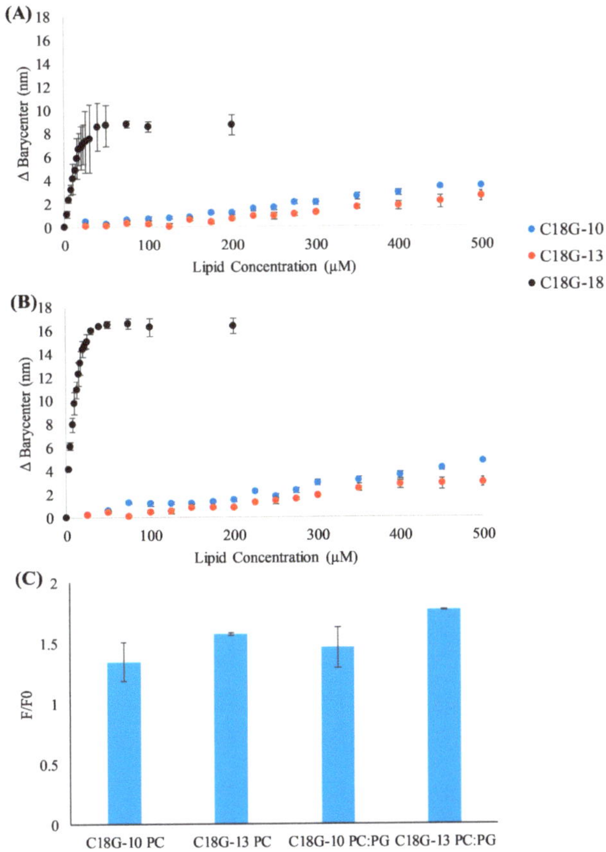

Figure 1. Vesicle binding assays. (**A**,**B**) Peptides C18G-18 (black), C18G-13 (red), or C18G-10 (blue) were titrated with lipid vesicles composed of (**A**) 100% DOPC or (**B**) 75% DOPC 25% DOPG. Trp fluorescence emission spectra were recorded after each addition, spectra were processed including barycenter calculation, and the final change in barycenter was calculated from the difference between the initial, 0 lipid spectra and that from each titration point. (**C**) Fluorescence intensity changes upon interaction with lipid bilayers. F/F_0 represents the ratio of the fluorescence intensity after binding (F) compared to the intensity in the absence of vesicles (F_0). The ratio was taken at λ = 355 nm and lipid concentration of 500 μM. All data are the averages of 2–3 independent samples and the error bars represent the standard deviations.

As C18G-10 and C18G-13 did not exhibit significant spectral shifts, the fluorescence intensity changes upon addition of vesicles were also examined. In many cases, environmentally sensitive fluorophores will exhibit an increased fluorescence emission intensity upon binding to the lipid bilayer or other hydrophobic structures [50,51]. The emission intensity changes for C18G-13 and C18G-13, represented as F/F_0 or the final fluorescence divided by the initial fluorescence, can be seen in Figure 1C. Notably, both peptides exhibited intensity increases upon titration with lipid vesicles 1.5–1.75-fold, indicating binding to the bilayers. Consistent with the barycenter analysis, the peptides did show a slightly enhanced fluorescence increase when interacting with anionic vesicles compared to zwitterionic vesicles, although this difference may not be significant.

2.4. Bacterial Membrane Permeabilization

The mechanism of action of many AMPs has been demonstrated to include bacteri membrane destabilization or disruption, including C18G. There have been numero reported approaches to monitor membrane disruption in live bacterial cells, includi measurement of membrane potential, leakage of DNA-binding dyes into the cell, ar through chromogenic substrate–enzyme pairs [52].

The ability of the peptides to per"eabi'Ize the *E. coli* inner membrane was assessed ing the cytoplasmic enzyme β-galactosidase and a chromogenic substrate ortho-Nitropher β-galactoside (ONPG). Under normal conditions, the bacterial inner membrane is relative impermeable to the ONPG substrate; however, if it is disrupted by peptides or oth molecules, the ONPG can more readily cross the membrane, resulting in an increas degree of substrate conversion. As shown in Figure 2A–C, the peptides induced varyi degrees of leakage across the *E. coli* inner membrane, and all acted in a dose-depende manner. Consistent with the MIC results, the full-length peptide was the most effective permeabilization while C18G-10 was the least effective. Results from the control expe ments using a membrane-solubilizing detergent can be found in Supplemental Figure S

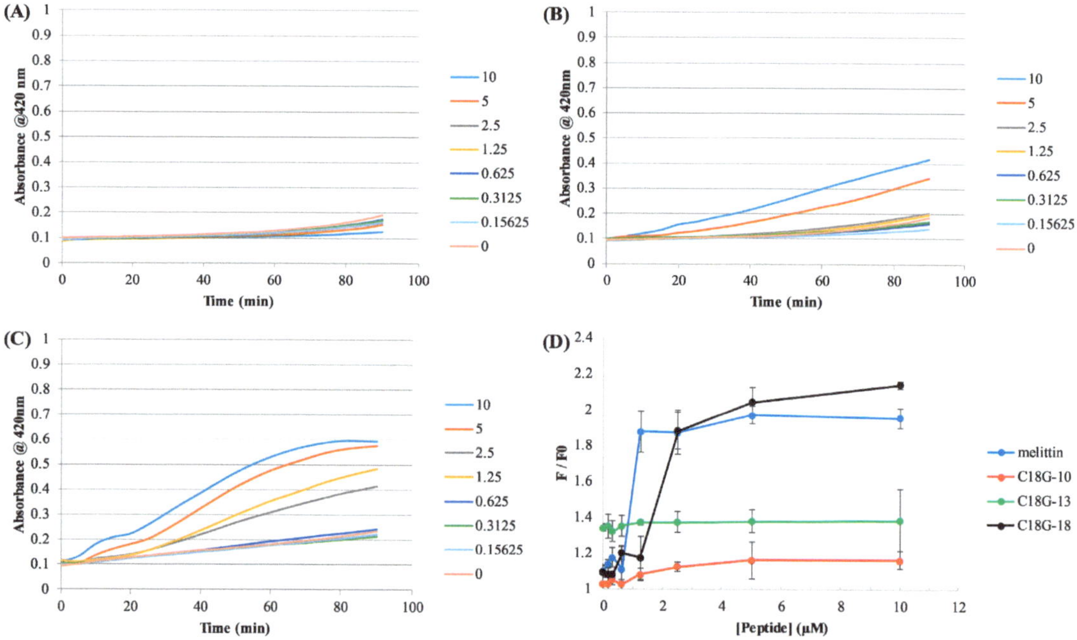

Figure 2. Bacterial membrane permeabilization. *E. coli* inner membrane permeabilization is show in panels (**A–C**) where peptide concentrations and corresponding colors are shown in the legend on the right. Panels represent (**A**) C18G-10, (**B**) C18G-13, and (**C**) C18G-18. Data were recorded 5 min intervals from triplicate samples and error bars represent the standard deviation. Positi controls using the detergent CTAB can be found in the supplemental information. *S. aureus* membra permeabilization is shown in (**D**). The colors associated with each peptide and the control melitti are shown in the legend. F/F_0 is calculated by the ratio of DAPI fluorescence at a given peptid concentration to the fluorescence prior to peptide addition. Data are from 3–5 independent sample and error bars represent the standard deviation.

The ability of the peptides to disrupt the membrane of Gram-positive *S. aureus* cells shown in Figure 2D. This assay relies on the passage of the DNA-binding dye DAPI acros the bacterial cell membrane. Under normal conditions, DAPI is minimally permeable acros the membrane and exhibits very low to negligible fluorescence emission when in aqueous environments. Upon membrane permeabilization, the DAPI can cross the membrane and interact with cellular DNA, inducing a dramatic increase in fluorescence emission intensity. The results in Figure 2D parallel those for *E. coli*, with both peptide length and dose dependence being linked to membrane disruption. The bee venom peptide Melittin was used as a positive control.

2.5. Vesicle Permeabilization

In light of the results of the bacterial membrane permeabilization studies, there appears to be a link between peptide length and ability to disrupt the bilayer. In an attempt to gain more insight into this relationship, a series of dye leakage experiments were carried out in which lipid vesicles were created using lipids with varying acyl chain lengths, thus varying the thickness of the bilayer. These vesicles were created with the self-quenching dye calcein trapped in the vesicle lumen which, upon leakage from the vesicle interior, is diluted, relieves the self-quenching, and results in a large increase in fluorescence intensity. Leakage was normalized by comparing the intensity before addition to peptide as the zero value, and after the vesicles were permeabilized with the detergent Triton X-100 as the complete or 100% leakage value [53].

Here, four different lipids were used to create vesicles: 1,2-dimyristoleoyl-sn-glycero-3-phosphocholine(14:1 (Δ9-Cis) PC; dMoPC), 1,2-dioleoyl-sn-glycero-3-phosphocholine (18:1 (Δ9-Cis) PC; DOPC), 1-palmitoyl-2-oleoyl-glycero-3-phosphocholine (16:0–18:1 PC; POPC), and 1,2-dierucoyl-sn-glycero-3-phosphocholine (22:1 (Δ13-Cis) PC; dEuPC). The acyl chain lengths and approximate resultant hydrophobic thickness of the bilayers can be found in Figure 3A, based on the measurements in references [54,55]. The results of dye leakage can be found in Figure 3B–D. These data show that a given peptide's ability to cause membrane permeabilization is linked both to bilayer thickness and peptide length, indicating a hydrophobic matching effect. The shortest peptide, C18G-10, was able to permeabilize the thinnest bilayers tested to some extent; however, this ability was lost when moving to thicker bilayers. C18G-13 displayed significantly enhanced leakage compared to C18G-10 in all bilayer thicknesses, but still incomplete permeabilization of the thickest bilayers tested. Finally, C18G-18 was able to disrupt the vesicles of all thicknesses tested, with a nearly complete disruption of the vesicles at the highest concentrations tested. Taken together, these data indicate that hydrophobic matching of the peptide to the target bilayer is an important consideration in the mechanism of AMPs, with shorter sequences being more susceptible to losing activity.

(A)

Lipid	Acyl Chain Length (# C)	Approx Bilayer Thickness (Å)
DMoPC	14:1(Δ9)	22.5
DOPC	18:1(Δ9)	29.5
POPC	16:0;18:1(Δ9)	28.8
DEuPC	22:1(Δ13)	36.5

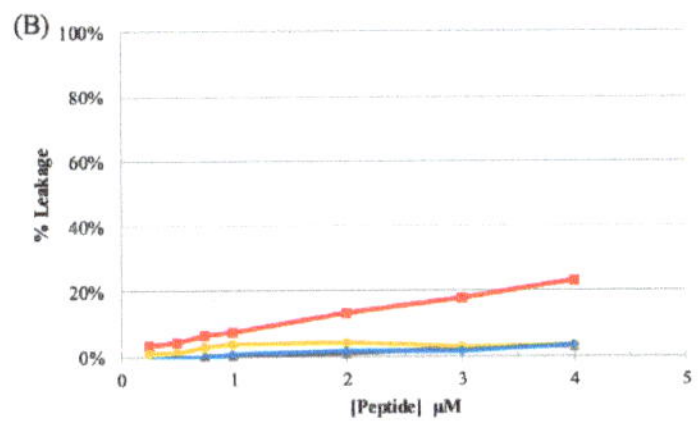

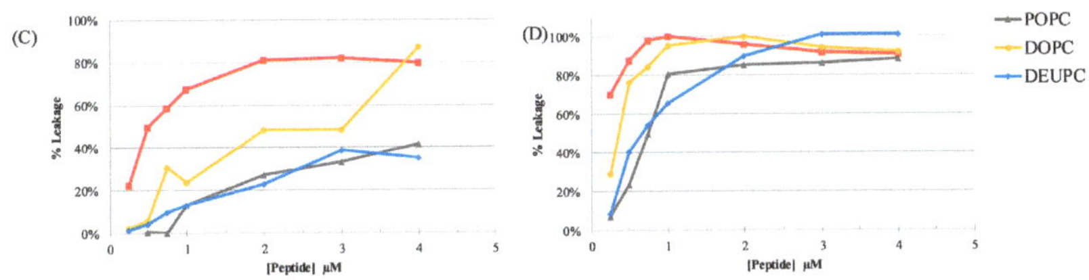

Figure 3. Vesicle leakage. (**A**) Synthetic lipids used to make vesicles with varied bilayer hydrophobic thickness and the approximate resultant bilayer thickness from each lipid type, based on data from references [54,55]. Leakage from vesicles of different bilayer thickness caused by (**B**) C18G-10, (**C**) C18G-13,

and (**D**) C18G-18. Lipid concentration was 200 μM. Leakage percentage was determined by measuri calcein fluorescence prior to addition of peptides (zero or baseline leakage), after 60 min of incubatic and after addition of the detergent Triton X-100 (100% leakage). Data are representative samples fro individual paired experiments.

3. Discussion

Hydrophobic mismatch has been an area of study for many years, primarily focuse on the function of transmembrane proteins and peptides. The concept of hydrophob matching proposes that protein transmembrane domains have evolved such that th length of the transmembrane segments "match" the hydrophobic thickness of the me brane in which they are functionally incorporated [56]. Hydrophobic matching has bee linked to numerous proteins and proper function [57–59], sorting and location in th membrane [60–64], or transport across the membrane [65–67]. Overall, this phenomenc is important in normal cellular function for a variety of protein systems.

The effect of hydrophobic mismatch on AMPs has been investigated in some mod systems, but this area is still largely unexplored. Much of the early work on hydroph bic matching with AMPs was focused on ion channel-forming peptides. These system such as gramicidin and alamethicin, adopt stable transmembrane orientations in order functionally transport ions across the bilayer [4,68,69]. These experimental investigatio were corroborated by simulations which showed similar mismatch dependence [70,7 These molecules function similarly to traditional transmembrane proteins, and thus it not surprising that they are impacted by hydrophobic mismatch. Ulrich and cowork used model AMPs based on a repeating sequence of amino acids to investigate the ro of hydrophobic mismatch by varying the number of repeats in the final sequence, an thus the length of the peptide. Using a combination of leakage experiments and NM approaches, they demonstrated that the model AMP peptides modulated the tilt angle which they were imbedded in the bilayer in response to hydrophobic mismatch, similar the paradigm in transmembrane helices [72–74].

The results from the work presented in this paper parallel the results in Grau-Campistar et al.'s work, with decreased bilayer permeabilization in the case of negative hydrophob mismatch (when the protein segment is shorter than the thickness of the bilayer) [7 While the structures of these peptides were unable to be experimentally determined, thre independent secondary structure prediction algorithms indicate that all three sequence are likely to adopt α-helical conformations (Supplemental Figure S4) [75–77]. In th conformation, the length of the helices formed by the peptides would be 15 Å, 19.5 and 27 Å, respectively. The permeabilization of the vesicles is completely lost for C18G-1 when the mismatch is greater than 7.5 Å, while for C18G-18 the ability to significantl destabilize bilayers when the mismatch corresponds to ~10 Å in the dEuPC bilayers maintained. There have been reports that some AMPs can induce modest bilayer thinnin (1–2 Å); however, it is unclear how overall peptide mass would impact this phenomeno Additionally, the peptide sequences were analyzed using the iTasser algorithm for protei structure prediction [78–80]. This algorithm also predicted that all three sequences woul adopt helical conformations, and the best fit models are shown in Supplemental Figure S The helical length measurements from these models are 14.3 Å (C18G-10), 18.4 Å (C18G-13 and 26.2 Å (C18G-18), measured from the Cα of the first helical residue to the Cα of the las helical residue. Overall, the data indicate that negative mismatch (helices shorter than th hydrophobic thickness of the bilayer) is more detrimental to the AMP permeabilization c membranes compared to positive mismatch.

In the context of the AMP's mechanism of action, the results indicate that peptid length and subsequent hydrophobic matching are important to consider for the evolutio and design of membrane-disrupting AMPs. Importantly, while the mechanism of actio of many AMPs involved membrane destabilization, there are several physical models b which this can occur: the barrel-stave pore model, the toroidal pore model, and the carpe model [81,82]. However, all these models involve AMPs crossing the bilayer in some sort of transmembrane architecture, either stably or transiently. While the relationship is clear between hydrophobic mismatch and stable transmembrane structures, the transient transmembrane conformations are also impacted. In the case of a peptide acting by the carpet model, a key proposed component of the mechanism is the transient membrane crossing and concomitant pore formation in the bacterial membrane. If the peptide physically cannot transit across the membrane due to any combination of physicochemical limitations, then the membrane disruption is compromised, likely leading to decreased antimicrobial activity. While these results can help guide the design of new AMPs, the key factor in activity is the interaction with bacterial membranes. Bacterial membranes are inherently more complex than the model vesicles used in this and many other studies, and they do not modulate hydrophobic thickness as dramatically as shown using synthetic lipids; thus, any matching or mismatch would be a result of changing the length of the peptide. Additionally, the inherent changes in overall peptide hydrophobicity are likely to impact the cytotoxicity of the molecules. Numerous previous studies have linked net hydrophobicity to the hemolytic activity of peptides and peptidomimetic polymers, with increased hydrophobic character resulting in increased cytotoxicity and/or hemolysis [83–87]. Thus, there will be a necessary interplay between sufficient length to permeabilize membranes while trying to optimize the necessary hydrophobicity to allow for membrane interaction while minimizing cytotoxic effects.

4. Materials and Methods

4.1. Materials

All chemicals and supplies were purchased from VWR (Radnor PA, USA) unless otherwise noted. The lipids 1,2-dimyristoleoyl-sn-glycero-3-phosphocholine(14:1 (Δ9-Cis) PC; dMoPC), 1,2-dioleoyl-sn-glycero-3-phosphocholine (18:1 (Δ9-Cis) PC; DOPC), 1-palmitoyl-2-oleoyl-glycero-3-phosphocholine (16:0–18:1 PC; POPC), and 1,2-dierucoyl-sn-glycero-3-phosphocholine (22:1 (Δ13-Cis) PC; dEuPC) were purchased from Avanti Polar Lipids (Alabaster AL, USA) and were stored at −20 °C as stock solutions dissolved in chloroform. All samples were measured in sodium phosphate buffer (150 mM NaCl, 50 mM NaH_2PO_4/Na_2HPO_4, pH 7.0) unless specifically indicated.

All peptides were synthesized using FMOC solid-phase synthetic methods. A rink-amide resin was used as the solid support, DMF as the primary solvent, and 20% piperidine in DMF (*v:v*) was used for FMOC deprotection. The removal of peptides from the rink-amide support was achieved by mixing resin with a cleavage "cocktai" of 92.5:2.5:2.5:2.5 trifluoroacetic acid (TFA):water (H_2O):triisopropylsilane (TIPS):ethanedithiol (EDT). The cleaved peptides were isolated from the spent resin by gravity filtration through glass wool and were subsequently precipitated by dropwise addition into cold diethyl ether ($(C_2H_5)_2O$). Reversed-phase HPLC (RP-HPLC), using a Jupiter 300 C4 column (Phenomenex, Torrance CA, USA) with the mobile phase being a linear gradient of acetonitrile and water containing 0.1% TFA, changing from 0–40% acetonitrile over 40 min, was used to purify the peptides. Confirmation of the peptide identities was performed by ESI-MS in negative ion mode.

4.2. Bacterial Strains and Culture

The bacterial species used were *E. coli* D31 [88], *B. subtilis* ATCC: 6633, and *S. aureus* ATCC: 27660. Beginning with a stock culture preserve at −80 °C with glycerol, the bacteria were inoculated and spread on LB–Miller agar (BD-Difco, Franklin Lakes, NJ, USA) plates. These plates were grown overnight at 37 °C to allow for the growth of single, isolated colonies. A single colony from the plates was subcultured into approximately 3 mL of fresh LB or Mueller Hinton (MH) media (BD-Difco, Franklin Lakes, NJ, USA) and grown overnight in a 37 °C shaking incubator @ ~250 rpm. After incubation, an aliquot of the overnight culture was diluted 1:200 in fresh LB or MH media and allowed to grow at 37 °C with shaking until the culture density reached an OD_{600} of approximately 0.5.

4.3. Minimal Inhibitory Concentration Assay

Antimicrobial activity was determined using the broth microdilution minimal hibitory concentration (MIC) assay [89]. Briefly, cultures of bacteria were grown as c scribed and then diluted to ~10^5 cfu/mL in fresh MH broth. Next, 90 μL of this dilut culture was added to each well of a sterile 96-well plate containing 10 μL each of seria diluted aliquots of the peptides. The plate was covered and incubated at 37 °C for 18 After the 18 h incubation, optical density at 600 nm was measured using a Spectramax M multimode plate reader (Molecular Devices, San Jose CA, USA). Optical density readin were compared to untreated controls and sterile media to determine MIC.

4.4. Lipid Binding Assays

Lipid vesicles for the binding assays were created by sonication of multilamell vesicles (MLVs). Briefly, the appropriate volumes of lipids in chloroform were dried und a gentle flow of N_2 gas and further dried by incubation in a vacuum dessicator for at least 1 The resultant lipid film was rehydrated by vigorous vortexing immediately upon additi of the appropriate volume of sodium phosphate buffer to create MLVs. The MLV soluti was then subjected to sonication in a high-power bath sonicator (Avanti) to produce sma unilamellar vesicles (SUVs). Fluorescence experiments were performed using a JY-Horil Fluoromax4 (JY Horiba, Edison, NJ, USA) with the emission and excitation slit widths s to 2.5 nm. The samples were prepared by mixing 2 mM peptides with the phosphate buff and then the initial fluorescence spectrum was collected. Samples were excited at 280 n with emission recorded over the range of 300–400 nm with 1 nm increments betwe measurements in the spectrum. After the spectrum was collected, the appropriate volun of lipid vesicles was added to the sample, mixed by pipetting, and allowed to incubate f 5 min at room temperature before the next measurement was taken. The spectral barycent and Δbarycenter calculations were performed as previously described [44]. All spect were corrected for background and dilution before barycenter calculations were performe Data are the averages of 2–3 samples and error bars represent the standard deviations.

4.5. Bacterial Membrane Permeabilization

Evaluation of peptide-induced permeabilization of the *E. coli* and *S. aureus* membran was carried out as described previously [44,45,52,90]. The permeabilization of the *E. c* inner membrane used the chomogenic substrate ortho-Nitrophenyl-β-galactoside (ONP which is broken down by the cytoplasmic enzyme β-galactosidase. A sample containi *E. coli* D31 in Z-buffer 0.1 M Na_2HPO_4/NaH_2PO_4, 10 mM KCl, 1 mM $MgSO_4$, 0.05 β-mercaptoethanol, pH 7.0) T was added to a 96-well plate containing serially dilute peptides, with the cationic detergent cetyltrimethylammonium bromide (CTAB) serving a positive control (see Supplemental Figure S3). ONPG was added immediately prior the first measurement, and subsequent measurements were taken every 5 min for 90 m total. Data are the average of 3–5 independent samples.

The permeabilization of the *S. aureus* membrane using the membrane impermeab DNA-binding fluorophore 4′,6-diamidino-2-phenylindole (DAPI). DAPI (final concentr tion in 85.7 nM) was added to 100 μL of bacterial cells resuspended in HBS in a 96-we plate. Peptides were added after initial background readings stabilized. As a contr only buffer with no peptide was added to control wells. The fluorescence was recorde immediately before (time 0) and after addition (time 1), at 10, 30, and 60 min using th excitation wavelength = 358 nm and emission wavelength = 461 nm. The pore-formin peptide melittin was used as a positive control. Data are the average of 3–5 independer samples and error bars represent the standard deviations.

4.6. Vesicle Leakage Assays

Vesicles containing calcein were prepared as noted above, with 75mM calcein solutio dissolved in HBS as the solvent. Calcein-loaded vesicles were then subjected to five round of freeze–thaw by alternating the sample between a liquid N_2 bath and a 37 °C water bat

Vesicles were then extruded 21 times through a 200 nm polycarbonate filter using a syringe extruder (Avanti). Loaded vesicles were separated from untrapped calcein by passage over a G25 Sephadex column equilibrated in HBS. Fractions containing vesicles were used directly on the same day. Lipid concentration was estimated using a ratiometric method in which identical preparations containing a fluorescently labeled lipid was used to determine the dilution factor from the column separation. Final lipid concentration in the samples was 200 μM. Fluorescence measurements were taken on the Spectramax M5 plate reader using excitation 495 nm and emission 520 nm. Normalization of leakage was determined by adding 20 uL of Triton X-100 to each well, incubating in the dark for 60 min, and then remeasuring fluorescence which was used as the 100% leakage for each sample.

5. Conclusions

Taken together, the results from this study demonstrate that hydrophobic matching between AMPs and target membranes is an important component of membranolytic activity. The ability of AMPs to permeabilize target membranes is at the core of many AMPs' mechanism of action, and thus critical to maintain when modifying the sequence or structure of AMPs in the development pipeline. Additionally, the relationship between AMP and membrane in hydrophobic matching may be a useful tool in the determination of mechanisms for novel and uncharacterized AMPs.

Supplementary Materials: The following supporting information can be downloaded at: https://www.mdpi.com/article/10.3390/antibiotics12111624/s1, Figure S1: Helical Wheel models; Figure S2: Representative Trp emission spectra; Figure S3: Detergent control for bacterial membrane permeabilization; Figure S4: helical propensity scores; Figure S5: Structural prediction for peptides (A) C18G-10, (B) C18G-13, (C) C18G-18. Structural predictions were generated using the iTasser server and software suite which is based on homology.

Author Contributions: Conceptualization, G.A.C.; Funding acquisition, G.A.C.; Investigation, S.M., Z.M.R. and A.L.P.; Supervision, G.A.C.; Writing—original draft, G.A.C.; Writing—review and editing, S.M., Z.M.R. and A.L.P. All authors have read and agreed to the published version of the manuscript.

Funding: This work was supported by NIH R15 GM094330 to Gregory Caputo.

Institutional Review Board Statement: Not applicable.

Informed Consent Statement: Not applicable.

Data Availability Statement: The Data are contained within the article and supplementary materials.

Conflicts of Interest: The authors declare no conflict of interest.

References

1. CDC. *Antibiotic Resistance Threats in the United States, 2019*; U.S. Department of Health and Human Services, CDC: Washington, DC, USA, 2019.
2. Ahmad, M.; Khan, A.U. Global economic impact of antibiotic resistance: A review. *J. Glob. Antimicrob. Resist.* **2019**, *19*, 313–316. [CrossRef]
3. Dadgostar, P. Antimicrobial Resistance: Implications and Costs. *Infect. Drug Resist.* **2019**, *12*, 3903–3910. [CrossRef]
4. Antimicrobial Resistance, C. Global burden of bacterial antimicrobial resistance in 2019: A systematic analysis. *Lancet* **2022**, *399*, 629–655. [CrossRef]
5. Seethaler, M.; Hertlein, T.; Wecklein, B.; Ymeraj, A.; Ohlsen, K.; Lalk, M.; Hilgeroth, A. Novel Small-molecule Antibacterials against Gram-positive Pathogens of Staphylococcus and Enterococcus Species. *Antibiotics* **2019**, *8*, 210. [CrossRef]
6. Iacopetta, D.; Ceramella, J.; Catalano, A.; D'Amato, A.; Lauria, G.; Saturnino, C.; Andreu, I.; Longo, P.; Sinicropi, M.S. Diarylureas: New Promising Small Molecules against Streptococcus mutans for the Treatment of Dental Caries. *Antibiotics* **2023**, *12*, 112. [CrossRef]
7. Magana, M.; Pushpanathan, M.; Santos, A.L.; Leanse, L.; Fernandez, M.; Ioannidis, A.; Giulianotti, M.A.; Apidianakis, Y.; Bradfute, S.; Ferguson, A.L.; et al. The value of antimicrobial peptides in the age of resistance. *Lancet Infect. Dis.* **2020**, *20*, e216–e230. [CrossRef]
8. Zhang, C.; Yang, M. Antimicrobial Peptides: From Design to Clinical Application. *Antibiotics* **2022**, *11*, 349. [CrossRef]
9. Makhlynets, O.V.; Caputo, G.A. Characteristics and therapeutic applications of antimicrobial peptides. *Biophys. Rev.* **2021**, *2*, 011301. [CrossRef]

10. Teng, P.; Shao, H.; Huang, B.; Xie, J.; Cui, S.; Wang, K.; Cai, J. Small Molecular Mimetics of Antimicrobial Peptides as a Promisi Therapy To Combat Bacterial Resistance. *J. Med. Chem.* **2023**, *66*, 2211–2234. [CrossRef]
11. Yasuhara, K.; Tsukamoto, M.; Kikuchi, J.I.; Kuroda, K. An Antimicrobial Peptide-Mimetic Methacrylate Random Copolym Induces Domain Formation in a Model Bacterial Membrane. *J. Membr. Biol.* **2022**, *255*, 513–521. [CrossRef]
12. Panjla, A.; Kaul, G.; Chopra, S.; Titz, A.; Verma, S. Short Peptides and Their Mimetics as Potent Antibacterial Agents ar Antibiotic Adjuvants. *ACS Chem. Biol.* **2021**, *16*, 2731–2745. [CrossRef]
13. Mankoci, S.; Ewing, J.; Dalai, P.; Sahai, N.; Barton, H.A.; Joy, A. Bacterial Membrane Selective Antimicrobial Peptide-Mimet Polyurethanes: Structure-Property Correlations and Mechanisms of Action. *Biomacromolecules* **2019**, *20*, 4096–4106. [CrossRef]
14. Dos Reis, T.F.; de Castro, P.A.; Bastos, R.W.; Pinzan, C.F.; Souza, P.F.N.; Ackloo, S.; Hossain, M.A.; Drewry, D.H.; Alkhazra S.; Ibrahim, A.S.; et al. A host defense peptide mimetic, brilacidin, potentiates caspofungin antifungal activity against huma pathogenic fungi. *Nat. Commun.* **2023**, *14*, 2052. [CrossRef]
15. D'Souza, A.; Yoon, J.H.; Beaman, H.; Gosavi, P.; Lengyel-Zhand, Z.; Sternisha, A.; Centola, G.; Marshall, L.R.; Wehrman, M.I Schultz, K.M.; et al. Nine-Residue Peptide Self-Assembles in the Presence of Silver to Produce a Self-Healing, Cytocompatib Antimicrobial Hydrogel. *ACS Appl. Mater. Interfaces* **2020**, *12*, 17091–17099. [CrossRef]
16. Edirisinghe, D.I.U.; D'Souza, A.; Ramezani, M.; Carroll, R.J.; Chicon, Q.; Muenzel, C.L.; Soule, J.; Monroe, M.B.B.; Patteson, A. Makhlynets, O.V. Antibacterial and Cytocompatible pH-Responsive Peptide Hydrogel. *Molecules* **2023**, *28*, 4390. [CrossRef]
17. Palermo, E.F.; Kuroda, K. Structural determinants of antimicrobial activity in polymers which mimic host defense peptides. *Ap Microbiol. Biotechnol.* **2010**, *87*, 1605–1615. [CrossRef]
18. Jones, J.B.; Liu, L.; Rank, L.A.; Wetzel, D.; Woods, E.C.; Biok, N.; Anderson, S.E.; Lee, M.R.; Liu, R.; Huth, S.; et al. Cation Homopolymers Inhibit Spore and Vegetative Cell Growth of Clostridioides difficile. *ACS Infect. Dis.* **2021**, *7*, 1236–1247. [CrossRe
19. Brackman, G.; Coenye, T. Quorum sensing inhibitors as anti-biofilm agents. *Curr. Pharm. Des.* **2015**, *21*, 5–11. [CrossRef]
20. Wu, B.; Capilato, J.; Pham, M.P.; Walker, J.; Spur, B.; Rodriguez, A.; Perez, L.J.; Yin, K. Lipoxin A4 augments host defense in seps and reduces Pseudomonas aeruginosa virulence through quorum sensing inhibition. *FASEB J.* **2016**, *30*, 2400–2410. [CrossRef]
21. Goderecci, S.S.; Kaiser, E.; Yanakas, M.; Norris, Z.; Scaturro, J.; Oszust, R.; Medina, C.D.; Waechter, F.; Heon, M.; Krchnavek, R.I et al. Silver Oxide Coatings with High Silver-Ion Elution Rates and Characterization of Bactericidal Activity. *Molecules* **2017**, 2 1487. [CrossRef]
22. Grass, G.; Rensing, C.; Solioz, M. Metallic copper as an antimicrobial surface. *Appl. Environ. Microbiol.* **2011**, *77*, 1541–154 [CrossRef]
23. Pasquet, J.; Chevalier, Y.; Pelletier, J.; Couval, E.; Bouvier, D.; Bolzinger, M.-A. The contribution of zinc ions to the antimicrobi activity of zinc oxide. *Colloids Surf. A Physicochem. Eng. Asp.* **2014**, *457*, 263–274. [CrossRef]
24. Sharmin, S.; Rahaman, M.M.; Sarkar, C.; Atolani, O.; Islam, M.T.; Adeyemi, O.S. Nanoparticles as antimicrobial and antivir agents: A literature-based perspective study. *Heliyon* **2021**, *7*, e06456. [CrossRef]
25. Urnukhsaikhan, E.; Bold, B.E.; Gunbileg, A.; Sukhbaatar, N.; Mishig-Ochir, T. Antibacterial activity and characteristics of silv nanoparticles biosynthesized from Carduus crispus. *Sci. Rep.* **2021**, *11*, 21047. [CrossRef]
26. MacNair, C.R.; Tsai, C.N.; Rutherford, S.T.; Tan, M.W. Returning to Nature for the Next Generation of Antimicrobial Therapeuti *Antibiotics* **2023**, *12*, 1267. [CrossRef]
27. Palla, F.; Bucchini, A.E.A.; Giamperi, L.; Marino, P.; Raimondo, F.M. Plant Extracts as Antimicrobial Agents in Sustainab Conservation of Erythrina caffra (Fabaceae) Historical Trees. *Antibiotics* **2023**, *12*, 1098. [CrossRef]
28. Kvich, L.; Christensen, M.H.; Pierchala, M.K.; Astafiev, K.; Lou-Moeller, R.; Bjarnsholt, T. The Combination of Low-Frequenc Ultrasound and Antibiotics Improves the Killing of In Vitro Staphylococcus aureus and Pseudomonas aeruginosa Biofilm *Antibiotics* **2022**, *11*, 1494. [CrossRef]
29. Sousa, M.; Afonso, A.C.; Teixeira, L.S.; Borges, A.; Saavedra, M.J.; Simoes, L.C.; Simoes, M. Hydrocinnamic Acid and Perill Alcohol Potentiate the Action of Antibiotics against Escherichia coli. *Antibiotics* **2023**, *12*, 360. [CrossRef]
30. Yang, D.D.; Paterna, N.J.; Senetra, A.S.; Casey, K.R.; Trieu, P.D.; Caputo, G.A.; Vaden, T.D.; Carone, B.R. Synergistic interactions ionic liquids and antimicrobials improve drug efficacy. *iScience* **2021**, *24*, 101853. [CrossRef]
31. Tossi, A.; Sandri, L.; Giangaspero, A. Amphipathic, alpha-helical antimicrobial peptides. *Biopolymers* **2000**, *55*, 4–30. [CrossRef
32. Lu, W. Antimicrobial peptides. *Semin. Cell Dev. Biol.* **2019**, *88*, 105–106. [CrossRef]
33. Luo, Y.; Song, Y. Mechanism of Antimicrobial Peptides: Antimicrobial, Anti-Inflammatory and Antibiofilm Activities. *Int. J. Mc Sci.* **2021**, *22*, 11401. [CrossRef]
34. Yeaman, M.R.; Yount, N.Y. Mechanisms of antimicrobial peptide action and resistance. *Pharmacol. Rev.* **2003**, *55*, 27–55. [CrossRe
35. Kumar, P.; Kizhakkedathu, J.N.; Straus, S.K. Antimicrobial Peptides: Diversity, Mechanism of Action and Strategies to Improv the Activity and Biocompatibility In Vivo. *Biomolecules* **2018**, *8*, 4. [CrossRef]
36. Zhang, R.; Xu, L.; Dong, C. Antimicrobial Peptides: An Overview of their Structure, Function and Mechanism of Action. *Protei Pept. Lett.* **2022**, *29*, 641–650. [CrossRef]
37. Dijksteel, G.S.; Ulrich, M.M.W.; Middelkoop, E.; Boekema, B. Review: Lessons Learned From Clinical Trials Using Antimicrobi Peptides (AMPs). *Front. Microbiol.* **2021**, *12*, 616979. [CrossRef]
38. Darveau, R.P.; Blake, J.; Seachord, C.L.; Cosand, W.L.; Cunningham, M.D.; Cassiano-Clough, L.; Maloney, G. Peptides related t the carboxyl terminus of human platelet factor IV with antibacterial activity. *J. Clin. Investig.* **1992**, *90*, 447–455. [CrossRef]

Peck-Miller, K.A.; Blake, J.; Cosand, W.L.; Darveau, R.P.; Fell, H.P. Structure-activity analysis of the antitumor and hemolytic properties of the amphiphilic alpha-helical peptide, C18G. *Int. J. Pept. Protein. Res.* **1994**, *44*, 143–151. [CrossRef]

Yadavalli, S.S.; Carey, J.N.; Leibman, R.S.; Chen, A.I.; Stern, A.M.; Roggiani, M.; Lippa, A.M.; Goulian, M. Antimicrobial peptides trigger a division block in Escherichia coli through stimulation of a signalling system. *Nat. Commun.* **2016**, *7*, 12340. [CrossRef]

Choi, J.; Groisman, E.A. Activation of master virulence regulator PhoP in acidic pH requires the Salmonella-specific protein UgtL. *Sci. Signal.* **2017**, *10*, eaan6284. [CrossRef]

Moskowitz, S.M.; Ernst, R.K.; Miller, S.I. PmrAB, a two-component regulatory system of Pseudomonas aeruginosa that modulates resistance to cationic antimicrobial peptides and addition of aminoarabinose to lipid A. *J. Bacteriol.* **2004**, *186*, 575–579. [CrossRef]

Hitchner, M.A.; Necelis, M.R.; Shirley, D.; Caputo, G.A. Effect of Non-natural Hydrophobic Amino Acids on the Efficacy and Properties of the Antimicrobial Peptide C18G. *Probiotics Antimicrob. Proteins* **2020**, *13*, 527–541. [CrossRef]

Kohn, E.M.; Shirley, D.J.; Arotsky, L.; Picciano, A.M.; Ridgway, Z.; Urban, M.W.; Carone, B.R.; Caputo, G.A. Role of Cationic Side Chains in the Antimicrobial Activity of C18G. *Molecules* **2018**, *23*, 329. [CrossRef]

5. Saint Jean, K.D.; Henderson, K.D.; Chrom, C.L.; Abiuso, L.E.; Renn, L.M.; Caputo, G.A. Effects of Hydrophobic Amino Acid Substitutions on Antimicrobial Peptide Behavior. *Probiotics Antimicrob. Proteins* **2018**, *10*, 408–419. [CrossRef]

Wilkins, M.R.; Gasteiger, E.; Bairoch, A.; Sanchez, J.C.; Williams, K.L.; Appel, R.D.; Hochstrasser, D.F. Protein identification and analysis tools in the ExPASy server. *Methods Mol. Biol.* **1999**, *112*, 531–552. [CrossRef]

7. Hristova, K.; White, S.H. An experiment-based algorithm for predicting the partitioning of unfolded peptides into phosphatidylcholine bilayer interfaces. *Biochemistry* **2005**, *44*, 12614–12619. [CrossRef]

3. Walker, J.M. *The Proteomics Protocols Handbook*; Springer: Berlin/Heidelberg, Germany, 2005.

Ramsey, J.; Rasche, H.; Maughmer, C.; Criscione, A.; Mijalis, E.; Liu, M.; Hu, J.C.; Young, R.; Gill, J.J. Galaxy and Apollo as a biologist-friendly interface for high-quality cooperative phage genome annotation. *PLoS Comput. Biol.* **2020**, *16*, e1008214. [CrossRef]

Raghuraman, H.; Chatterjee, S.; Das, A. Site-Directed Fluorescence Approaches for Dynamic Structural Biology of Membrane Peptides and Proteins. *Front. Mol. Biosci.* **2019**, *6*, 96. [CrossRef]

Raghuraman, H.; Chattopadhyay, A. Interaction of melittin with membrane cholesterol: A fluorescence approach. *Biophys. J.* **2004**, *87*, 2419–2432. [CrossRef]

2. Ridgway, Z.; Picciano, A.L.; Gosavi, P.M.; Moroz, Y.S.; Angevine, C.E.; Chavis, A.E.; Reiner, J.E.; Korendovych, I.V.; Caputo, G.A. Functional characterization of a melittin analog containing a non-natural tryptophan analog. *Biopolymers* **2015**, *104*, 384–394. [CrossRef]

3. Sovadinova, I.; Palermo, E.F.; Huang, R.; Thoma, L.M.; Kuroda, K. Mechanism of polymer-induced hemolysis: Nanosized pore formation and osmotic lysis. *Biomacromolecules* **2011**, *12*, 260–268. [CrossRef]

4. Hills, R.D., Jr.; McGlinchey, N. Model parameters for simulation of physiological lipids. *J. Comput. Chem.* **2016**, *37*, 1112–1118. [CrossRef]

5. Ridder, A.N.; van de Hoef, W.; Stam, J.; Kuhn, A.; de Kruijff, B.; Killian, J.A. Importance of hydrophobic matching for spontaneous insertion of a single-spanning membrane protein. *Biochemistry* **2002**, *41*, 4946–4952. [CrossRef]

6. Killian, J.A. Hydrophobic mismatch between proteins and lipids in membranes. *Biochim. Biophys. Acta* **1998**, *1376*, 401–415. [CrossRef] [PubMed]

7. Foo, A.C.; Harvey, B.G.; Metz, J.J.; Goto, N.K. Influence of hydrophobic mismatch on the catalytic activity of Escherichia coli GlpG rhomboid protease. *Protein Sci.* **2015**, *24*, 464–473. [CrossRef] [PubMed]

8. Soubias, O.; Niu, S.L.; Mitchell, D.C.; Gawrisch, K. Lipid-rhodopsin hydrophobic mismatch alters rhodopsin helical content. *J. Am. Chem. Soc.* **2008**, *130*, 12465–12471. [CrossRef]

9. Windisch, D.; Ziegler, C.; Grage, S.L.; Burck, J.; Zeitler, M.; Gor'kov, P.L.; Ulrich, A.S. Hydrophobic Mismatch Drives the Interaction of E5 with the Transmembrane Segment of PDGF Receptor. *Biophys. J.* **2015**, *109*, 737–749. [CrossRef]

0. Caputo, G.A.; London, E. Cumulative effects of amino acid substitutions and hydrophobic mismatch upon the transmembrane stability and conformation of hydrophobic alpha-helices. *Biochemistry* **2003**, *42*, 3275–3285. [CrossRef]

1. Lin, Q.; London, E. Altering hydrophobic sequence lengths shows that hydrophobic mismatch controls affinity for ordered lipid domains (rafts) in the multitransmembrane strand protein perfringolysin O. *J. Biol. Chem.* **2013**, *288*, 1340–1352. [CrossRef]

2. Milovanovic, D.; Honigmann, A.; Koike, S.; Gottfert, F.; Pahler, G.; Junius, M.; Mullar, S.; Diederichsen, U.; Janshoff, A.; Grubmuller, H.; et al. Hydrophobic mismatch sorts SNARE proteins into distinct membrane domains. *Nat. Commun.* **2015**, *6*, 5984. [CrossRef] [PubMed]

3. Monne, M.; von Heijne, G. Effects of 'hydrophobic mismatch' on the location of transmembrane helices in the ER membrane. *FEBS Lett.* **2001**, *496*, 96–100. [CrossRef]

4. Schmidt, U.; Weiss, M. Hydrophobic mismatch-induced clustering as a primer for protein sorting in the secretory pathway. *Biophys. Chem.* **2010**, *151*, 34–38. [CrossRef]

5. Basu, I.; Chattopadhyay, A.; Mukhopadhyay, C. Ion channel stability of Gramicidin A in lipid bilayers: Effect of hydrophobic mismatch. *Biochim. Biophys. Acta* **2014**, *1838*, 328–338. [CrossRef]

6. Hao, B.; Zhou, W.; Theg, S.M. Hydrophobic mismatch is a key factor in protein transport across lipid bilayer membranes via the Tat pathway. *J. Biol. Chem.* **2022**, *298*, 101991. [CrossRef]

67. Mehner-Breitfeld, D.; Ringel, M.T.; Tichy, D.A.; Endter, L.J.; Stroh, K.S.; Lunsdorf, H.; Risselada, H.J.; Bruser, T. TatA and Ta generate a hydrophobic mismatch important for the function and assembly of the Tat translocon in Escherichia coli. *J. Biol. Che* **2022**, *298*, 102236. [CrossRef] [PubMed]
68. Kelkar, D.A.; Chattopadhyay, A. Modulation of gramicidin channel conformation and organization by hydrophobic mismatch saturated phosphatidylcholine bilayers. *Biochim. Biophys. Acta* **2007**, *1768*, 1103–1113. [CrossRef] [PubMed]
69. Pan, J.; Tristram-Nagle, S.; Nagle, J.F. Alamethicin aggregation in lipid membranes. *J. Membr. Biol.* **2009**, *231*, 11–27. [CrossRe [PubMed]
70. Kim, T.; Lee, K.I.; Morris, P.; Pastor, R.W.; Andersen, O.S.; Im, W. Influence of hydrophobic mismatch on structures and dynami of gramicidin a and lipid bilayers. *Biophys. J.* **2012**, *102*, 1551–1560. [CrossRef]
71. Rui, H.; Im, W. Protegrin-1 orientation and physicochemical properties in membrane bilayers studied by potential of mean fo calculations. *J. Comput. Chem.* **2010**, *31*, 2859–2867. [CrossRef] [PubMed]
72. Gagnon, M.C.; Strandberg, E.; Grau-Campistany, A.; Wadhwani, P.; Reichert, J.; Burck, J.; Rabanal, F.; Auger, M.; Paquin, J. Ulrich, A.S. Influence of the Length and Charge on the Activity of alpha-Helical Amphipathic Antimicrobial Peptides. *Biochemis* **2017**, *56*, 1680–1695. [CrossRef] [PubMed]
73. Grau-Campistany, A.; Strandberg, E.; Wadhwani, P.; Rabanal, F.; Ulrich, A.S. Extending the Hydrophobic Mismatch Concept Amphiphilic Membranolytic Peptides. *J. Phys. Chem. Lett.* **2016**, *7*, 1116–1120. [CrossRef] [PubMed]
74. Grau-Campistany, A.; Strandberg, E.; Wadhwani, P.; Reichert, J.; Burck, J.; Rabanal, F.; Ulrich, A.S. Hydrophobic mismat demonstrated for membranolytic peptides, and their use as molecular rulers to measure bilayer thickness in native cells. *Sci. R* **2015**, *5*, 9388. [CrossRef]
75. Chou, P.Y.; Fasman, G.D. Prediction of the secondary structure of proteins from their amino acid sequence. *Adv. Enzymol. Rel Areas Mol. Biol.* **1978**, *47*, 45–148. [CrossRef]
76. Deleage, G.; Roux, B. An algorithm for protein secondary structure prediction based on class prediction. *Protein. Eng.* **1987**, 289–294. [CrossRef]
77. Levitt, M. Conformational preferences of amino acids in globular proteins. *Biochemistry* **1978**, *17*, 4277–4285. [CrossRef] [PubMe
78. Yang, J.; Yan, R.; Roy, A.; Xu, D.; Poisson, J.; Zhang, Y. The I-TASSER Suite: Protein structure and function prediction. *Nat. Metho* **2015**, *12*, 7–8. [CrossRef] [PubMed]
79. Yang, J.; Zhang, Y. Protein Structure and Function Prediction Using I-TASSER. *Curr. Protoc. Bioinformatics* **2015**, *52*, 5–8. [CrossRe
80. Zhang, Y. I-TASSER server for protein 3D structure prediction. *BMC Bioinform.* **2008**, *9*, 40. [CrossRef]
81. Jenssen, H.; Hamill, P.; Hancock, R.E. Peptide antimicrobial agents. *Clin. Microbiol. Rev.* **2006**, *19*, 491–511. [CrossRef] [PubMe
82. Wimley, W.C. Describing the mechanism of antimicrobial peptide action with the interfacial activity model. *ACS Chem. Biol.* **201** 5, 905–917. [CrossRef]
83. Kuroda, K.; Caputo, G.A.; DeGrado, W.F. The role of hydrophobicity in the antimicrobial and hemolytic activities of polymethac late derivatives. *Chemistry* **2009**, *15*, 1123–1133. [CrossRef] [PubMed]
84. Takahashi, H.; Sovadinova, I.; Yasuhara, K.; Vemparala, S.; Caputo, G.A.; Kuroda, K. Biomimetic antimicrobial polymers-Desig characterization, antimicrobial, and novel applications. *Wiley Interdiscip. Rev. Nanomed. Nanobiotechnol.* **2023**, *15*, e1866. [CrossRe [PubMed]
85. Cuervo-Rodriguez, R.; Munoz-Bonilla, A.; Lopez-Fabal, F.; Fernandez-Garcia, M. Hemolytic and Antimicrobial Activities of Series of Cationic Amphiphilic Copolymers Comprised of Same Centered Comonomers with Thiazole Moieties and Polyethyle Glycol Derivatives. *Polymers* **2020**, *12*, 972. [CrossRef]
86. Hollmann, A.; Martínez, M.; Noguera, M.E.; Augusto, M.T.; Disalvo, A.; Santos, N.C.; Semorile, L.; Maffía, P.C. Role amphipathicity and hydrophobicity in the balance between hemolysis and peptide–membrane interactions of three relate antimicrobial peptides. *Colloids Surf. B Biointerfaces* **2016**, *141*, 528–536. [CrossRef] [PubMed]
87. Phuong, P.T.; Oliver, S.; He, J.; Wong, E.H.H.; Mathers, R.T.; Boyer, C. Effect of Hydrophobic Groups on Antimicrobial an Hemolytic Activity: Developing a Predictive Tool for Ternary Antimicrobial Polymers. *Biomacromolecules* **2020**, *21*, 5241–525 [CrossRef] [PubMed]
88. Burman, L.G.; Nordstrom, K.; Boman, H.G. Resistance of Escherichia coli to penicillins. V. Physiological comparison of tw isogenic strains, one with chromosomally and one with episomally mediated ampicillin resistance. *J. Bacteriol.* **1968**, *96*, 438–44 [CrossRef] [PubMed]
89. Wiegand, I.; Hilpert, K.; Hancock, R.E. Agar and broth dilution methods to determine the minimal inhibitory concentration (MI of antimicrobial substances. *Nat. Protoc.* **2008**, *3*, 163–175. [CrossRef] [PubMed]
90. Shirley, D.J.; Chrom, C.L.; Richards, E.A.; Carone, B.R.; Caputo, G.A. Antimicrobial activity of a porphyrin binding peptide. *Pe Sci.* **2018**, *110*, e24074. [CrossRef] [PubMed]

antibiotics

rticle

Evidence of Antibiotic Resistance and Virulence Factors in Environmental Isolates of *Vibrio* Species

ajkishor Pandey [1,2,*], Simran Sharma [3] and Kislay Kumar Sinha [1,*]

1 Department of Biotechnology, National Institute of Pharmaceutical Education and Research, Hajipur 844102, Bihar, India
2 School of Medicine, University of Missouri, Columbia, MO 65211, USA
3 Department of Basic and Applied Sciences, National Institute of Food Technology Entrepreneurship & Management (NIFTEM), Kundli, Sonipat 131028, Haryana, India; ssharma1650@gmail.com
* Correspondence: rajpandeydops@gmail.com (R.P.); kksinha@gmail.com (K.K.S.)

Abstract: The outbreak of waterborne diseases such as cholera and non-cholera (vibriosis) is continuously increasing in the environment due to fecal and sewage discharge in water sources. Cholera and vibriosis are caused by different species of *Vibrio* genus which are responsible for acute diarrheal disease and soft tissue damage. Although incidences of cholera and vibriosis have been reported from the Vaishali district of Bihar, India, clinical or environmental strains have not been characterized in this region. Out of fifty environmental water samples, twelve different biochemical test results confirmed the presence of twenty *Vibrio* isolates. The isolates were found to belong to five different *Vibrio* species, namely *V. proteolyticus*, *V. campbellii*, *V. nereis*, *V. cincinnatiensis*, and *V. harveyi*. From the identified isolates, 65% and 45% isolates were found to be resistant to ampicillin and cephalexin, respectively. Additionally, two isolates were found to be resistant against six and four separately selected antibiotics. Furthermore, virulent *hlyA* and *ompW* genes were detected by PCR in two different isolates. Additionally, phage induction was also noticed in two different isolates which carry lysogenic phage in their genome. Overall, the results reported the identification of five different *Vibrio* species in environmental water samples. The isolates showed multiple antibacterial resistance, phage induction, and virulence gene profile in their genome.

Keywords: *Vibrio* species; environmental isolates; antimicrobial resistance; virulence; vibriophage

itation: Pandey, R.; Sharma, S.; nha, K.K. Evidence of Antibiotic esistance and Virulence Factors in nvironmental Isolates of *Vibrio* pecies. *Antibiotics* **2023**, *12*, 1062. tps://doi.org/10.3390/ ntibiotics12061062

cademic Editors: Piyush Baindara, larisa Di Pietro and Jie Fu

eceived: 1 May 2023
evised: 3 June 2023
ccepted: 15 June 2023
ublished: 16 June 2023

1. Introduction

In India, many people have limited access to safe drinking water, especially in rural areas where the water sources are often untreated and contaminated with sewage/feces. Contaminated water and food are major risk factors for the spread of waterborne infectious diseases such as cholera and vibriosis. Cholera is a severe acute watery diarrheal disease caused by the O1 and O139 serogroup of *Vibrio cholerae* [1]. Apart from *V. cholerae*, many non-cholera pathogenic species of the genus *Vibrio* which cause vibriosis have been discovered in America, Europe, Asia, and other low-income countries [2–4]. *Vibrio* species are naturally found in freshwater, brackish water, and marine water. Generally, vibriosis infection is acquired by the ingestion of contaminated water or seafood [5]. It has been observed that environmental factors affect both eukaryotic and prokaryotic life through direct or indirect interaction. Jamie et al. mentioned in their study that warm sea surface influences the growth of pathogenic *Vibrio* species which increases the occurrence of infections in humans as well as marine animals [6]. Tropical countries such as India and other Caribbean countries are the most favorable place for the growth of toxigenic *Vibrio* species [7].

The pathogenicity of *Vibrio* species is determined by the virulent factors encoded by virulent genes. Virulence factors influence the severity of infection and drug resistance [8,9]. It has been reported that *Vibrio* species acquire external genetic material from environmental sources or other bacteria by horizontal gene transfer (HGT) [10]. The shared genetic material might be encoding virulence factors that help to enhance *Vibrio* species' adaptability diverse environments [11]. Virulence-associated elements such as toxin production, quoru sensing (cell to cell communication), presence of lysogenic phage, hemolysin, protease etc., are major pathogenic factors involved in the pathogenicity of *Vibrio* species [12,1 Therefore, assessment of virulence factors in environmental *Vibrio* isolates provides the ro of pathogenesis and helps to begin a path for treatment.

It has been reported that resistance developed by *Vibrio* species against antibiotics associated with virulent factors [8,9]. Over time, due to the immoderate use of antibioti in human disease, agriculture, and aquaculture, the *Vibrio* species developed antimicr bial resistance [14]. In another study, it has been characterized that the unique genet makeup and competency of *Vibrio* species help them in adapting to adverse environment conditions and resisting the antibacterial agent [15]. Due to the resistance developed b *Vibrio* species, they adversely affect marine, terrestrial animals, as well as human life. A reported by the Centers for Disease Control and Prevention (CDC), in mild infection *Vibrio* species, treatment is not compulsory, but a sufficient amount of liquid should b consumed by the patient to substitute the fluid that was lost in diarrhea. In the case mild to moderate infection of *V. cholerae*, it can be reversed by the administration of a oral rehydration solution (ORS) that helps in rehydration. For severe cholera condition antibacterial agents such as tetracycline, fluoroquinolones, ceftriaxone, cefotaxime, an macrolides are most effective and are used for the treatment of the disease. However, i severe vibriosis conditions, no such evidence was found that antibiotics reduce the severi of illness [16–18].

In earlier studies on the hotspot of cholera, most of the *Vibrio* species have been cha acterized in marine or brackish water. In the present study, environmental water sampl in the Vaishali district of Bihar, India, have been investigated for *Vibrio* strains. Therefor taking this into consideration, this study was performed with the objective of evaluatin the presence of virulence-associated factors and antibiotic resistance in the environment isolates of different *Vibrio* species. Additionally, it involved the characterization of phag induction in the identified isolates with the lysogenic phage-inducing agent mitomycin (MMC).

2. Results

2.1. Isolation and Biochemical Identification of Vibrio Isolates

Out of 50 environmental water samples, 20 bacterial isolates colony (River-1, Pond-Stagnant water-5, Sevage-8) were picked from the TCBS plate (Figure 1A) and confirme to be *Vibrio* positive by the 12 biochemical tests (Figure 1 and Supplementary Table S1 The biochemical test strip changed color after the addition of *Vibrio* culture, and as per th guidelines of manufacturer kit instruction (HiMedia, Mumbai, India), five different *Vibr* species (*V. proteolyticus*, *V. campbellii*, *V. harveyi*, *V. cincinnatiensis*, *V. nereis*) were identifie (Table 1).

Table 1. Different identified *Vibrio* species from different environmental water resources.

S/N	Isolate Code	Identified Species
1	V-Gan	*V. proteolyticus*
2	VR	
3	VH-I4	
4	VH-II4	
5	VP-IA	
6	VP-IB	
7	VM-IIA	
8	VHVB-I	

Table 1. *Cont.*

S/N	Isolate Code	Identified Species
9	VH-I3	*V. campbellii*
10	VH-II1	
11	VHMC-A	
12	VM-IB	
13	VHMC-C	*V. cincinnatiensis*
14	VH-II2	
15	VHVB-II	
16	VHMC-B	*V. nereis*
17	VHMC-D	
18	VM-IA	
19	VH-II3	*V. harveyi*
20	VM-IIB	Not identifiable

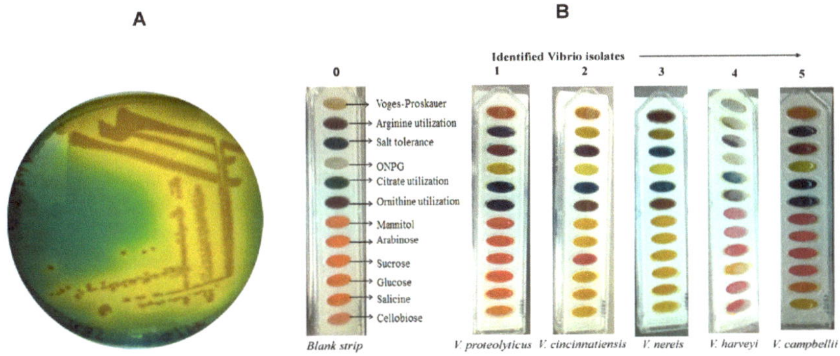

Figure 1. Isolation and confirmation of *Vibrio* isolates: (**A**) Image of presumed grown *Vibrio* colonies on selective TCBS agar plate. (**B**) Biochemical identification of five different *Vibrio* species. Zero indicates blank strip (without the addition of *Vibrio* culture) of 12 different biochemical tests. Numbers one, two, three, four, and five represent five different *Vibrio* species namely, *V. proteolyticus*, *V. cincinnatiensis*, *V. nereis*, *V. harveyi*, and *V. campbellii*, respectively.

2.2. Antibiotic Resistance and Susceptibility Assay

The environmental *Vibrio* isolates were checked for resistance against 14 commonly used antibiotics by disk diffusion methods (Figure 2A). Out of twenty isolates, thirteen isolates (65%) were found ampicillin resistant, and nine isolates (45%) were cephalexin resistant (Figure 2B and Supplementary Table S2). Co-trimoxazole resistance was observed in seven isolates (35%) followed by nalidixic acid in six resistant isolates (30%), as depicted in Figure 2B. Resistance against streptomycin, neomycin, cefotaxime, and furazolidone was found in one isolate each (Figure 2 and Supplementary Table S2). Antibiotics such as ciprofloxacin, gentamycin, norfloxacin, tetracycline, chloramphenicol, and polymyxin-B showed sensitivity to all isolates. Isolate VHMC-D was found to be resistant against six antibiotics namely ampicillin, cephalexin, nalidixic acid, cefotaxime, co-trimoxazole, and furazolidone. Another isolate, VHMC-B was resistant to ampicillin, cephalexin, nalidixic acid, and co-trimoxazole (Supplementary Table S2). Another seven isolates were found to be resistant to more than one antibacterial agent. Two isolates were found to be sensitive to all the antibiotics checked in the study (Supplementary Table S2).

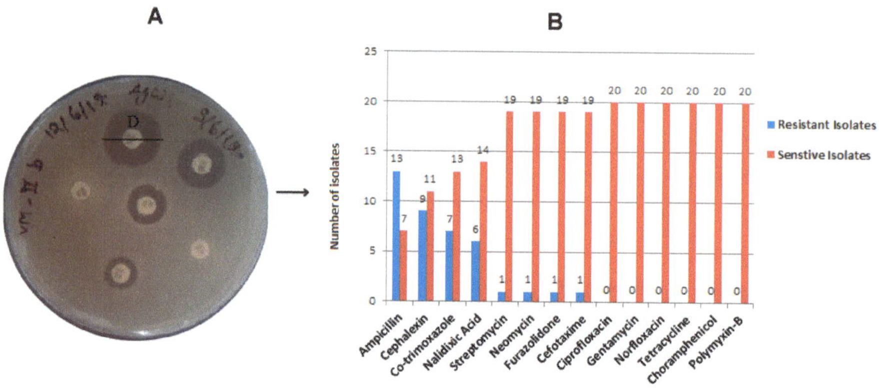

Figure 2. Antimicrobial susceptibility testing of identified different *Vibrio* isolates: (**A**) A represen tive plate of antibiotic susceptibility test assay of different *Vibrio* isolates by diffusion disk methc showing the diameter of zone of inhibition. (**B**) Antimicrobial resistance and susceptibility pattern each confirmed *Vibrio* isolates against selected antimicrobials.

2.3. Virulence Profile of Vibrio Isolates

Out of twenty *Vibrio* isolates, six isolates were chosen for the amplification of *Vibr* genes primers as mentioned in Table 2. All the mentioned genes primers were tested f amplification with the six selected isolates. The virulence gene primer *hlyA El Tor* was four to be amplified in isolate VH-I4 (*V. proteolyticus*) at 56 °C (Figure 3A), which represen hemolysin, which is an extracellular pore-forming hemolytic toxin [19]. Another *omp* gene primer amplified in VHMC-A (*V. campbellii*) isolate at 56 °C (Figure 3B). *ompW* a major outer membrane protein in *Vibrio* and is involved in salt tolerance as well as i the transferring of hydrophobic small molecules [20]. However, other selected *Vibrio* ger primers were not amplified in the selected isolates.

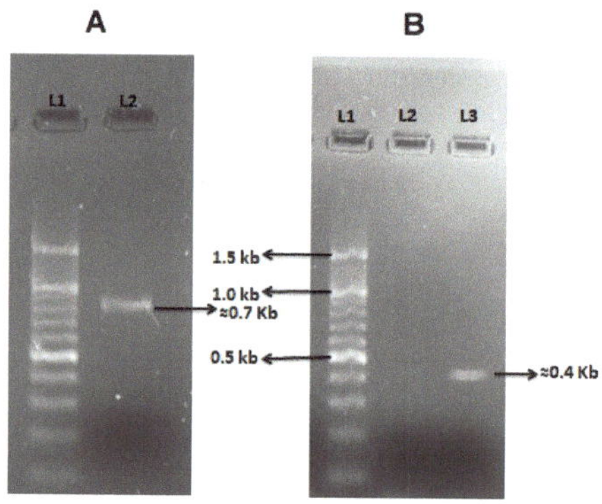

Figure 3. Amplification of *Vibrio* isolates' virulent genes: (**A**) Image of 1.5% agarose gel run wher Lane 1 (L1) represents 1 Kb marker and Lane 2 (L2) represents virulent gene primer hly-A (El To amplification by PCR with genomic DNA of VH-I4 isolate. (**B**) View of 1.5% agarose gel run showin Lane 1 (L1): 1 kb marker, Lane 2 (L2): unamplified gene primers with selected other isolates, an Lane 3 (L3) indicating virulent gene primer *ompW* amplification by PCR with genomic DNA of isola VHMC-A.

Table 2. Primers for amplification of *Vibrio* virulent genes.

S/N	Primers	Nucleotide Sequences (5′−3′)	Amplicon Size (bp) and Annealing Temperature in °C	Reference
1	*ctx*A-F	CTCAGACGGGATTTGTTAGGCACG	302 (64 °C)	[21]
	*ctx*A-R	TCTATCTCTGTAGCCCCTATTACG		
2	ctxB-F	GATACACATAATAGAATTAAGGATG	460 (60 °C)	[22]
	ctxB-R	GGTTGCTTCTCATCATCGAACCAC		
3	O1 *rfb*F	GTTTCACTGAACAGATGGG	192 (57 °C)	[23]
	O1 *rfb*R	GGTCATCTGTAAGTACAAC		
4	O139 *rfb*F	AGCCTCTTTATTACGGGTGG	449 (57 °C)	[23]
	O139 *rfb*R	GTCAAACCCGATCGTAAAGG		
5	*omp*W-F	CACCAAGAAGGTGACTTTATTGTG	304 (56 °C)	[24]
	*omp*W-R	GGTTTGTCGAATTAGCTTCACC		
6	tcpA-F	CACGATAAGAAAACCGGTCAAGAG	451 (El Tor) (60 °C)	[25]
	tcpA-R	CGAAAGCACCTTCTTTCACACGTTG		
	tcpA-R	TTACCAAATGCAACGCCGAATG	620 (Class)	
7	zot-F	TCGCTTAACGATGGCGCGTTTT	947 (60 °C)	[24]
	zot-R	AACCCCGTTTCACTTCTACCCA		
8	hlyA-F	GGCAAACAGCGAAACAAATACC	481 (El Tor)	[24]
	hlyA-F	GAGCCGGCATTCATCTGAAT	738/727 (ET and Class) (56 °C)	
	hlyA-R	CTCAGCGGGCTAATACGGTTTA		
9	toxR-F	CCTTCGATCCCCTAAGCAATAC	779 (60 °C)	[24]
	toxR-R	AGGGTTAGCAACGATGCGTAAG		
10	ompU-F	ACGCTGACGGAATCAACCAAA	869 (62 °C)	[26]
	ompU-R	GCGGAAGTTTGGCTTGAAGTAG		

2.4. Phage Induction Assessment by Mitomycin C

In the present study, it was found that after induction most of the isolates (both mitomycin C treated as well as untreated control culture) grew exponentially at comparable rates. Approximately, 2.5 h after the addition of mitomycin C (MMC+), the OD_{550} value of the induced culture of *Vibrio* isolate VH-II1 (*V. campbellii)* drastically decreased from 1.45 to 0.3 (Figure 4A). Similarly, 1.5 h after the addition of mitomycin C, the OD_{550} value of the induced culture of *Vibrio* isolate VHMC-A (*V. campbellii)* dramatically decreased from 2.05 to 0.1 (Figure 4B), whereas the control sample without mitomycin C (MMC−) continued to grow in both of the isolates (Figure 4A,B). In both isolates, typical fibers such as particles were observed in the induced culture. The significant reduction in OD at 550 nm indicated that both isolates carried lysogenic phages in their genome.

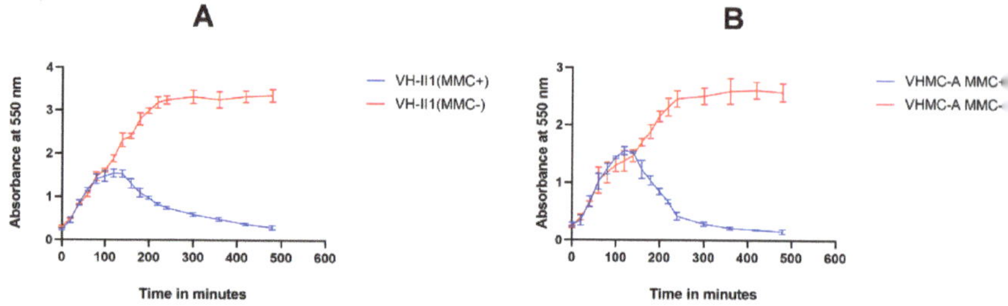

Figure 4. Vibriophage induction: Growth curve of vibriophage in (**A**) VH-II1 isolate and (**B**) VHMC- isolate, after Mitomycin C induction (MMC+). The absorbance was monitored over time at 550 n and confirmed the presence of vibriophage in VH-II1 and VHMC-A isolates.

3. Discussion

Organic waste such as foods, sewage, fertilizers, and animal and human feces co taminate water sources and play an important role in providing the most favorable plac for the growth of microbes including *Vibrio* species [27]. Further climatic changes and th enormous use of antimicrobial agents potentiate prokaryotic population and their surviv in harsh environmental conditions [28].

In the present study, five different *Vibrio* species have been isolated from fresh env ronmental water resources. The results indicated that out of twenty isolates, eight isolate (V-Gan, VR, VM-IIA, VP-IA, VP-IB, VH-I4, VH-II4, and VHVB-I) (40%) had characte istics of *V. proteolyticus*; four isolates (VHMC-A, VM-IB, VH-I3, and VH-II1) (20%) ha characteristics of *V. campbellii*; three isolates (VHMC-C, VHVB-II, and VH-II2) (15%) ha characteristics of *V. cincinnatiensis*; and another three isolates (VHMC-B, VHMC-D, an VM-IA) (15%) showed *V. nereis* characteristics. There was one isolate, VH-II3,identified a *Vibrio harveyi*. A fairly mixed population of different species of *Vibrio* was observed in th region (as shown in Figure 1, Table 1 and Supplementary Table S1). In the previous stud these *Vibrio* species have been characterized as potential pathogens for humans as we as aquatic animals [1,2,5]. In an in vitro study, it was found that *V. proteolyticus* produce virulent factors cytotoxic to eukaryotic cell lines such as macrophages and HeLa cells [29 In the year of 2017, Paek, Jayoung et al. isolated *V. cincinnatiensis* from the clinical specime sample of a patient having symptoms of *Vibrio*-associated disease, such as watery diarrhe and soft tissue injury [30]. Another *Vibrio* species mentioned in the Harveyi Clade wa considered to be the most severely pathogenic *Vibrio* cluster for aquatic animals, capable c generating more than 50 different *Vibrio* diseases [31].

Globally, antimicrobial resistance is a big challenge for the health authority. Microbe develop anti-microbial resistant genes in their genome for their protection [32]. Antim crobial susceptibility assay of *Vibrio* isolates revealed that out of twenty isolates, thirtee (65%) and nine (45%) isolates were found to be ampicillin-resistant and cephalexin-resistan respectively (Figure 2B and Supplementary Table S2). Those thirteen isolates were fron *V. proteolyticus* [6], *V. campbellii* [2], *V. cincinnatiensis* [3], and *V. nereis* [2]. The nine isolate were from *V. proteolyticus* [5], *V. campbellii* [2], *V. cincinnatiensis* [1], and *V. nereis* [1]. Co trimoxazole resistance was observed in seven isolates (35%) from all the isolated *Vibri* species except *V. harveyi,* followed by nalidixic acid for which six isolates (30%) of *V. prote olyticus* and *V. nereis* were found to be resistant. Resistance against streptomycin, neomycir cefotaxime, and furazolidone was found in one isolate each (Supplementary Table S2 The identified *Vibrio* isolates exhibited a high incidence of resistance against selected ar tibacterial agents that are commonly used [33]. Additionally, two isolates, VHMC-D anc VHMC-B, of *V. nereis* showed resistance against six (ampicillin, cephalexin, nalidixic acic cefotaxime, co-trimoxazole, and furazolidone) and four (ampicillin, cephalexin, nalidixi acid, and co-trimoxazole) selected antibiotics, respectively. Another seven isolates fron four different *Vibrio* species except *V. harveyi* were found to be moderate to highly resistan to more than one selected antibacterial agent (Figure 2 and Supplementary Table S2). I was noted that out of twenty isolates, sixteen isolates (80%) showed resistance against beta-lactam and its derivatives. In the previous study, it was also indicated that *Vibrio* isolates had resistance against beta-lactam antibiotics and their derivatives derived from China, Italy, and the U.S. [34–36]. However, it was observed that two isolates, VHMC-A (*V. proteolyticus*) and VH-II3 (*V. harveyi*), were sensitive to all the selected antibiotics used in this study. Further, it was found that all the isolates were sensitive to ciprofloxacin, gentamycin, norfloxacin, tetracycline, chloramphenicol, and polymyxin-B. This finding of antibiotic resistance of the *Vibrio* isolates indicates the emergence of multidrug resistance that could be a health threat.

In general, microbes show virulence by the secretion of virulent factors which are primarily infectious to host cells. It has been observed that virulent factors of *Vibrio* species are responsible for the pathogenesis and cause diarrhea, gastroenteritis, cholera, tissue injury, and blood infections [37]. In the present study, 10 different virulent genes primers were selected for the amplification of genes in *Vibrio* isolates (Table 2). As shown in Figure 3A, the isolate (VHI4) of *V. proteolyticus* showed amplification of *hlyA* gene (hemolysin) which indicates the presence of the pore-forming toxin in the genome of the isolate. Hemolysin is a well-known pore-forming hemolytic bacterial toxin, encoded by the *hlyA gene*. This toxin is present in the genome of various pathogenic *Vibrio* species. Hemolysin is a member of the class leucocidin superfamily, it alters the host cell membrane permeability and leads to cell death by the activation of the inflammasome pathway [38]. Further, another virulent gene primer, *ompW* was found to be amplified with the genomic DNA of VHMV-A isolate (*V. campbellii*) (Figure 3B). The bacterial outer membrane protein is a major bacterial protein which is expressed on the surface of the bacterial cell membrane. It participates in the regulation of salt stress, substance transport, and osmoregulation in many *Vibrio* species [39]. OmpW expression enhances the growth of *V. cholerae* in high saline water by carnitine channel [40]. In a previous study, an increase in the OmpW expression in *V. alginolyticus* was observed in the presence of high NaCl concentration [41].

In previous studies, horizontal transfer of virulent factors has been observed in *Vibrio* through phages and other mobile genetic elements [42–44]. In another study, it has been mentioned that virulent genes and anti-microbial resistant genes encoded by prophage-like elements get exchanged between pathogenic and nonpathogenic *Vibrio* species [45]. In our study, interestingly, two isolates (VHMC-A and VH-II1, both were *V. campbellii*) showed a drastic decrease in the OD_{550} after 1.5 and 2.5 h of the addition of mitomycin C (MMC+), respectively, which was a typical indication of the presence of lysogenic virulent phage (Figure 4A,B). Additionally, characteristic fibers such as particles were seen in mitomycin C induced cultures (MMC+). However, without mitomycin C added, bacterial culture (MMC−) was grown constantly, and, at one point, bacterial stationary phase was established. These observations strongly suggest that both these isolates, VHMC-A and VH-II1, carry lysogenic phages in their genome. Supernatants from the induced cultures were spotted on all the isolates in order to find a suitable host that can support the entry and growth of these putative phages, but no clearing was seen which indicated that these isolates could not be infected with these phages.

4. Materials and Methods

4.1. Sample Collection and Isolation of Bacterial Colonies

A total of 50 water samples were collected in sterile 500 mL bottles from various sources such as rivers, ponds, sewage, and stagnant rainwater from different locations in the Vishal district of Bihar, India (duration: October to June). After sample collection, 50 mL of water sample was filtered through a 0.22 µm membrane filter (Millipore, Bethesda, MD, USA). The retained content was suspended in phosphate-buffered saline (PBS, pH 8.4)and then it was enriched in alkaline peptone water (APW, pH 8.4) and incubated at 37 °C for 6–8 h in an incubator shaker (ThermoFisher Scientific, Berkeley, MO, USA). Further presumptive bacterial colonies were isolated according to the method described by Mishra et al. [46], with slight modification. Briefly, the enriched subculture was streaked using a sterile inoculation loop on a thiosulfate citrate bile salts sucrose (TCBS) agar plate (Sigma Aldri Kenilworth, NJ, USA) and incubated at 37 °C for 18–24 h. The promising isolated smoo yellow colonies were selected and subjected to biochemical tests for further confirmation *Vibrio* species.

4.2. Identification of Vibrio Isolates by Biochemical Tests Assay

The preserved bacterial colonies on agar stabs (1.5% Agar, 5 g yeast extract, 10 tryptone, and 10 g NaCl in 1000 mL of nuclease free water, pH = 7.2) were picked u for biochemical tests. A total of 12 biochemical tests were performed for each samp These tests include Voges–Proskauer, arginine utilization, salt tolerance, ortho-nitrophen b-D-galactopyranoside (ONPG), citrate utilization, ornithine utilization, and differe carbohydrate tests as mentioned in manufacturer protocol (HiMedia, Mumbai, India).

The selected colony was inoculated in a conical flask containing 5 mL alkaline pepto water and incubated at 37 °C until the inoculum turbidity became ≥0.5 OD at 600 nm. that point, the sample was ready to be used for biochemical tests. A test strip (*Hivibrio*T Identification kit, HiMedia, Mumbai, India) with 12 wells, each designated for a differe biochemical test analysis, was used. The test strip was opened aseptically, then 50 µL test sample was inoculated in each well of the test strip by surface inoculation meth and it was incubated for 18–24 h at 37 °C. After incubation, the test strip was analyz for different species of *Vibrio* on the basis of color changes as per the manufacturer cha description (Hivibrio™ Identification kit, HiMedia, Mumbai, India).

4.3. Antibiotic Susceptibility Testing

Antibiotic susceptibility testing was implemented in accordance with the Kirby–Bau disk diffusion method following the Clinical and Laboratory Standards Institute (CLS guidelines [47]. In brief, 150 µL of overnight *Vibrio* culture was spread aseptically o LB agar plates. *E. coli* DH5α was used as the negative control (usually *E. coli* strain K1 MG1655 is used as a negative control). The antibiotics which are most commonly used f the treatment of cholera were chosen for antibiotic susceptibility testing [15,33]. Differe antibiotics discs [ampicillin (AMP, 10 µg), chloramphenicol (C, 30 µg), cefotaxime (C 30 µg), ciprofloxacin (Cf, 5 µg), co-trimoxazole (Co, 25 µg), polymyxin-B (PB, 300 uni furazolidone (FR, 100 µg), gentamycin (GEN, 10 µg), neomycin (N, 30 µg), norfloxaci (Nx, 10 µg), nalidixic acid (Na, 30 µg), tetracycline (T, 30 µg), cephalexin (CN, 30 µg streptomycin (S, 10 µg), trimethoprim (Tr, 5 µg)] (HiMedia, Mumbai, India) were placed certain distances on the surface of the agar plate. The plates were incubated at 37 °C f 24 h and then the diameter of the zone of inhibition of each antibiotic disc was measure

4.4. Virulent Genes Amplification by Polymerase Chain Reaction (PCR)

The genomic DNA of different *Vibrio* isolates was extracted as per manufactur protocol (QIAGEN, Germantown, MD, USA). The genomic DNA was used as a templa for *Vibrio* virulent gene amplification. In the present study, specific *Vibrio* species gen primers [*ctxA*, *ctxB*, *O1rfb*, *O139rfb*, *tcpA* (*El Tor* and *Classical*), *hlyA* (*El Tor* and *Classica* *ompW*, *ompU*, *toxR*, *zot*] were used for PCR amplification (Table 2). The PCR reaction wa performed in 25 µL solution containing 10 µL PCR buffer including $MgCl_2$, dNTP mix (10× (Promega, Madison, WI, USA), 1 µL Taq DNA Polymerase (5 U/µL) (Promega, Madiso WI, USA), 1 µL each primer (10 µM), 1 µL of template DNA, and 11 µL of nuclease fre water. The reaction mixture was set up on the thermal cycler (Bio-Rad, Hercules, CA, USA The reaction conditions involved one cycle of initial denaturation at 95 °C for 5 min an 35 cycles of denaturation at 95 °C for 30 s, annealing at the annealing temperature for 1 mi extension at 72 °C for 30 s, and one cycle of final extension at 72 °C for 7 min. The PC amplified product was run on 1% agarose gel electrophoresis for separation. The *E. co* template DNA and their gene primer were used as a positive control for the PCR reactio

4.5. Induction of Vibriophage by Mitomycin C

Induction of vibriophage by mitomycin C was performed according to the methodology mentioned by *Castillo* et al. [48], with slight modification. In brief, preserved *Vibrio* species in LB agar stab were aseptically transferred with a sterile inoculum loop to 10 mL of autoclaved LB media and incubated at 37 °C overnight at 200 rpm in an incubator shaker (ThermoFisher Scientific, Waltham, MA, USA). The overnight grown culture was sub-cultured in LB medium and incubated again at approximately 37 °C for 2 h at 200 rpm to reach the *Vibrio* culture optical density (OD_{550}) of $\geq$0.2 at 550 nm. Phage induction was initiated by the addition of mitomycin C (at a final concentration of 0.5 µg/mL) (ThermoFisher Scientific, MA, USA) in grown culture (MMC+), except for the control (MMC−) without mitomycin C. Then, the optical density (OD_{600}) of samples was measured at a time interval of 20 min for 8 h.

4.6. Statistics

Statistical analysis of Mitomycin-C (MMC+) induced and uninduced (MMC−) *vibrio* culture was expressed in mean ± SD (standard deviation). All statistically analyzed data were graphed using Graph Pad Prism version 8.0.2 and Microsoft 365 (2010).

5. Conclusions

The present study concluded that different environmental water resources of Vaishali district, Bihar, India have the presence of distinct *Vibrio* species. Most of the identified isolates showed potential drug resistance against different selected antibacterial agents which are commonly used for the treatment of cholera as well as other bacterial diseases. Additionally, the two identified isolates (VH-I4 and VHMC-A) showed virulence gene in their genome. In spite of these characterizations, two *Vibrio* isolates (VHMC-A and VH-II1) showed lysogenic phage induction as detected by the treatment of the inducing agent MMC. Furthermore, the characterization of more isolates would be helpful in identifying other virulent strains that could give more explanatory results regarding the virulent genes, phage induction, and other multidrug resistant *Vibrio* species. These findings suggest that this district might be the next hotspot of cholera and non-cholera associated diseases in the future.

Supplementary Materials: The following supporting information can be downloaded at: https://www.mdpi.com/article/10.3390/antibiotics12061062/s1, Figure S1: Result of biochemical test confirmation of each *Vibrio* isolates; Table S1: Results of biochemical test assay of environment water samples isolates; Table S2: Antibiotic resistance and susceptibility assessment of each isolate.

Author Contributions: R.P. collected the samples and performed all experiments, and wrote and revised the manuscript. S.S. helped in writing and revising this manuscript. K.K.S. devised the main conceptual idea, designed the research, and supervised this study as well as reviewed the manuscript. All authors have read and agreed to the published version of the manuscript.

Funding: This research received no external funding.

Institutional Review Board Statement: Not applicable.

Informed Consent Statement: Not applicable.

Data Availability Statement: All data generated or analyzed during this study are included in this published article in the main manuscript.

Acknowledgments: The authors are obligated to the National Institute of Pharmaceutical Education and Research (NIPER), Hajipur, Ministry of Chemical and Fertilizer, Govt. of India for the fellowship awarded to Rajkishor Pandey to pursue his research project work at NIPER, Hajipur.

Conflicts of Interest: The authors declare no conflict of interest.

References

1. Muzembo, B.A.; Kitahara, K.; Debnath, A.; Ohno, A.; Okamoto, K.; Miyoshi, S.I. Cholera Outbreaks in India, 2011–2020: Systematic Review. *Int. J. Environ. Res. Public Health* **2022**, *19*, 5738. [CrossRef] [PubMed]
2. Zhang, X.H.; He, X.; Austin, B. *Vibrio harveyi*: A serious pathogen of fish and invertebrates in mariculture. *Mar. Life Sci. Techn* **2020**, *2*, 231–245. [CrossRef] [PubMed]
3. Le Roux, F.; Wegner, K.M.; Baker-Austin, C.; Vezzulli, L.; Osorio, C.R.; Amaro, C.; Ritchie, J.M.; Defoirdt, T.; Destoumieux-Garzó D.; Blokesch, M. The emergence of *Vibrio* pathogens in Europe: Ecology, evolution, and pathogenesis (Paris, 11–12th March 201 *Front. Microbiol.* **2015**, *6*, 830. [PubMed]
4. Brehm, T.T.; Dupke, S.; Hauk, G.; Fickenscher, H.; Rohde, H.; Berneking, L. Non-cholera *Vibrio* species-currently still rare b growing danger of infection in the North Sea and the Baltic Sea. *Internist* **2021**, *62*, 876–886. [CrossRef] [PubMed]
5. Baker-Austin, C.; Oliver, J.D.; Alam, M.; Ali, A.; Waldor, M.K.; Qadri, F.; Martinez-Urtaza, J. *Vibrio* spp. infections. *Nat. Rev. D Prim.* **2018**, *4*, 8. [CrossRef] [PubMed]
6. Harrison, J.; Nelson, K.; Morcrette, H.; Morcrette, C.; Preston, J.; Helmer, L.; Titball, R.W.; Butler, C.S.; Wagley, S. The increase prevalence of *Vibrio* species and the first reporting of *Vibrio jasicida* and *Vibrio rotiferianus* at UK shellfish sites. *Water Res.* **2022**, *21* 117942. [CrossRef]
7. Letchumanan, V.; Chan, K.-G.; Lee, L.-H. *Vibrio parahaemolyticus*: A review on the pathogenesis, prevalence, and advan molecular identification techniques. *Front. Microbiol.* **2014**, *5*, 705. [CrossRef]
8. Gxalo, O.; Digban, T.O.; Igere, B.E.; Olapade, O.A.; Okoh, A.I.; Nwodo, U.U. Virulence and antibiotic resistance characteristics *Vibrio* isolates from rustic environmental freshwaters. *Front. Cell. Infect. Microbiol.* **2021**, *11*, 765. [CrossRef]
9. Schroeder, M.; Brooks, B.D.; Brooks, A.E. The complex relationship between virulence and antibiotic resistance. *Genes* **2017**, *8*, 3 [CrossRef]
10. Soucy, S.M.; Huang, J.; Gogarten, J.P. Horizontal gene transfer: Building the web of life. *Nat. Rev. Genet.* **2015**, *16*, 472–48 [CrossRef]
11. Hazen, T.H.; Pan, L.; Gu, J.-D.; Sobecky, P.A. The contribution of mobile genetic elements to the evolution and ecology of *Vibri FEMS Microbiol. Ecol.* **2010**, *74*, 485–499. [CrossRef]
12. Ruwandeepika, H.; Defoirdt, T.; Bhowmick, P.; Shekar, M.; Bossier, P.; Karunasagar, I. Presence of typical and atypical virulen genes in vibrio isolates belonging to the *Harveyi* clade. *J. Appl. Microbiol.* **2010**, *109*, 888–899. [CrossRef] [PubMed]
13. Xiao, Y.; Huang, Z.; Yu, K.; Wang, M.; Gao, H.; Bai, X.; Jiang, M.; Wang, D. Distribution and Molecular Characteristics of *Vibr* Species Isolated from Aquatic Environments in China, 2020. *Microorganisms* **2022**, *10*, 2007. [CrossRef]
14. Cabello, F.C.; Godfrey, H.; Tomova, A.; Ivanova, L.; Dölz, H.; Millanao, A.; Buschmann, A. Antimicrobial use in aquacultur re-examined: Its relevance to antimicrobial resistance and to animal and human health. *Environ. Microbiol.* **2013**, *15*, 1917–194 [CrossRef] [PubMed]
15. Das, B.; Verma, J.; Kumar, P.; Ghosh, A.; Ramamurthy, T. Antibiotic resistance in *Vibrio cholerae:* Understanding the ecology resistance genes and mechanisms. *Vaccine* **2020**, *38*, A83–A92. [CrossRef] [PubMed]
16. Centers for Disease Control and Prevention. Diagnosis and Treatment. 2023. Available online: https://www.cdc.gov/vibri diagnosis.html#:~:text=Treatment%20is%20not%20necessary%20in,in%20severe%20or%20prolonged%20illnesses (accessed o 8 January 2023).
17. World Health Organization (WHO). Cholera. 2023. Available online: https://www.who.int/news-room/fact-sheets detail/cholera?gclid=CjwKCAjwpuajBhBpEiwA_ZtfhRtJtsyyYxOn61c8unDkw_YvdjmCsJjbTBh_dC3OKsFrUUhC3CcFjhoC sQAvD_BwE (accessed on 8 January 2023).
18. Yuan, X.H.; Li, Y.M.; Vaziri, A.Z.; Kaviar, V.H.; Jin, Y.; Jin, Y.; Maleki, A.; Omidi, N.; Kouhsari, E. Global status of antimicrobi resistance among environmental isolates of *Vibrio cholerae* O1/O139: A systematic review and meta-analysis. *Antimicrob. Resis Infect. Control* **2022**, *11*, 62. [CrossRef]
19. Gao, H.; Xu, J.; Lu, X.; Li, J.; Lou, J.; Zhao, H.; Diao, B.; Shi, Q.; Zhang, Y.; Kan, B. Expression of hemolysin is regulated under th collective actions of HapR, Fur, and HlyU in *Vibrio cholerae* El Tor serogroup O1. *Front. Microbiol.* **2018**, *9*, 1310. [CrossRef]
20. Sharma, A.; Chaturvedi, A.N. Prevalence of virulence genes (ctxA, stn, OmpW and tcpA) among non-O1 *Vibrio cholerae* isolate from fresh water environment. *Int. J. Hyg. Environ. Health* **2006**, *209*, 521–526. [CrossRef]
21. Keasler, S.P.; Hall, R.H. Detecting and biotyping *Vibrio cholerae* O1 with multiplex polymerase chain reaction. *Lancet* **1993**, *34* 1661. [CrossRef]
22. Ogawa, A.; Takeda, T. The gene encoding the heat-stable enterotoxin of *Vibrio cholerae* is flanked by 123-base pair direct repeat *Microbiol. Immunol.* **1993**, *37*, 607–616. [CrossRef]
23. Madhusudana, R.B.; Surendran, P.K. Detection of ctx gene positive non-O1/non-O139 *V. cholerae* in shrimp aquaculture enviror ments. *J. Food Sci. Technol.* **2013**, *50*, 496–504. [CrossRef]
24. Nandi, B.; Nandy, R.K.; Mukhopadhyay, S.; Nair, G.B.; Shimada, T.; Ghose, A.C. Rapid method for species-specific identificatio of *Vibrio cholerae* using primers targeted to the gene of outer membrane protein OmpW. *J. Clin. Microbiol.* **2000**, *38*, 4145–415 [CrossRef]
25. Rivera, I.N.; Chun, J.; Huq, A.; Sack, R.B.; Colwell, R.R. Genotypes associated with virulence in environmental isolates of *Vibri cholerae*. *Appl. Environ. Microbiol.* **2001**, *67*, 2421–2429. [CrossRef]

Singh, D.V.; Isac, S.R.; Colwell, R.R. Development of a hexaplex PCR assay for rapid detection of virulence and regulatory genes in *Vibrio cholerae* and *Vibrio mimicus*. *J. Clin. Microbiol.* **2002**, *40*, 4321–4324. [CrossRef] [PubMed]

Adesiyan, I.M.; Bisi-Johnson, M.A.; Okoh, A.I. Incidence of antibiotic resistance genotypes of *Vibrio* species recovered from selected freshwaters in Southwest Nigeria. *Sci. Rep.* **2022**, *12*, 18912. [CrossRef] [PubMed]

Li, W.; Liu, C.; Ho, H.C.; Shi, L.; Zeng, Y.; Yang, X.; Huang, Q.; Pei, Y.; Huang, C.; Yang, L. Association between antibiotic resistance and increasing ambient temperature in China: An ecological study with nationwide panel data. *Lancet Reg. Health-West. Pac.* **2023**, *30*, 100628. [CrossRef] [PubMed]

Ray, A.; Kinch, L.N.; Santos, M.D.S.; Grishin, N.V.; Orth, K.; Salomon, D. Proteomics analysis reveals previously uncharacterized virulence factors in *Vibrio proteolyticus*. *mBio* **2016**, *7*, e01077-16. [CrossRef]

Paek, J.; Shin, J.H.; Shin, Y.; Park, I.-S.; Kim, H.; Kook, J.-K.; Kang, S.-S.; Kim, D.-S.; Park, K.-H.; Chang, Y.-H. *Vibrio injenensis* sp. nov., isolated from human clinical specimens. *Antonie Leeuwenhoek* **2017**, *110*, 145–152. [CrossRef]

Kang, H.; Yu, Y.; Liao, M.; Wang, Y.; Yang, G.; Zhang, Z.; Li, B.; Rong, X.; Wang, C. Physiology, metabolism, antibiotic resistance, and genetic diversity of *Harveyi* Glade bacteria isolated from coastal mariculture system in China in the last two decades. *Front. Mar. Sci.* **2022**, *9*, 932255. [CrossRef]

2. Hernando-Amado, S.; Coque, T.M.; Baquero, F.; Martínez, J.L. Antibiotic resistance: Moving from individual health norms to social norms in one health and global health. *Front. Microbiol.* **2020**, *11*, 1914. [CrossRef]

3. Heng, S.P.; Letchumanan, V.; Deng, C.Y.; Ab Mutalib, N.S.; Khan, T.M.; Chuah, L.H.; Chan, K.G.; Goh, B.H.; Pusparajah, P.; Lee, L.H. *Vibrio vulnificus*: An environmental and clinical burden. *Front. Microbiol.* **2017**, *8*, 997. [CrossRef] [PubMed]

4. Zhu, Z.M.; Dong, C.F.; Weng, S.P.; He, J.G. The high prevalence of pathogenic *Vibrio harveyi* with multiple antibiotic resistance in scale drop and muscle necrosis disease of the hybrid grouper, *Epinephelus fuscoguttatus* (♀)× *E. lanceolatus* (♂), in China. *J. Fish Dis.* **2018**, *41*, 589–601. [CrossRef]

5. Fernández-Delgado, M.; Suárez, P.; Giner, S.; Sanz, V.; Peña, J.; Sánchez, D.; García-Amado, M.A. Occurrence and virulence properties of *Vibrio* and *Salinivibrio* isolates from tropical lagoons of the southern Caribbean Sea. *Antonie Leeuwenhoek* **2017**, *110*, 833–841. [CrossRef]

6. Zanetti, S.; Spanu, T.; Deriu, A.; Romano, L.; Sechi, L.; Fadda, G. In vitro susceptibility of *Vibrio* spp. isolated from the environment. *Int. J. Antimicrob. Agents* **2001**, *17*, 407–409. [CrossRef]

7. Schirmeister, F.; Dieckmann, R.; Bechlars, S.; Bier, N.; Faruque, S.M.; Strauch, E. Genetic and phenotypic analysis of *Vibrio cholerae* non-O1, non-O139 isolated from German and Austrian patients. *Eur. J. Clin. Microbiol. Infect. Dis.* **2014**, *33*, 767–778. [CrossRef]

8. Cohen, H.; Baram, N.; Edry-Botzer, L.; Munitz, A.; Salomon, D.; Gerlic, M. *Vibrio* pore-forming leukocidin activates pyroptotic cell death via the NLRP3 inflammasome. *Emerg. Microbes Infect.* **2020**, *9*, 278–290. [CrossRef] [PubMed]

9. Jalajakumari, M.B.; Manning, P.A. Nucleotide sequence of the gene, ompW, encoding a 22kDa immunogenic outer membrane protein of *Vibrio cholerae*. *Nucleic Acids Res.* **1990**, *18*, 2180. [CrossRef] [PubMed]

0. Fu, X.; Zhang, J.; Li, T.; Zhang, M.; Li, J.; Kan, B. The outer membrane protein OmpW enhanced *V. cholerae* growth in hypersaline conditions by transporting carnitine. *Front. Microbiol.* **2018**, *8*, 2703. [CrossRef]

1. Xu, C.; Wang, S.; Ren, H.; Lin, X.; Wu, L.; Peng, X. Proteomic analysis on the expression of outer membrane proteins of *Vibrio alginolyticus* at different sodium concentrations. *Proteomics* **2005**, *5*, 3142–3152. [CrossRef]

2. Deng, Y.; Xu, H.; Su, Y.; Liu, S.; Xu, L.; Guo, Z.; Wu, J.; Cheng, C.; Feng, J. Horizontal gene transfer contributes to virulence and antibiotic resistance of *Vibrio harveyi* 345 based on complete genome sequence analysis. *BMC Genom.* **2019**, *20*, 761. [CrossRef]

3. Antonova, E.S.; Hammer, B.K. Genetics of natural competence in *Vibrio cholerae* and other *vibrios*. *Microbiol. Spectr.* **2015**, *3*, 3. [CrossRef]

4. Lo Scrudato, M.; Blokesch, M. The regulatory network of natural competence and transformation of *Vibrio cholerae*. *PLoS Genet.* **2012**, *8*, e1002778. [CrossRef]

5. Castillo, D.; Kauffman, K.; Hussain, F.; Kalatzis, P.; Rørbo, N.; Polz, M.F.; Middelboe, M. Widespread distribution of prophage-encoded virulence factors in marine *Vibrio* communities. *Sci. Rep.* **2018**, *8*, 9973. [CrossRef]

6. Mishra, A.; Taneja, N.; Sharma, M. Environmental and epidemiological surveillance of *Vibrio cholerae* in a cholera-endemic region in India with freshwater environs. *J. Appl. Microbiol.* **2012**, *112*, 225–237. [CrossRef] [PubMed]

7. Bauer, A.W. Antibiotic susceptibility testing by a standardized single disc method. *Am. J. Clin. Pathol.* **1966**, *45*, 149–158. [CrossRef]

8. Castillo, D.; Andersen, N.; Kalatzis, P.G.; Middelboe, M. Large phenotypic and genetic diversity of prophages induced from the fish pathogen *Vibrio anguillarum*. *Viruses* **2019**, *11*, 983. [CrossRef] [PubMed]

MDPI AG
Grosspeteranlage 5
4052 Basel
Switzerland
Tel.: +41 61 683 77 34

Antibiotics Editorial Office
E-mail: antibiotics@mdpi.com
www.mdpi.com/journal/antibiotics

www.ingramcontent.com/pod-product-compliance
Lightning Source LLC
Chambersburg PA
CBHW041355210926
51850CB00013B/3085